Political Geography

Territory, State, and Society

Kevin R. Cox

Blackwell
Publishers

First published 2002

2 4 6 8 10 9 7 5 3 1

Blackwell Publishers Ltd
108 Cowley Road
Oxford OX4 1JF
UK

Blackwell Publishers Inc.
350 Main Street
Malden, Massachusetts 02148
USA

British Library Cataloguing in Publication Data

A CIP catalogue record for this book is available from the
British Library.

Library of Congress Cataloging-in-Publication Data
Cox, Kevin R., 1939–
    Political geography : territory, state, and society / Kevin R. Cox.
       p.; cm.
    Includes bibliographical references and index.
    ISBN 0–631–22678–8 (hbk : alk. paper) — ISBN 0–631–22679–6 (pbk : alk. paper)
  1. Political geography.   2. Human territoriality.   3. Group identity.   4. Nationalism.
  5. Capitalism.   6. State, The.   I. Title.
    JC319 .C69 2002
    320.1′2—dc21

                                                                        2001002457

Typeset in 10.5 on 12 pt Palatino
by Best-set Typesetter Ltd., Hong Kong
Printed in Great Britain by TJ International Ltd, Padstow, Cornwall

This book is printed on acid-free paper.

Contents

Figures

Tables

Preface

Like many books, this one began as an attempt to develop some reading materials for a course: reading materials that would complement what I was trying to do in the lecture room. This has been going on for quite a few years, during which time the materials have been refined, filtered, and, in many cases, discarded wholesale. That the book I was aiming to write should shed light on the world as we experience it today should go without saying. But beyond that, two major criteria guided the writing process.

The first was that the book should be organized relatively tightly around a coherent framework of ideas. A major problem with political geography has always been lack of coherence in its subject matter, and to some degree that continues down to the present day. From the viewpoint of the student coming to political geography for the first time, I have never thought that this could possibly be satisfactory. We learn by constructing, by trial and error, networks of relations between different ideas and logics; and it is the job of the teacher to communicate such a framework and facilitate the process of making connections. In this book the focus of political geography is on the twin concepts of territory and territoriality: territory as the object and territoriality as the practice. Given that the objective is to illuminate contemporary political geographies, the state has also had to figure prominently in the argument. Some concept of social process has then given life to territory, territoriality and the principle vehicle through which they are mobilized today, the state.

The term social process is clearly a very abstract one which could cover a multitude of different logics. In recent work on human geography one of the emergent distinctions in approaches to this question has been between political economy and the cultural, as, for example, in the recent collection *Geographies of Economies*, edited by Roger Lee and Jane Wills. I think that this is a helpful distinction in thinking about social process as it relates to human geography, and it provides the windows on the subject matter of political geography which are presented in the first and second parts of what is a three-part book. The third part re-examines the subject matter from the standpoint of the state and brings the cultural, the economic, and the territorial together.

My second guiding principle in writing this book has been that it should embody an approach to the subject matter that is critical; critical not so much of other approaches in a direct sense but of the world those other approaches are supposed to shed light on: the world of public pronouncements, editorials, news reports, real historico-geographical events. Over many years of undergraduate teaching I have found that this is best achieved through an approach that foregrounds the structuring role of capitalism and of the logics of capitalist development and how they work themselves out over space. Capitalism is the fundamental structuring force in the world today. It is, therefore, insight into the logics of capitalism, and its characteristic ways and forms of development, that is crucial to making sense of those issues of territory which I believe to be central to an understanding of the political geography of the contemporary world.

This in turn has led to me to make copious use of sources that will be, at least in books oriented to an undergraduate market, unconventional to say the least. But I see no reason to apologize for the numerous references to articles in *New Left Review* or for references to those who have contributed so much to the tradition of thought, historical materialism, that that periodical represents: David Harvey, Eric Hobsbawm, Michael Mann, among many others. On the other hand, given this particular perspective, the informed reader may find equally surprising the use I have made of some of the more conservative financial publications, such as *The Wall Street Journal* or *The Economist*, and certainly their editorials and op-ed pages can leave much to be desired. But their news stories often stand in sharp counterpoint to that free market flim-flam, providing sober assessments, with a sharp eye for the significance of economic forces, of unfolding issues. So the critical approach embraced by this book comes far less from a self-conscious examination of other bodies of literature, and alternative theories, than it does from a perspective on territory, the state, and the social process underpinning them, that is itself critical. If students are slow to recognize this, then, the questions I pose to them throughout the text in the form of "Think and Learn" boxes should serve to get them back on the right track as well as helping to impart that overall sense of coherence which has been my other goal. On the other hand, for those instructors who feel that something is still missing, there is no reason why the book should not be used as a foil for exploring those alternative views.

As a result of my emphasis on capitalism and its logics some might expect my approach to be economistic. I do not think that they will find this to be borne out in the reading. The cultural turn in geography and earlier interests in post-Marxist social theory have provided a challenge to us all. But historical materialism is a living body of thought which is constantly being reworked not only in the light of the necessarily contingent element assumed by the course of development but also with respect to the numerous intellectual challenges it has had to confront. It will, however, be for the reader to judge how successful I have been in responding to the claims that have been made. Indeed, while the book is primarily intended as a text I hope it will also stimulate a wider audience interested in a fresh approach to political geography.

Finally we reach the point in the preface which typically begins "This book could not have been written without . . ." Certainly there are many people who have over the years stimulated me to think about the issues I have addressed in this book. Not least are former graduate students, including, and in no particular order, Kim England, Raju Das, Jeff McCarthy, John Agnew, Felicity Sutcliffe, Paul Herr, Karen Walby, Andy Jonas, Murray Low, Andy Wood, Mike Sutcliffe, Golden Mergler, and Andy Mair. They cannot know how important they have been in helping me reconsider and refine my ideas, and how as they learnt, I learnt too. Every academic should be so lucky! My son Gerard has also shown a growing interest in and appreciation of the sorts of argument embodied in this book. I am grateful to him for keeping me on my toes at the dinner table, as well as for joining me in my long-suffering support of the Cincinnati Bengals, and reminding me that there *is* modern jazz after the 1950s.

I should also pay tribute to a series of equally long-suffering geography editors at Blackwell. It was John Davy who originally suggested the project to me back in 1995. Jill Landeryou helped coax me along and brought me to the point of embarking on a first draft. And Sarah Falkus is the one who has presided over the final stages, with great intelligence and care. It was through her good offices that I had the benefit of two excellent readers in Jenny Robinson and Byron Miller.

I hope, however, that none of these will feel slighted if I express the burden of my debt as owing to David Harvey. His probing, critical intelligence, his creative reworking of the field of human geography, have been an inspiration not just to me, but to many of us, and for a long time. He is the one above all who has shown the way forward. And while I suspect he will not agree with everything written in this book or how I have gone about writing it, I hope he will appreciate it as a contribution to spreading the word about that historic-geographical materialism whose contours he has developed and explored. It is, therefore, to David Harvey, political geographer *sans pareil*, that I dedicate this book.

Kevin Cox
Ohio State University

Acknowledgments

The author and publishers gratefully acknowledge the following for permission to reproduce copyright material:

Figure 1.1, from Peter J. Taylor (1971) "Distances within Shapes," *Geografiska Annaler B*, 53, 43, © Swedish Society for Anthropology and Geography.

Figure 1.2, from R. A. Dodgson and R. A. Butlin (eds) (1993) *An Historical Geography of England and Wales*, Academic Press.

Figure 1.4, reprinted from *Political Geography*, volume 19, number 4, B. Giordano, "Italian Regionalism or 'Padanian' Nationalism – The Political Project of the Lega Norda in Italian Politics," pp. 445–471, copyright (2000), with permission from Elsevier Science.

Figure 2.2, from H. C. Darby (ed.) *New Historical Geography of England*, p. 393, figure 83, and p. 452, figure 94. Reprinted with the permission of Cambridge University Press.

Figure 3.3, from *Atlas of Industrialising Britain 1780–1914*, edited by John Langton and R. J. Morris (1986), Routledge.

Figure 3.5, reprinted, with permission, from *The Columbus (Ohio) Dispatch*.

Figure 3.6, from *Regional Policy and Planning in Europe* by Paul Balchin, Ludek Sykora and Gregory Bull (1999), Routledge.

Figure 3.10 reprinted by permission of Paul Chapman Publishing from P. Dicken, *Global Shift*, © Paul Chapman Publishing (1998).

Figure 4.1, from H. V. Savitch and R. K. Vogel (eds) *Regional Politics: America in a Post-city Age*, figure 4.2, copyright © Sage Publications, Inc. Reprinted by Permission of Sage Publications, Inc.

Figure 4.4, from *Maps of Meaning*, by Peter Jackson (1989), Routledge.

Figure 6.3 to 6.6, from Eugen Weber, *Peasants into Frenchmen: The Modernization of Rural France, 1870–1914*, Copyright © 1976 by the Board of Trustees of the Leland Stanford Junior University. With the permission of Stanford University Press, www.sup.org.

Figure 6.10, from *The Atlas of Apartheid*, by A. J. Christopher (1994), Routledge.

Figure 9.8 and 9.9, from John Allen, Doreen Massey, and Allan Cochrane (1998) *Rethinking the Region*, reprinted with the kind permission of the authors.

Figure 9.11, from P. Dicken, A. Tickell, and H. Yeung (1997) "Putting Japanese Investment in Its Place," *Area*, 29, 208, © Royal Geographical Society (with the Institute of British Geographers).

Figure 9.13, from Wolf Roder (1964) "Division of Land Resources in Southern Rhodesia," *Annals of the Association of American Geographers*, 54, 47, © Association of American Geographers.

Figure 10.3, reprinted from figure 4 of "The New Geography of Automobile Production: Japanese Transplants in North America," by A. Mair, R. Florida, and M. Kenney, *Economic Geography* (Vol. 64 No. 4, 1988), with the permission of *Economic Geography*.

The publishers apologize for any errors or omissions in the above list and would be grateful to be notified of any corrections that should be incorporated in the next edition or reprint of this book.

Chapter 1

Fundamental Concepts of Political Geography: An Introduction

Introduction

The simple answer to the question "what is political geography about?" is what it says it is about: politics and geography. But that is altogether *too* simple. Political geography is by no means the sum of its two parts. In political geography, "geography" is drawn on in selective ways: in ways which illumine the political. By the same token, "politics" is drawn on in ways which shed light on the geographic. Above all, political geography focuses on the twin ideas of territory and territoriality.

Territory and territoriality are the defining concepts of political geography in that they bring together the ideas of power and space: territories as spaces that are defended, contested, claimed against the claims of others; in short, through territoriality. Territory and territoriality mutually presuppose one another. There can't be one without the other. Territoriality is activity: the activity of defending, controlling, excluding, including; territory is the area whose content one seeks to control in these ways.[1]

But again, that only takes us so far. To understand territory and territoriality as opposed to describing what they are about, we need understandings of space relations and politics. As geographic concepts territory and territoriality have their roots, their conditions, in other spatial practices; in particular those relating to movement and those that have to do with the embedding of people and their activities in particular places – ideas that are fundamental to contemporary human geography. Likewise, in order to understand the political in political geography we need to come to terms with *the* central concept

---

1 Consider in this regard the definitions given by *The Dictionary of Human Geography* (1986). Territory: "A general term used to describe areas of land or sea over which states and other political entities claim to exercise some form of control" (p. 483); territoriality: "The attempt by an individual or group to influence or establish control over a clearly demarcated territory" (p. 482).

of modern political science, the state. The state is itself an expression of territorial power: it has an area over which it claims jurisdiction, it has boundaries and it has powers to influence movement and what goes on in any part of its jurisdiction. For any territorial strategy, any expression of territoriality advanced by a neighborhood organization, a business or ethnic group, or whatever, the state is, accordingly, of crucial significance.

This begs the question, however, of what motivates people to defend particular areas and so to seek out the help of the state. It also begs the question of why the state might be responsive. Territory itself has no substance and what motivate people are interests which are, by definition, substantive in character: they refer to things, perhaps symbols, that people want. In short we need some concept of what it is that drives people in their territorial activities and what produces conflict over territory. Ultimately it has to do with our relationship to the material world: our need to relate to that world if we are to survive. But that relationship is always socially mediated. It is always in and through others that we appropriate and transform aspects of that material world into forms which we can use. Concepts of social process, therefore, are central to understanding territory and territoriality. But specifically *what* social process are we talking about? In human history there has been a succession of highly diverse social formations. This book, however, has to do with the political geography of the specifically *contemporary* world. Accordingly our focus here has to be that highly dynamic force that we know as capitalism.

Now, this may sound as if the treatment is to be economically deterministic. This is far from my aim. Rather I recognize that social life is highly diverse; that it consists of many different conditions, without which it could not function. There is something that I will call the social process that is separate from capitalism. But capitalism is the energizing moment of that process and continually strives to mobilize those other conditions for its own purposes. And in this it is no different from previous forms of social life. Production is always the central pivot around which social life is continually being organized and shaped.

In the first major section of this opening chapter, therefore, the three principal ideas around which the argument in this book is organized are introduced: territory, the state, and the social process. The second part of the chapter is devoted to a consideration of some case studies through which I want to illustrate how these fundamental ideas can be applied. In a brief closing section I will then outline how the book as a whole is organized.

Fundamental Concepts

Territory

The core concepts of political geography can be stated quite simply: they are *territory* and *territoriality*. These ideas are inextricably interrelated. Territory is to be understood through its relations to those activities we define as territorial: the exercise of territoriality, in other words. Robert Sack (1983) has defined it as activity aimed at influencing the content of an area. This means that activities of an exclusionary or, alternatively, of an inclusionary nature would be regarded as territorial and the area the content of which one wants to influence as the territory in question. This means that in addition to territory having associations of area and boundary it also has ones of defense: territories are spaces which people defend by excluding some activities and by including those which will enhance more precisely what it is in the territory that they want to defend.

In these terms examples of territorial activity are legion. Import quotas and tariffs are obvious cases in point as are restrictions on immigration. Sometimes the products whose movement is being regulated have a strong cultural content: the French government has tried to limit the amount of non-French programming shown on French television. This is not to say that exclusionary processes are limited to the level of the nation state so that the territory that political geographers focus on is that of the state's jurisdiction. Examples can be found at all manner of scales: the gated communities that have become common in the suburbs of many American cities, for example; or the greenbelts which surround every British city of any size and which limit new residential development within their boundaries. And the latter example reminds us that *any* form of land use zoning is a territorial form of activity.

There are also activities or processes of a more inclusionary nature. People and organizations try to regulate the content of geographic areas by attracting in certain sorts of people or activity. The constitution of the state of Israel mandates that all Jews should be accorded full rights of residency in Israel if they should request it. A different sort of example has to do with the channeling of investment flows. For many years in the United States local and State[2] governments have implemented a variety of policies the goal of which has been to attract new investment inside their boundaries: investment that will, among other things, generate employment and add to the local tax base. This sort of activity is now becoming more common in Western Europe. The member states of the European Union have been especially active in competing for choice investments like those of the Japanese auto companies.

This is not to say that exclusionary and inclusionary forms of policy are unrelated. What is inclusionary for some may be exclusionary for others, and that may be the point of the exercise. Gentrification has been a common housing market process in neighborhoods close to the downtowns of major cities in both North America and Western Europe. As wealthier people move into an area so rents and housing prices tend to increase. This results in the exclusion of long-term, low-income residents who can no longer afford the rents. But this is a process the gentrifiers promote through trying to secure for the area various local government expenditures and regulatory policies that will make the area more attractive to the well heeled buyer. And one of the purposes of that is, through the medium of increasing real estate

---

2   Throughout this book I use State with a capital "S" to indicate US (and Mexican) States, and state with a lower case "s" to indicate the state as a universal concept.

values, to drive out the poor, who for various reasons are regarded as lowering the tone of the area, perhaps introducing a criminal element into the neighborhood.

The idea of territoriality is derivative of other concepts absolutely crucial to contemporary human geography. These are the related ones of mobility and immobility. Geography, bear in mind, is the study of objects, activities, institutions from the standpoint of their space relations (both internal and external), what we might call their various where-nesses. These include their accessibility relations with respect to one another, and their distributions.

One way of studying human geography is in terms of movements. This was a dominant theme in the spatial analysis school which dominated human geography for much of the sixties and which is still influential today. The point is that the reproduction of a particular distribution of objects – factories, houses, highways, airports, the people themselves – depends on various sorts of flow: movements of raw materials for the factories, movements of money with which to buy the raw materials, movements of labor among others. To the extent that the geography of movement changes then so will the distribution of houses, factories, and the like. As investment moves out to the suburbs, for example, so the form of the city changes: housing is added on the edge but we often find housing towards the center of the city being deleted. The shift of investment to the suburbs is a major reason for the fact of housing abandonment that is so apparent in some American central cities, like Detroit and Chicago.

But more recently, the converse of movement, the idea of settlement, of immobilization or embedding in a particular place, has come to be recognized as of immense significance. This is particularly so from the standpoint of understanding territoriality. It is certainly true that people move around. Residential mobility within cities is a fact of life and without it realtors would go out of business. And people also move over much longer distances, retiring from, say, New York or Montreal to Florida or from the United Kingdom to the Costa Blanca in Spain. In similar fashion firms move. They close or sell factories in one location and shift their operations elsewhere. But there are contrary tendencies as well. People, firms, organizations of all types get embedded in particular places: embedded in the sense that other places become costly substitutes for their current locations. People put down what are often referred to as "roots." They buy houses in neighborhoods, and raise families. Their children marry and some, at least, will live in the same city. People also get locked into particular careers with particular firms: they develop skills which are appropriate to their particular employer but which have limited portability. So leaving the area, moving elsewhere, can mean a serious diminution of life chances, a deep sense of loss as one moves away from one's loved ones and the familiar, or both. Even owning a house is a source of geographic inertia since buying and selling is such a protracted and time-consuming process.

In similar fashion firms develop collaborative relations with other firms in the same locality and these can be a source of competitive advantage.

Firms may share the same labor supply. A virtue of being located next to other firms manufacturing similar products is that when one of them is releasing workers another is likely to be hiring. So labor shortage is unlikely to be a problem in the area whereas moving to a city where the firm is the only one that has those sorts of skill demands is. In short, firms can get locked into areas not just through the productive relations they enter into with other firms but also through the way they may share with those firms labor reserves or suppliers.[3]

This means that people, firms, organizations may be very dependent on what happens in the area they happen to be located in. People buy houses in neighborhoods and see the house, to some degree at least, as an investment: an asset like stocks or bonds or a savings account on each of which they expect a return. In the case of investment in the house you live in the return is in the form of an increase in its value. But neighborhoods can change as some people leave and others move in, as undeveloped land is rezoned for gas stations or bars. In short, movements in and out can threaten investments in homes. Money has been invested in something which is difficult to move, which is literally embedded in the ground. If values are to be maintained let alone increase, territorial strategies have to be deployed: attempts to structure movements into the area by (e.g.) opposing the rezonings that will allow gas stations or bars or the conversion of existing owner-occupied housing into apartments.

As we have seen, firms likewise get immobilized, dependent on particular localities or those in them, and the continual flow of value through them. But the arrival of new firms in the area can threaten that flow of value and hence their profitability. The increased demand for labor that comes about can result in increased wage levels, particularly if the new arrivals are the branch plants of unionized firms. To the extent that labor shortages are moderated by in-migration then pressure may be transferred to the housing market, and as housing prices increase this too can exercise upward pressure on wages. Yet relocation by the firms so affected to areas where lower wages prevail will be difficult. It may be hard to persuade the workers on whose skills the firms depend to move with them, and training new workers will be a protracted and costly process. And it will certainly be hard to reconstitute elsewhere the collaborative relations with other firms so important to competitiveness.

As a result they can be expected to organize to defend their territory and the advantages it provides them. They may, for example, pressure city government to ease bottlenecks in housing supply so that the upward shift in housing prices can be contained: facilitate the speedier rezoning of

---

3   Firms may have specialized transport or marketing needs. If located alongside firms producing similar sorts of product demand may be such that these activities can be subcontracted to specialized suppliers able to operate at lower cost. Whereas if the firm is only one in the area with those particular needs it may have to (e.g.) purchase its own trucks, even though it may not need them on a continual basis.

land to higher densities, eliminate delays in servicing raw land with water and sewerage. And the policies that are bringing new firms into the area will also come under review: should city government be so aggressively courting firms to locate their branch plants there, for instance?

We will see in what ensues that territory and territoriality can assume more complex forms: that what is a territorial strategy for some is a threat to the territorial strategies of others. But the relation between mobility and immobility, of movement and embeddedness, is central to the emergence of territory as an issue: to the desire to influence the content of an area. And as we have seen from the examples above, territorial strategies typically draw in some way on the power of the state: its power over rezonings, over local economic development policy, for example. It is therefore entirely appropriate that the state should be our next focus of concern.

The state

For a start, notice how important the state and its various agencies are in regulating geographies: in structuring movements, in defending the interests of the more immobilized, the more embedded. Central governments everywhere regulate movements across their boundaries: movements of people, of commodities and of money. They may restrict imports in order to protect particular industries, their workers and the cities in which they are located from foreign competition. They may also restrict exports for a similar purpose: a duty on exports of American leather protects the shoe making industry by driving up its price to overseas producers at the same time as it lowers it for the American producer. Limits on immigration on the part of the more developed countries[4] are the norm and so too is the regulation of foreign investment. In the latter regard there are often laws governing the takeover of firms by foreign corporations or foreign investment in certain sensitive industries like arms firms.

Likewise there are things that local government can do that impact on geographic change through their effects on movement. This is despite the fact that central branches of the state protect the freedom of movement of labor

4   Use of the term "more developed countries" raises an important issue for this text. The problem is one of differentiating between countries but not in a way that implies value judgments. The "First World/Third World" distinction clearly implies hierarchy and will not be used. Alternatives to "more developed"/"less developed" are "rich"/"poor" and "North"/"South". I am deterred from using the latter by virtue of its transparent inaccuracy. There are more developed countries in the South, like Australia and New Zealand, and less developed countries in the North, like Egypt or Pakistan. The "rich country"/"poor country" distinction seems no improvement on the "more developed"/"less developed" distinction since to define someone or someplace as poor is often (not always) to imply some sort of lack on their or its part. At least the term "less developed" implies a process of change towards the more developed pole. That gets us into the problem of whether development is a good or a bad thing but it nevertheless softens the sense of invidious distinction between countries. It is, therefore, the term that I will use in the remainder of this book.

and of commodities within national boundaries and so local governments cannot try to achieve their ends by interfering with them: protecting a major local employer by imposing restrictions on the sale of goods from competing firms elsewhere in the country, say. Rather there are other means of structuring location choice. Urban development, the siting of new housing developments, new industrial estates, and the location of new highways must invariably run the gauntlet of a local permitting process: public hearings, rezoning hearings, objections from national public health authorities, and so on.

Nevertheless, the relation between the state on the one hand, and power in society on the other, including power over geography, is not straightforward. Power comes in different forms. Immensely important in contemporary social life is the power of money. This is not something which is foreign to the state. This is because it itself draws on that power in persuading others to do what it wants: tax concessions, subsidies, various forms of duty, the threat of fines. But it is also a power that anyone participating in a market, or for that matter trying to purchase the favors of a legislator, draws on. The power of money is expressed among other things in what urban analysts call the competitive bidding process. The wealthy, by and large, live in the more desirable neighborhoods because they can afford to: they have the money to outbid other would-be purchasers.

Likewise there is the power of the normative. Norms are important in regulating family life and much else besides. It isn't just the power of money that makes us punctual for work; the fear that we will be fired if we don't turn up on time. We have been socialized into it from early childhood on: "do not be late for meals," "do not be late for school," "hurry, or you'll miss the bus." Again, this is something that the state can turn to its own advantage. It is a form of power that it employs through the schools. It is through the educational system, both state schools and the private schools – that are always regulated by the state – that certain rules of good citizenship are imparted. And through its public statements, if not always through its actions, it advocates the ideal of equality as a principle of social justice.

Yet in talking about the state and its relation to various forms of social power we need to bear in mind that the state form is not a universal. There have been societies which lacked states. Some of these exist at the present time in, among other places, the jungles of Amazonia or Borneo. And in many other so-called states, particularly in less developed countries, the power of the state, its ability to penetrate and regulate social life, is weak indeed.

But having said that, a case can be made for some sort of regulation in *all* societies. Government with the intent of harmonizing the activities of different people one with another has been an omnipresent feature of all social life: the household, kinship, and the various norms accompanying them, for example. And indeed today these regulatory mechanisms continue to play a role alongside more historically recent ones like the market. But what is characteristic of the present era is the role of the state as, in effect, the regulator of regulators: as the ultimate guarantor – and limiter – through the law, of the social power of others, whether that of capitalists, husbands, and parents, or that of money in the abstract. In other words, there can be government without states; but states always entail government.

Territorial strategies are *always* exercises of power. To some degree they may depend on the direct exercise of state power: redrawing the catchment districts of schools so as to simultaneously include some and exclude others; or assigning additional police patrols to a neighborhood. Sometimes, on the other hand, strategies appear to be more private in character. This would apply to the gated community or the private school, both of which can have exclusionary intent. But ultimately they both depend on the state. Gated communities have to be legal, as do private schools. And even if private schools are legal the state can take steps to make them more or perhaps less attractive as territorializing options through the sorts of tax concessions it makes to parents (i.e. whether or not school fees are tax deductible).

But what is attractive about the state as a means of regulating space relations, as a vehicle for the various exclusionary and inclusionary policies different organizations, firms, political parties, residents' organizations push for, is its own territorial character. Consider the variety of possibilities here. Imagine, for example, a state whose power was not territorial in the sense of areal and bounded. What if (e.g.) people who were the citizens of different states were not as they are now, geographically segregated one from another, but geographically *integrated*? Imagine a situation, in other words, in which your next door neighbors, other people living in the same city or region as you belonged not to the same state but to different states: that American citizens lived in the same neighborhood alongside French, German, British, Mexican, Australian, Nigerian citizens *and* they were all subject to the laws of their respective countries.

While on the one hand this might have its advantages – it would make warfare a very difficult enterprise, for example, since "friendly fire" victims would be at least as numerous as enemy dead – it would also make the implementation of other, less lethal, territorial strategies highly problematic. An interest in remedying something like acid rain in response to the demands of people downwind of factories and power stations with high sulfur emissions would be extremely difficult to bring about. This is because it would involve so many independent sovereign powers in multiple, many sided, negotiations with one another: a high level of geographic fragmentation of power where what is needed to remedy the situation is a spatial centralization of power. In other words, what is required is states that respectively enjoy uninterrupted sovereign power over large, continuous areas that in terms of their shape are relatively compact: neither punctured, highly elongated, fragmented, nor indented (figure 1.1). And of course it is precisely towards the latter compact form that states in their jurisdictional geography tend. This is what makes them so appealing to those promoting territorial strategies of various sorts: it promises some sort of resolution of conflicts, though not necessarily in favor of them or their particular territorial projects as opposed to those of others.

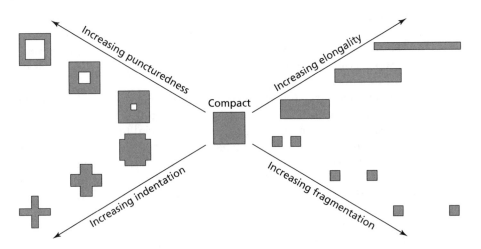

Figure 1.1 Deviations from compactness along four dimensions. Consider the compactness or otherwise of state forms and those jurisdictional subdivisions like the Canadian provinces, British counties, French *départements* and US states in terms of these different dimensions.

Source: P. J. Taylor (1971) "Distances within Shapes: An Introduction to a Family of Finite Frequency Distributions." *Geografiska Annaler*, 53B(1), 43. © Swedish Society for Anthropology and Geography.

Think and Learn

In talking about the compactness of state jurisdictions I used the term "tend." Think of exceptions to the compactness rule. What states *are* elongated, punctured, indented, or fragmented? How would you judge the US or Canada in these regards? Peruse a world atlas in order to identify these deviations.

Significantly, the territorial principle is writ large in the geographic structures of states. The internal organization of the state includes a division into local and central branches and sometimes branches at a more intermediate level (regional or provincial governments, for instance) and these all tend to the same compact form, as a scrutiny of the geometry of the States of the US, the counties of the United Kingdom or the *départements* of France would quickly confirm. The territorial principle likewise extends to representation and to many state policies. The constituencies or Congressional districts that legislators represent are discrete, bounded, relatively compact areas. Compactness is viewed as a virtue to the extent that any serious departure from it is likely to be viewed with suspicion: as signifying, that is, some attempt to manipulate boundaries in order to guarantee a particular electoral outcome.[5]

5 So-called "gerrymandering".

Within state jurisdictions there are yet other partitions that relate not to representation but to actual policies: the land use zones of local governments; the Special Areas of the United Kingdom designated for assistance in attracting new employment; conservation areas, historical districts, urban renewal districts, Areas of Outstanding Natural Beauty, etc.

None of this is accidental. One can say that it is this which makes the state so important to those with territorial interests. But it also reflects the significance of territoriality as an organizing principle of social life. People have territorial interests that they share with at least some people in the same area and which bring them into competition and conflict with those elsewhere. If these interests are to be expressed then it makes sense to organize elections through territorially defined voting districts.[6] And if they are to be satisfied, then some policies at least should be territorially differentiating.

So it is important that the state's organization be through and through territorial: that there be local as well as central branches; that legislators represent geographically discrete districts; and that there be, for some policies at least, ways of making their incidence geographically differentiated in some way. This is a state in short that is appropriate to the expression and realization of interests of a territorial nature.

But a territorial form that facilitates the expression and realization of one territorial interest may be less satisfactory from the standpoint of others. Just as state policy is a stake, therefore, as people, firms, labor organizations, and so on struggle for policy outcomes enhancing to their neighborhoods, regions, industrial districts, and countries, so too is *the structure* of the state itself. We will see later that a major issue dividing people, firms, and other organizations has been the internal organization of the state in its territorial aspects: the degree to which, that is, the state should be a highly centralized one, one which reserves few powers and responsibilities for more local branches, as opposed to one that decentralizes a good deal of its power to more local or regional levels. Recently this has come to the fore in the UK with the implementation of some devolution of power to Scotland, Wales, and Northern Ireland. It is also of ongoing significance in debate about the future form of the European Union. But it is not just the territorial organization of the state that is contested. Modes of representation, how territorial they should be, have often surged to the fore as an issue. In the US Senate each State has two Senators regardless of population; so representation is by State rather than proportional to State population. But in Canada there is no such equality between the provinces. Not surprisingly, perhaps, the less populous provinces there are pushing for a US-style Senate.

The social process and political geography

What is lacking from this picture is some sense of what energizes the political process in a geographic context. It is not enough to refer to territorial inter-

---

6   Though we will see later that this is actually not a universal in democracies.

ests and projects in the abstract. They always have some substantive content. They are interests in particular things, practices, relations. Ultimately, as I remarked earlier, our interest has to be in relating to the material world: in harnessing its naturally occurring substances and forces in order to realize our changing needs for sustenance, shelter, affection, creative expression, etc. This is why no human science can ignore the relationship to nature, including, of course, our own nature. But this relationship is always socially mediated. It is always in and through our relations with others that we relate to nature (as in production, narrowly conceived) and our own nature (as in the socialization process). So our needs assume socially mediated forms. In the advanced industrial societies of today they become interests in profits, wages, property values, trade, labor, and housing markets: in other words interests in categories that only make sense given the existence of a capitalist society, and that are *entailed* by it.

Other stakes are less obviously related to capitalist development and the material objectives of those participating in it. These include demands as diverse as upholding the national honor, protecting particular landscapes from development, recognizing favorite daughters or sons by creating national holidays in their honor, or controlling the activities of white policemen in black neighborhoods. All these seem a little remote from money making and distributing among various claimants the wealth so produced. What ties them together is in part the symbolic: actions that recognize, accord respect (or disrespect, perhaps, in the case of the white policemen). What are at stake are less objectives of an instrumental nature (achieving them as a means to an end) but ones that are more consummatory in character, that by their very writing into law perform an important symbolic role for some people.

On the other hand, these different types of demand are not unrelated either. Struggles for recognition are often prosecuted through mobilizing the power of money. The recognition of Martin Luther King Day has been an issue in a number of the American States. One of the ways in which blacks and white liberals have sought to achieve their ends has been through influencing the location of national conventions. In other words, if Arizona refused to recognize Martin Luther King Day then various professional associations threatened to move their conventions, with all their implications for local hotel and restaurant trades, to cities in other States. Similarly, in South Africa the black boycott of white stores became a favored tactic in the dying days of apartheid.

Conversely, in more clearly economic struggles, struggles in which the ultimate stakes are ones of wages, welfare benefits, etc., leverage of a more moral sort may be resorted to: "our rights as British citizens" fairly cries out for recognition not on instrumental grounds but as an end in itself.[7] In the United States blacks struggle for improved life chances through the educational system. One of their arguments is that they are disadvantaged through the

---

7   In other words, policies should be such as to recognize certain claims because to do so imparts respect and dignity to the people making them – as if the increased public benefits are a mere unintended consequence!

cultural bias of educational testing instruments, i.e. that those who are different are being marginalized, being treated unfairly. So cultural struggles are often conducted, at least in part, through the exercise of economic leverage and vice versa.

But what does this imply for how we approach questions in political geography? Does it mean that in thinking about the economic and the cultural, the material and the symbolic, and how they articulate with the politics of space, we should see them as *independent* sources of social power, as substitutable one for another depending on circumstance? There are several quite crucial points to bear in mind here.

First, any social process, indeed any action we perform in society, has a diversity of aspects. It is, for example, both material and ideal. A necessity of our existence is that we have to relate to the material world. We have to transform it into usable forms and then consume or experience the product. But in order to relate to the material world, to produce, to consume, or whatever the material practice is, we have to have some idea of what we are doing: how to cultivate, how to operate, how to cook, how to assemble. On the other hand, practice is a precondition for our ideas. It is in terms of those material practices that our ideas about them change and, for example, new technologies are developed.

Likewise, action is invariably both individual and social. People are irremediably social creatures. They depend on others for (e.g.) the systems of communication like language through which they acquire ideas about nature and how to appropriate useful things from it; they depend on others through a division of labor. This socialized nature of what we do does not mean to say that we can read off individual thoughts and actions from a knowledge of forms of communication and the division of labor. These change and it is people who do the changing. They develop new modes of communication, new metaphors, perhaps, new roles in the division of labor. But they always do these things using the raw materials provided by the existing division of labor and existing forms of communication. Nothing is totally novel. So while people are indeed creative and can change things, can make a difference, if often only to infinitesimally slight degrees, they do not do it out of nothing. The resources they draw on are social in character and available to others, though perhaps not in the same sort of mix.

Finally, the social process is always cultural, always political and, one might add, always spatial. Culture enters in the form of the meaning systems through which we are able to interact meaningfully with others and with the material world in general. It is always political because some invariably have power over others by virtue of (e.g.) some skill or knowledge lacking but important to others. And it is always spatial because it requires connections over space with others and (again) the material world in its entirety. If we want to interact with others we have to get close to them. If we need water we need to move in the direction of the tap or the water fountain.

We can, in short, think of social processes in terms of mutually presupposing parts, though without consigning those processes to stasis, to stagnation, to always reproducing what was there before and in the same forms. There *is* change. People are inventive, they come up with new ways of doing things; but only through a contact with the material world that is invariably socially mediated. Likewise this idea of mutually presupposing parts, how the material entails the ideal and vice versa, how the material practice entails the social and vice versa, should not lead us to the view that that is all there is to their interrelations. Some things are more fundamental than others, some aspects are more conditions than they are conditioned. As Marx and Engels (1845–6, p. 48) famously remarked:

> we must begin by stating the first premise of all human existence and, therefore, of all history, the premise, namely, that men must be in a position to live in order to be able to "make history." But life involves before everything else eating and drinking, a habitation, clothing and many other things. The first historical act is thus the production of the means to satisfy these needs, the production of material life itself. And indeed this is an historical act, a fundamental condition of all history, which today, as thousands of years ago, must daily and hourly be fulfilled merely in order to sustain human life.

The development of social and intellectual life as we know it today would have been impossible without a relation to the material world of a particular sort: one of control and the harnessing of natural forces to productive purposes, that enables people to be productive on a virtually heroic scale. Without it there would be no schools, universities, opera, libraries, foreign holidays, modern medicine, pensions, and so on.

One can, of course, retort that that development of productive abilities has in turn depended on particular social configurations, particular ideas and insights, and that is true. But not *any* meanings, social relations, power relations are effective in this regard. In organizing a hunt we would not give the role of coordinator to someone who had never hunted before. And mobilizing the power of steam or aerodynamics to productive purposes depends on getting the equations right.

The fact is, the material world – the world of physical, chemical, and biological objects and forces – has its own ways of acting. Material objects, as diverse as pylons and people, have their own powers and limitations. As far as people are concerned, it is by virtue of our own nature, our own material nature, that we can develop ideas,[8] new social forms, and have, unlike other organisms, social and technical revolutions. But our nature is also limiting. These powers have to be deployed, as the quote above indicates, towards satisfying our material needs. As we develop new material needs so this necessity reasserts itself in new forms. And how we go about satisfying those material needs in turn depends on the nature of the material world outside us.

8 If you disagree with this emphasis on the material conditioning of thought consider what happens to a person's powers of cognition and of thinking when the material character of the brain changes, as with Alzheimer's or a tumor.

But this is to talk in very general terms. It applies to social life anywhere and everywhere. While the material is primary quite how it all works out depends on more concrete forms of social life and these change over time and space. Again, this is not to argue that all social forms are possible. They have to be such that material needs can be satisfied. Of those concrete social forms capitalism is one. We live in a capitalist society and this has distinct implications for the social process and therefore for the political geography of the contemporary world. Capitalism is, in fact, thoroughly consequential.

In the first place under capitalism the different aspects of the social process, the material and the ideal, the cultural, the political, etc., are separated out and *seemingly* take on lives of their own as independent forces. But only apparently. So, for example, some of our material relations, in particular those that require commodity exchange, are reconstituted as something that we start calling "the economic." One important consequence of this is that what goes on in the household is not defined as "economic." Housewives work – they cook, make beds, launder and a whole variety of other material practices – but since they don't get a wage for their work they are not defined as part of the economy, except, of course, when they make forays to the supermarket and purchase things, i.e. enter into commodity exchange.

Alongside the idea of a distinct economic sphere arise notions about the political, the cultural, the spatial as independent areas of social life with their own distinct logics. Likewise we come to see the material as separate from the ideal, as in books with titles like Great Ideas that Changed the World.[9] In part this is a consequence of the division of labor subsequent to capitalist development. There are, for instance, not only assembly line workers who supposedly work only with their hands but also scientists who, it is believed, do their work with their heads. There are captains of industry who are classified as part of the economy but also politicians whose specialty is power and using it. Likewise the cultural appears in a (again, seemingly) separate form as art museums, folkways, ethnic groups with their own languages and practices, newspapers and the media, literature, and so on.

This appearance of separation, however, is misleading: everything we do involves both a material practice and some idea of what we are doing whether we work on an assembly line or in a research laboratory. Likewise art museums have their politics as much as corporations do and they also depend for their continued existence on a healthy economy. But things do *seem* to take on a life of their own and give credence to the view that there is a culture separate from an economy which is separate from the state which is separate from technology and other material practices and so forth. Indeed, the state may well act as if the economy didn't matter and the economy as if space relations were of no consequence. But the unity of these different aspects of the social process will – necessarily – reassert itself: states will go bankrupt as will firms in the "wrong" locations.

This suggests that the active, structuring process, what holds things together, what integrates, what drives the social process forward is capitalism

9   For example, Robert B. Downs (ed.) (1956) *Books that Changed the World*. New York: Mentor Books.

and its agents. In order to reproduce itself, to endure, capitalism requires, for example:

1   A state that works for it rather than against it: legislating a body of labor law that allows profits to be made, facilitating the provision of a physical infrastructure of highways, railroads, cities, airports, and the like.
2   A set of cultural understandings celebrating the virtues of money making, hard work, private property, and "progress," and denigrating improvidence, idleness, and lack of (a particular sort of) ambition.
3   A geography which enhances productivity through speeding up the circulation of capital through its various phases, bringing together those who need to work together, minimizing the time in which materials are being transformed into useful states.

Capitalists may not take the lead directly in structuring the world thus, in attempting to reduce everything to its money-making logic. Indeed the state may take the lead. But the state can only act within the constraints defined by capitalism as a particular form of social life. The state needs money to do what it does but it can only mobilize that particular form of social power to the extent that it promotes capitalist development. Thus in capitalist societies it is only through capitalist development that the state can appropriate its revenues.

This is not to say that the agents of capitalist development, the investors, and those state officials who work alongside them, create symbolic and cultural worlds, distributions of power, and the like as they see fit. Capitalism emerges in a world that is already differentiated in many and diverse ways: culturally, geographically, historically, politically. Its agents have to work with what is available as they try to accomplish their ends: what antagonisms they can exploit, what alliances they can form, who can be seduced in their cultural battles through the power of money, what forms of organization they can orchestrate in such a way as to give them competitive advantages vis-à-vis firms elsewhere. And so it goes too for their working class antagonists. As they (e.g.) struggle for a larger share of the product or even for an alternative way of organizing production, they try to mobilize the symbolic on *their* side appealing to the need to protect "American jobs" or the dignity of the working *man*.

The position of this book, therefore, is that the logic of capitalist development, its attempt to subordinate everything else to its purposes and logics, including culture and the state, is central to understanding the political geography of the contemporary world. It is around these attempts that struggles over space, over the territorial, ultimately revolve. It may not always appear that way. It may, rather, seem that struggles around (e.g.) the symbolic character of particular spaces have an autonomy. But that autonomy is always limited. It has to be consistent with the logic of making money to make more money. Spaces can be set aside as (e.g.) Arctic Wildlife Refuges or National Parks. But if there is a sense that oil lurks underneath then nothing will be sacred.

Any approach to political geography that has aspirations to balance has to consider the cultural alongside the economic, the moral alongside the material: struggles that are seemingly more cultural, like the women's movement – struggling against the marginalization of women in social life – and struggles that are seemingly more economic, like the labor movement. But the relation between the two always has to be borne in mind. They never exist independent of one another. They are intertwined. But it is through understanding the logic of capitalist development, its attempt to subordinate everything to its logic, how it exploits particular configurations of power that may appear *either* cultural *or* economic, that we can ultimately bring the two into a fruitful and illuminating relationship.

Case Studies

We have seen that movement and fixity are central to political geography. As far as the fixed are concerned movement can be both fact and possibility. As such it can be both threatening and enabling. And it can be threatening to some and enabling to others: in which case the precise form that territoriality will take will depend on who is able to prevail, who is able to mobilize the powers of the state on their respective behalves. If it is those for whom movements are enabling, then the attempts to influence what happens in particular areas will be more inclusionary than exclusionary and vice versa. Should, for example, national policy be one that dismantles tariff barriers and immigration controls? Should a local government change its zoning policy so as to shift the balance of land uses in the direction of more rather than fewer apartments? This is not to argue that the movements are necessarily coming from "outside" as these examples might suggest. They can also be coming from "inside." A brain drain can be a threat to a national economy. The same applies to the movement of money "offshore" subsequent to an economic crisis.

Movement has always been with us, as has the fact of settlement and fixity. It has accordingly elicited various forms of territorial response. But over the past 200 years or so many of the movements that affect us in our daily lives have increased, and in at least two senses. They have tended to extend their geographic reach; and they have also grown in their magnitude. The labels on the products that we habitually consume are enormously expressive of the way in which we are now connected to people and places scattered across the globe. The wines from Chile and Australia, the shoes from China, the French cheeses, the tuna from Thailand. Movement moreover is lubricated by the shrinkage of space brought about by increasingly speeedy forms of transportation and communication (figure 1.2); and that speed has also brought down the real cost of movement.[10] The increasing distances over which people

10 Consider, for example, the sheer cost of the meals and accommodations required on an ocean-going liner compared with the needs of people traveling on jet planes; little wonder that trans-Atlantic travel has increased so dramatically over the past 40 years.

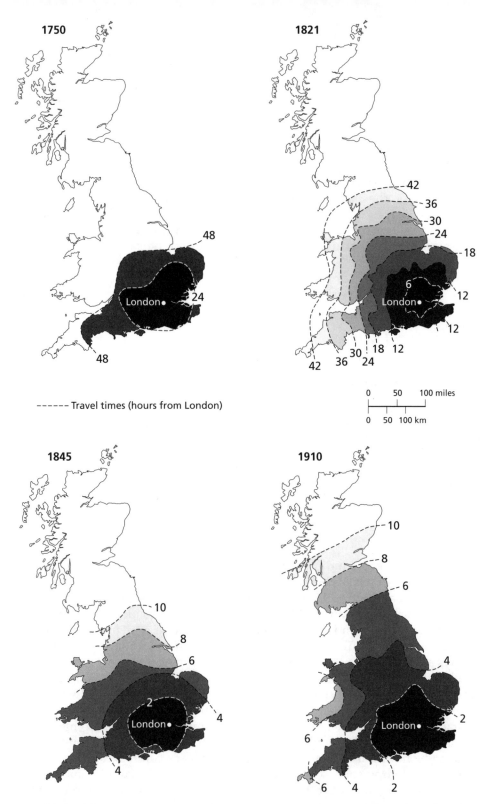

Figure 1.2 The shrinkage of travel times in Britain, 1750–1910. The two maps in the upper panels refer to stage-coach journey times; those in the lower panels are for railroad journey times. Note the dramatic shrinkage of times, allowing increasing movement over longer distances, that occurred.

Source: Figure 16.2 (p. 456) and Figure 16.8 (p. 463) in N. J. Thrift (1993) "Transport and Communication 1730–1914." In R. A. Dodgson and R. A. Butlin (eds) *An Historical Geography of England and Wales.* London and New York: Academic Press.

go on vacation, as well as their increasing numbers, tell a similar story. Thirty years ago the only cities in the United States with direct airline service to London were Chicago, Los Angeles, and New York, but now there are at least a dozen cities with such links.

There are other indicators with which we will be familiar from everyday life: the proliferation of Chinese and Indian restaurants in the cities of North America and Western Europe; the increasing diversity of their populations; the spread of the fast-food chain McDonald's throughout the world; the variety of images of other places and other peoples to which people are exposed through the media, especially television; the increase in marriage across international boundaries. There are also ecological effects: the emergence of a "dead zone" offshore from the delta of the Mississippi is testimony to the huge amounts of fertilizer that run off farmland throughout the river basin and are transported downstream; the appearance of air pollution downwind from large metropolitan areas. And then, of course, there is global warming.

Think and Learn

We talked earlier about the importance of capitalism and capitalist development to an understanding of the contemporary world and its political geography. The increasing reach of movements and their increasing magnitude coincide with the rise of capitalism as *the* way of organizing production. Do you think that this is mere coincidence? How do you think that the increasing reach and magnitude of movements might be related to the capitalist form of development?

As these examples indicate, this increasing geographic reach, the increasing magnitude of what is being conveyed/moved over space is something that is apparent at all geographical scales. The current interest in globalization has tended to focus attention on the growth of trade worldwide, the expansion of foreign investment, and the movements of people in search of jobs, though the novelty of this can surely be exaggerated (Hirst and Thompson, 1996). But always accompanying these changes have been changes at other scales. At the level of the nation state, for instance, one can point to the increasing penetration of the state into everyday lives: the way in which government within the household is displaced by government through the state. The rise of child welfare officers, the redefinition of the disciplining of children as child abuse and as something to be regulated by the state, the recognition of spousal abuse as a problem and a similar tightening up of state controls, and the earlier growth of compulsory schooling are all indicative. Likewise more local branches of the state have tended to see their power shift to more central branches as the latter, for example, become more responsible for providing the money. There are yet other changes of an economic and cultural sort: the displacement of the local provider by chains (the fast-food chain versus the local hamburger joint, the chain hotels), the decline of minority languages and even dialects; and the increasing distances over which people have been able to move in their daily lives, using the bicycle, the train, the bus and car, the airplane.

Think and Learn

We have been talking about the increasing geographical reach of movements and their increasing magnitude by reference to more global and national scales. How do you think these arguments apply to metropolitan areas? Would you expect there to have been similar changes there? What sorts of movements might exemplify this?

In terms of political geography, and in particular the geographic structure of the state, the effects of this have been contradictory. There has been territorial *integration* and there have been tendencies also towards *disintegration*. On the one hand we can point to the extension of jurisdictional boundaries, the emergence of new territorial structures at larger scales which come into being in order to facilitate movement and the advantages it can bring. The most striking recent example of this has been the European Union (EU). The control that member states once had over trade regulations has been ceded to the European Commission in Brussels, creating an area within which commodities are free to move from one country to another. The justification for this was the classic free trade argument: that it would induce increased competition, heightened specialization, and therefore increased efficiency, lower prices, greater prosperity. The recent adoption of a common currency (the euro) by most of the members has promoted this goal still further by eliminating the currency risk that exporters typically face.

There are other inter-nation arrangements around the globe also aiming to dismantle trade barriers among members – the North American Free Trade Area or NAFTA uniting Canada, Mexico, and the US is one example, and the free trade area linking Australia and New Zealand is another. But none have such far reaching goals as the EU. One should also point to the European empires as earlier attempts to capture the advantages of moving commodities over long distances; though clearly the distribution of those advantages was geographically highly uneven between imperial power on the one hand and the colonies on the other.

There have also been disintegrating effects, however. Decolonization and the breakup of the Belgian, British, Dutch, French, and Portuguese empires produced a massive increase in the number of individual states during the period from about 1950 to 1980. Since the ending of the Cold War there has been another burst of territorial fragmentation. The most obvious expression of this has been the breakup of the Soviet Union, creating the independent states of Ukraine, the Baltic states of Estonia, Latvia, and Lithuania, and those of the Caucasus (e.g. Armenia, Azerbaijan) and former Soviet Central Asia (Turkmenistan, Uzbekistan, etc.). But the former state of Yugoslavia has given way to five independent states. In Italy the northern region has threatened to break away and form its own state. In the UK substantial devolution to Wales, Northern Ireland, and Scotland is under way, raising questions as to how long the United Kingdom will indeed remain "united."

All of these partitions and fragmentations have effects on movements; and that, of course, is the point. They can, for example, lock up fiscal resources: prevent the leakage of taxes to populations elsewhere. The drive for an independent Scotland, for example, gains energy from the prospect of diverting the highly lucrative severance taxes imposed on the North Sea oil industry centered in Aberdeen to exclusively Scottish use. They can alter the geography of civil service appointments so that cultural minorities are no longer administered by those defined as alien: Croats by Serbs, Estonians by Russians, and so on. And they can protect infant industry from "foreign" competition and so nurture a native capitalist class.

These are tendencies, moreover, that apply at many different geographical scales: not just the more global with its nation states, empires, and free trade areas but also the metropolitan. Within every metropolitan area there are competing tendencies towards greater integration, greater centralization of power, and tendencies that would enhance the power of constituent local governments or even neighborhoods. So at the same time that metropolitan areas move towards a common water and sewerage system, a metropolitan airport authority and transport authority, for example, there may be counter movements: demands from neighborhoods that decisions on land use rezonings be delegated to them. And there will be resistances: resistance to the creation of larger, metropolitan-scale school districts that would change pupil compositions in ways seen as threatening by some.

So whatever the outcome – integrating or disintegrating, unifying or fragmenting over space – there is always a struggle over what is good for "our neighborhood/city/region/country" since some are likely to benefit from particular territorial projects and others will lose. This is because of what those projects will do to various movements of people, commodities, tax resources, school pupils, etc., and what the implications of those movements are for different social groups: for employers as opposed to workers, for black parents as opposed to whites, for the middle class as compared with those of lesser means, for those who see themselves as paying more in taxes than they receive in government benefits versus those who see themselves as net beneficiaries, and so on.

To illustrate these points I now want to move to some examples. These are very different and have been selected in order to illuminate different aspects of the argument set forth above. We will look, for instance, at the struggle around acid rain in the US because of the way it underlines the importance to politics of our relation to nature. The notion of a separate state of Padania in Northern Italy is important because of the way it shows how what is disintegrating at one scale (that of Italy) has integrating effects at a larger scale, since it was seen as facilitating the fuller integration of Northern Italy into the European Union. The discussion of Chinese migrants to Vancouver and the subsequent politics, on the other hand, brings together the cultural and the urban.

Case study 1   The breakup of Italy?

The notion of an independent state of Padania in the north of what is presently Italy, leaving a rump Italian state in the south, is an idea that has generated increasing attention over the past ten years or so. It is the brainchild of a political party known as the Northern League. The forerunner of the Northern League was the Lombard League, which was founded in 1981 and which changed its name in 1992. Both the Lombard League and the Northern League have drawn their support from what is the most modern and prosperous part of Italy. And that support is not insignificant. In 1992 they got 25–30 percent of the vote in Lombardia and landslides in neighboring Piemonte and Veneto where it became the second largest party. Even so, and despite secessionist demands, it is not clear that this is the ultimate goal of the movement. Another of its proposals, for example, has been for greatly enhanced regional autonomy. This idea envisages dividing Italy into three regions, North, Central, and South. Each would have its own parliament with responsibility for income tax, health care, and education. The suspicion is that if the Italian state could be restructured to the advantage of the North, then that would be the end of the matter.

The appeal of the Northern League is to a hostility towards three closely intermeshed objects: Southerners, the South (figure 1.3), and the Italian state. The South of Italy is the backward part of the country and has been for a long time. In 1988 unemployment was a relatively meager 7.7 percent in the North but 21 percent in the South. Gross regional product per capita also varies. That in the South is barely two-thirds of that in Italy as a whole but as a result of transfer payments from the rest of the country – largely the North – the standard of living enjoyed by Southerners was much closer to the national average (83 percent in 1987). This uneven development has been the condition for two processes impacting on Northerners.

In the first place there has been considerable migration from the South to northern industrial cities like Turin, Bologna, and Milan. This has in turn engendered hostility on the part of Northerners, not so much through fear of job competition but more because of perceived cultural differences. There is, for example, a fear of mafia influence and concern about the intrusion of the "kidnapping industry," based in Sardinia and Calabria, into the North. There is also a broader sense of difference rooted in the urban–rural contrast between Northerner and immigrant. The fact that they occupy the lower levels of the occupational hierarchy in the North and so are more vulnerable to layoffs has also made them an easy target as lazy and welfare-dependent.

Second, uneven development has been the motivation for various programs aimed at improving living standards in the South: programs aimed at enhancing economic development there and, if that should not suffice, redistributing

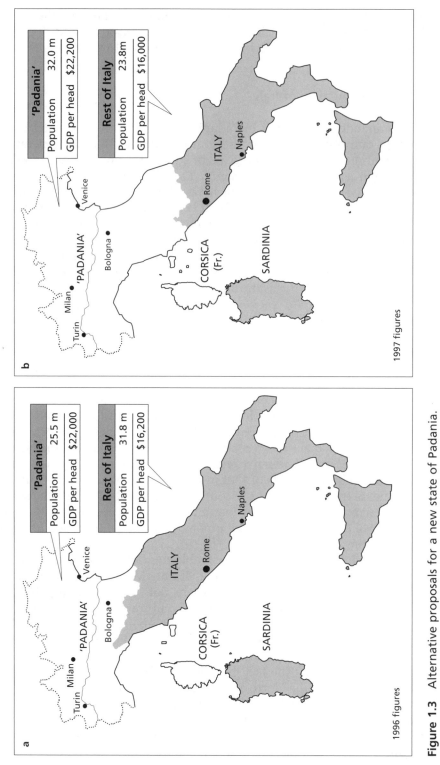

Figure 1.3 Alternative proposals for a new state of Padania.

Source: **a**, after *The Economist*, March 29, 1997, p. 58; **b**, after *The Wall Street Journal*, September 13, 1996, p. A8.

Figure 1.4 An example of the Northern League's anti-Southern propaganda. Note the sense of territorial exploitation depicted by this cartoon: the Northern hen is exerting considerable energy in laying eggs that go to benefit the South, as represented by the peasant woman. The inclusion of the word "Roma" underneath her figure identifies the Roman bureaucracy as part of the problem as well.

Source: B. Giordano (2000) "Italian Regionalism or 'Padanian' Nationalism – The Political Project of the Lega Nord in Italian Politics." *Political Geography*, 19(4), 462. Copyright (2000) with permission from Elsevier Science.

income. The perception of many Northerners is that those programs have been at their expense (figure 1.4). This is part of the reason for hostility to the Italian state since it is seen as the mediating force in these efforts at Southern uplift. But it is not the only reason. The Roman bureaucracy is also despised for its inefficiency and corruption. This in turn is linked to the bogy of the Southerner since the Italian civil service, particularly in the lower echelons where it comes into contact with the Italian public – postal workers and railroad employees, for example – is manned disproportionately by Southerners (up to 80 percent according to some accounts). Accordingly the Northern League is characterized by strong hostility not only to the South and to Southerners but to the central state itself, which is seen as fiscally oppressive, corrupt, and inefficient. Small businesses complain about the stifling bureaucracy and bad public services while tax rates on individual and corporate incomes are indeed high by international standards. The dominant feeling is one of exploitation to the advantage of the Roman bureaucracy and the South.

On the other hand, there are also important conjunctural elements that help to explain the appeal of the Northern League. In the first place the end of the Cold War and the Soviet threat has meant changes in the Italian party system. The Communist Party has lost some of its grip on the Italian working class; while the party that for a long time dominated the Italian right wing, the Christian Democrats, has lost its credibility as a shield against communism. This has created an opening for the Northern League.

In the second place there has been the attraction of the European Union. Italy is a member of the EU but membership of the latest phase of its development, European Monetary Union (EMU), required meeting criteria that business people in Northern Italy feared would be difficult to achieve given the Southern incubus. An attraction of the EMU is that while Italian firms can borrow more cheaply within the EU than they can in Italy, membership of the EMU eliminates the currency risk.[11] In the event Italy did indeed meet the criteria through reducing its budget deficit, but with the South, and the demands made by the South on the national budget, it looked for a while as if this might be difficult.

As a national movement the drive for an independent state of Padania is unusual. This is because typically national movements are not just driven by motivations of the clearly material sort that are driving the Northern League. Rather there is commonly a strong sense of difference with respect to non-nationals: a sense of difference in terms of culture, history, senses of belonging and solidarity that go to create a distinct national identity. But there is little or no Padanian identity, apart from some minor differences of dialect, and little hint of how it might be constructed: no distinct history, no political entity such as a provincial subdivision of Italy, for instance. On the other hand, in material terms many in Northern Italy clearly believe they would be better off with a state of their own. And the fact that Italian nationalism is weak is also a factor that works in their favor.[12]

Case study 2   States and the federal government: the acid rain issue

In Italy the response to what is perceived to be some regional disadvantage has been to propose a restructuring of the territorial structure of the state,

---

11   The risk, that is, of (e.g.) borrowing German marks but having to pay back at a rate of exchange adverse to the borrower owing to inflation and a devaluation of the Italian lira.
12   There is an old saying that little holds Italy together apart perhaps from its soccer team (*The Economist*, November 8, 1997, p. 13 of Special Supplement on Italy).

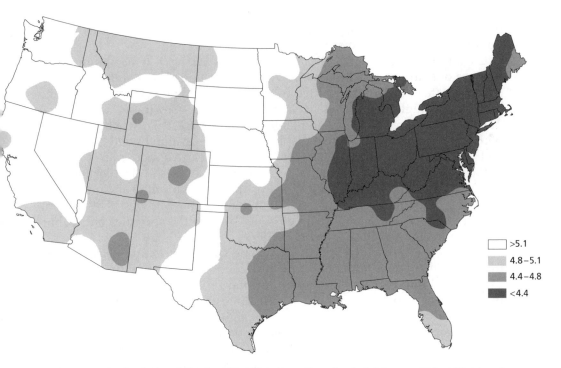

Figure 1.5   Acid rain levels in the US, 1994. Unpolluted rainfall has a pH (acidity) level of 5.6. Readings below that indicate the presence of acid rain, and the lower the figure, the greater its intensity. Note the relatively low readings across the Middle Atlantic and Northeastern States of the US. These were the areas from which greatest pressure was brought to bear on the federal government to do something about the problem. The result was the Clean Air Act of 1993. And indeed if you check the same map for 1999 (http://nadp.sws.uiuc.edu) you will see that there have been reductions in intensity in that area. Note also that one of the biggest offenders in the production of acid rain, Ohio, had very low pH readings, presumably as a result of the burning of high sulfur coal in States further west like Illinois and Indiana.

Source: National Atmospheric Deposition Program/National Trends Network (http://nadp.sws.uiuc.edu).

though whether that restructuring should take the form of fragmentation or federalism is undecided. In the case to which we now turn the territorial structure of the state, through its capacities for controlling movements, has also been to the forefront. But here the movements to be controlled are not those of people or tax resources but of air pollutants. And for the affected regions to have broken away from the United States, in the way that some in Italy support the idea of an independent Padania, would have served no useful purpose. This is because it would have impeded their ability to control the acid rain at its source: a source that lay elsewhere.

Acid rain is something that seriously afflicts areas in the northeastern part of the US, including New England and the States of the mid-Atlantic seaboard, including New York and Pennsylvania (see figure 1.5). It has a number of dele-

terious effects on ecosystems significant to people. It tends to kill off fresh-
water fish, for instance, and can also kill forests. It has implications for the
hunting, fishing, and timber industries therefore.

Chemically it is a highly dilute form of sulfuric acid. It results from the
mixture of sulfurous particles with water vapor aloft. These sulfurous particles originate in the smoke emitted from factories and power stations burning
fuels with a high sulfur content. In the American cases these factories and
power stations are found, by and large, in the Midwest. Their effluent mixes
with water vapor which is then displaced by the prevailing winds to the east
and northeast where it condenses and falls to the earth as acid rain. The view
of the environmental and economic development lobbies in the Northeastern
and Middle Atlantic States, therefore, was that the solution to acid rain lay in
some sort of regulation of smokestack emissions in the Midwest. This became
the core of various policy initiatives going back to the seventies and culmi-
nating in the federal Clean Air Act of 1993.

The earlier approach had been to try to persuade the Midwestern States to
undertake their own initiatives aimed at controlling high sulfur emissions.
This proved impossible. State Environmental Protection Agencies, recognizing the substantial costs it would impose on respective State economic bases,
simply dragged their feet. Attention then shifted to the federal arena where
those States could be forced to act. In short, the anti-acid rain forces mobilized
those branches of the American state where they would have most leverage.
And this in turn was to result in a further centralization of environmental
policy at the federal level in the form of the Clean Air Act.

The Clean Air Act calls on the major sources of high sulfur emissions to
simply reduce them. The major sources targeted have been the power stations
burning high sulfur coal. The possible solutions are twofold. The first is that
the electric utilities install so-called scrubbers in their smokestacks: these take
most of the sulfurous particles out of the smoke prior to it leaving the smokestack. The second solution is to burn low sulfur coal. Which solution to adopt
has been a major source of contention in the States most affected: Illinois,
Indiana, and Ohio. The difficulties have been several-fold.

The scrubbers are expensive and this would result in higher electricity
prices in the States affected. This in turn would have impacts on major electricity users and on the ability of State development departments to attract in
new businesses. Major electricity consumers in particular, like the automobile
industry, have been important sources of pressure for the power companies
to adopt the low sulfur coal solution. The problem with this, however, is that
the high sulfur coal that is burnt comes from the Midwestern States in question. To cut back the consumption of high sulfur coal in favor of the low sulfur
variety would result in serious unemployment in respective coal mining
industries. As a result the coal mining companies have lobbied in favor of the
scrubber solution and against low sulfur coal.

There are other pressure groups involved from outside the Midwest and
quite apart from the Northeastern and Middle Atlantic States affected by acid
rain. Low sulfur coal is an important export for some Western States, in particular Colorado, Montana, and Wyoming. They all supported the Clean Air

Act of 1993 because they believed it would increase the demand for their low sulfur coal. At the same time they have opposed any policy that would increase the likelihood of the power companies installing scrubbers.

The Midwestern States affected, like Ohio, pushed for federal subsidies to defray the cost of the scrubbers, i.e. a federalization of the cost of the legislation alongside a federalization of regulation. This was vigorously opposed by the low sulfur coal producers and respective States which benefit from severance taxes on the coal extracted. In this, moreover, there has been some overlap with the interests of other Western States, like California. Many of these States have relatively high electricity rates. The high sulfur coal burning States of the Midwest, on the other hand, have often enjoyed relatively low electricity rates and this has redounded to the benefit of their economic development initiatives: it has made them more attractive for some firms (than, say, California) as places in which to invest. Federal subsidies for the installation of scrubbers would do nothing from their viewpoint to "level the playing field." They too, therefore, have opposed a federalization of defraying the costs of clean air.

Another approach by the high sulfur coal-producing States has been to give financial incentives to the utilities if they continue to buy in-State coal. This does not mean that the utilities can ignore the Clean Air Act; only that the scrubber option becomes financially more palatable. But this too has been opposed by the coal producers of the States of Colorado, Montana, Utah, and Wyoming operating through their lobbying association, the so-called Alliance for Clean Coal. The basis of their case is that this interferes with the commerce clause of the US Constitution which proscribes barriers to trade between the States.

In this case study ecological issues mingle with the economic, all within the framework of strong territorial interests. Movements are at stake and they are not all channeled by those natural pathways that convey the acid rain. Rather they are structured by market gradients in contrast to those of the world's atmospheric pressure differentials. So alongside attempts to block the effluents causing acid rain at their source there have also been initiatives designed to exploit the new markets that were seemingly in prospect, to the extent that the effluent producing States actually opted for the low sulfur coal alternative. And of course, as we have seen, the States producing low sulfur coal took the necessary legislative steps, or more accurately legislation-blocking steps, that that would require. Clearly in this instance, at least, and in contrast to the common image, environmental action may have its market appeals!

Case study 3 The monster houses of Vancouver[13]

If ecological effects are increasingly felt "at a distance" and to a substantial degree, so too might it be said of cultural effects. The movement of people around the world creates juxtapositions that can be the source of strong

---

13   This discussion is based on Mitchell (1993).

exclusionary sentiment as cultural prerogatives, deeply embedded senses of superiority and power, get challenged.

To some degree these effects are closely bound up with movements more of an economic character. Countries, regions, and cities compete for capital on what seems like an increasingly global stage. And in some instances those who own the capital move with it. Such was the case of the wealthy Chinese who invested in, and set up residence in, Vancouver in the late eighties and early nineties.

Anticipating an increased anxiety level on the part of the Hong Kong Chinese as reunification with China in 1997 loomed closer, Canada set out to attract the wealthy as immigrants. Starting in 1984 a business immigration program aimed at this group was initiated. In exchange for a higher process-ing priority for immigration they were required to bring in a certain amount of money and commit to investing some of it. In 1991, for those moving to British Columbia (and usually therefore to Vancouver), they had to have a minimum personal worth of C$500,000 and promise to invest C$350,000 in a Canadian business over a three-year period. In Vancouver this legislation was acted on with vigor by a mix of local government officials and businesses that stood to benefit – like developers and banks – as a means of boosting the local economy. The result was a considerable inflow of wealthy Chinese into the city. In the single year of 1988, for example, British Columbia was the pre-ferred destination for over 300 immigrants, and the vast majority of these would have ended up in Vancouver.

But although they were eagerly solicited by local business interests and local government in Vancouver there have been tensions, particularly with more elite elements of the Anglo-Canadian mainstream in the city. Most of the tensions have focused on the housing market, though the initial stimulus for disquiet was neighborhood change. There is, for example, a historic Chinese presence on Canada's Pacific coast and one which has always been marked by separate areas of residence. The new immigrants have not observed this norm and this has been an affront to the identities of the old Anglo elite. Historically racial separation has been seen as part of what it means to be a member of the Anglo ruling class. Adding to the sense of cultural threat have been the "monster houses" favored by many of the immigrants. Often with large extended families, they have sought appropriately large houses. But the way in which they have gone about this has been to purchase property in the more desirable neighborhoods, demolish the existing structure, and replace it with something that consumes a much larger proportion of the lot's surface area: hence the sobriquet "monster houses." This has served to further disrupt the sense of cultural integrity of the Anglo elite as their architecutural tastes have been quite different: mock Tudor, for example, in an ample garden with numerous trees. And in addition to neighborhood change there have been wider housing market effects. Housing prices in the Vancouver area have accelerated considerably and this has sparked concern on the part of residents that their children would not be able to afford to live in the city.

The response of the Anglo elite to these threats has been various. One has been to call for new restrictions on the proportion of the lot area that a house can consume, in order, one assumes, to inhibit the purchase of property by the Chinese in choice neighborhoods. The other, of course, is to stop encouraging the Chinese to come and live in Vancouver. But both of these actions are problematic from the standpoint of those who want to see the Chinese comfortable about living in Vancouver and continuing to bring their money and invest it in the city. In other words a struggle over the nature of Vancouver, what and who it should consist of, has been joined between the old elite concerned about their neighborhoods and those business elements who want to see the city's economy grow.

The local growth lobby, for instance, has been anxious to see the Hong Kong link preserved. It has therefore been closely associated with think tank research into the housing market demonstrating that the housing price rise has nothing to do with the arrival of the Chinese but that it is the baby boom that is responsible. The other part of the counter-offensive has been to claim the moral high ground. Accordingly the opposition has been branded as racist. There has also been an attempt to align their position on the Chinese with the idea of multiculturalism which had been embraced earlier by the Canadian state and which was designed for purposes quite other than encouraging the immigration of the wealthy with a view to boosting local economies. This is the idea of equal rights under law but respecting the fundamental differences of individuals that stem from diverse cultural and "racial" backgrounds. But as Mitchell (1993, p. 265) comments, "the attempt to shape multiculturalism can be seen as an attempt to gain hegemonic control over concepts of race and nation in order to further expedite Vancouver's integration into the international networks of global capitalism."

Summary

The central focus of political geography, the point from which it starts and to which it returns, is defined by the twin concepts of territory and territoriality. Neither of these can be understood apart from each other. In order to talk of territory one must talk of territoriality and vice versa. Territoriality refers to actions designed to exercise control over some area: the territory. Territory and territoriality, therefore, bring together the two concepts of space and power: geography and the political, as in political geography. Accordingly, in order to understand territory and territoriality we need understandings of relations over space and of politics.

Territoriality is rooted in the contradiction between movement and fixity. In order to carry on their various activities people seek some fixity in their lives. They "settle" in particular places, become embedded in them through (e.g.) the relatively permanent transformations they make to the immediate environment (draining the land, cutting down the forests, building houses, creating tracks) and through the relations they develop with other people: relations of kinship, friendship, cooperation. But there are wider movements which either underpin or threaten these place-bound activities. These include natural movements like those that convey acid rain and socially mediated ones like those of trade. There are also displacements of population which can result in the threat of invasion and dislodgement, and threats to a place-based identity as we discussed in the case of the Chinese immigrants to Vancouver. To protect the place-bound relations that they have created, therefore, people in particular areas seek to control the movements in and out of them by defending, excluding, including; in short by regulating this wider set of movements to local advantage.

The notion of power, on the other hand, is closely bound up today with that of the state. Most of what we talk about in this book will have to do with the state for in the contemporary world the state is an extremely important regulatory agent. This is not to say that it has been a universal of human existence. There have been stateless societies. But there have been no societies that lacked means of regulating their activities. Even today, regulation cannot be reduced to the state. But the state is now the ultimate regulator which either regulates directly or regulates the regulations of others.

The particular forms of regulation that we are most interested in in this book are those of a territorial sort. We look to the state to control the content of those areas important to us. In this regard the state is a highly appropriate vehicle, not just because of its regulatory power, but because of the territorial form of its jurisdiction: the tendency for it to regulate within relatively compact spaces so that it can indeed arbitrate between (e.g.) neighbors, or the movements coming from some place within its jurisdiction and impacting on others elsewhere. This is not to say that its territorial form is uncontested. Those who feel oppressed by the state may want to see it divided so that they can take control of one of its fragments: this is the goal of the Padanian project we reviewed in the case studies.

Rooting territoriality in the contradiction of fixity and movement, of course, serves to underline what territoriality is ultimately about: maintaining a relation to the material environment that will facilitate the realization of human needs. But we never deal with human needs in the abstract, but always with human needs as they are socially mediated. And so too is it with the activities through which we relate to the material world: they also are socially mediated since it is only in and through our relations with others that we can appropriate naturally occuring substances or forces. In brief, our powers and our needs with respect to that material world are *social* powers and needs. Our forms of production are social forms and our needs are socially defined. In today's world people's activities are coordinated through markets and they seek wages, profits and rents: the categories, in other words, of a capitalist society. Central to the notion of social process that we will draw upon in this book, therefore, is capitalism and capitalist development.

All of the case studies reflect these different concepts and how they interact one with another: the concepts of territory and territoriality, as structured by the tension between movement and fixity, the state and capitalism. The movements are diverse: acid rain in one instance, tax resources and people in another, and in the case of Vancouver, wealthy immigrants. In each case people have stakes in particular places that they seek to defend against the threats implied by these movements: investments in forests in the case of acid rain, a local status for the Anglos of Vancouver, businesses and jobs in Northern Italy. In every case the movements are mediated in some way by capitalist forms of development – the desire of the booster lobby to attract inward investment, the need of the Midwestern power utilities to minimize their costs, the importance of being part of European Monetary Union for the businesses of Northern Italy. And finally, of course, the state is central to the action as it unfolds.

The Organization of the Book

The remainder of this book is divided into three major sections. Parts I and II take up the distinction made earlier in this chapter about the economic and the cultural, matters material and matters having to do with identity and feelings of significance. Part I addresses the economic, or more accurately the politico-economic: why, by whom, how, and where governments are mobilized in order to intervene in the production of economic geographies. The first chapter of this section, chapter 2, provides a general background of political economy as context for what is to come. Chapter 3 focuses on workplace issues, in particular those of economic development. Chapter 4, on the other hand, is concerned with issues we encounter in the living place: issues of schools, home values, housing availability.

Part II takes up the issue of Difference: how it is that we come to differentiate Others, how we define each other socially, how the state is implicated in this process of definition, how in short it is a state for Some rather than for Others, and the implications that has for identity and struggles around identity. Chapter 5 is the counterpart for this part of the book to chapter 2: it addresses the question of social definition and the politics of Difference in general terms, trying to establish some principles that can be applied in the two subsequent chapters. Chapter 6 applies these ideas to the formation of nations and nationalism, and why a sense of nationhood is so important to people. The final chapter of this section then examines from the same viewpoint some Differences that have become quite central to politics more recently – those of race and gender.

Part III examines political geography more explicitly from the standpoint of the state. For the most part, the first two sections of the book take the state for granted. It is part of the background. It is an organization that various interest groups and social movements mobilize in order to secure their ends. The state also has its own effects on those struggles. It endorses particular social orders. It redistributes, not least geographically. But quite why states should exist is bracketed. In chapter 8 we address the nature of the modern state and why it has the features it does: features that make it attractive to groups struggling to achieve ends of a territorial character. What we find is that the modern state is a necessary condition for that development, and its territorial character facilitates solutions to the territorial dilemmas that capitalist development confronts. Chapters 9 and 10 then explore two particular dilemmas confronted by states when they are placed in a geographical context.

Chapter 9 examines the politics of geographically uneven development from this standpoint. Chapter 10 then looks at the politics of scale, a politics that is expressed most clearly today in what has become known as the politics of globalization but which, as I hope to show, has also been generous in the illusions which it has spawned.

REFERENCES

Hirst, P. and Thompson, G. (1996) *Globalization in Question*. Cambridge: Polity Press.
Marx, K. and Engels, F. (1845–6) *The German Ideology*.
Mitchell, K. (1993) "Multiculturalism, or the United Colors of Capitalism." *Antipode*, 25(4), 263–94.
Sack, R. (1983) "Human Territoriality: A Theory." *Annals, Association of American Geographers*, 73(1), 55–74.

FURTHER READING

An excellent reading for a person coming to political geography for the first time is Doreen Massey's (1995) "Making Spaces: Or, Geography Is Political Too." *Soundings*, 1 (Autumn), 193–208. The essential source on the questions of territory and territoriality is Robert Sack. Read in particular his paper referred to in the references. Writings on the state are voluminous and none of them especially accessible to the novice. Especially interesting to geographers, however, because of the links he makes with territory, is the work of Michael Mann. See in particular his provocative 1984 statement "The Autonomous Power of the State: Its Origins, Mechanisms and Results." *Archives Européennes de Sociologie*, 25, 185–213.

For further reading on the Northern League and Padania consult: B. Giordano (2000) "Italian Regionalism or 'Padanian' Nationalism – The Political Project of the Lega Nord in Italian Politics." *Political Geography*, 19(4), 445–72; and J. A. Agnew (1995) "The Rhetoric of Regionalism: The Northern League in Italian Politics." *Transactions, Institute of British Geographers* NS, 20(2), 156–72. On the monster houses the paper by Katharyne Mitchell listed in the references is well worth reading.

Part I

Territory and Political Economy

Chapter 2

The Political Economy of the Contemporary World: Fundamental Considerations

Introduction

No discussion of geography and political economy that aspires to coherence and comprehensiveness can afford to ignore one crucial fact about the times in which we live: the dominant role of capitalism as a way of organizing production. It has proved itself to be *the* great motor of development. Nothing so far encountered has shown itself nearly as capable of harnessing the forces of nature, raising the productivity of workers, and improving material standards of living, though that does not mean to say that it might not, and nor does it mean that its outcomes have not been highly uneven, including geographically. The ending of the Cold War has led to a retreat from socialist economic strategies around the globe and a search for so-called market (i.e. capitalist) solutions but that does not license a belief in "the end of history" (Fukuyama, 1992). For as well as promoting the growth of productivity and raising material standards of living capitalism is also conflict-ridden. Tensions are omnipresent, particularly between workers and owners. The search for alternatives is far from over. The struggle to define the meaning of life and modes of social organization appropriate to realizing that meaning continues. This drama of development and conflict is played out on a stage substantially differentiated geographically. This geographical differentiation structures both the course of development and the conflicts that subsequently arise. Geography is used as a weapon in the pursuit of conflicting goals and is itself shaped so as to provide advantage in those conflicts.

In this chapter I start out by a discussion of what capitalism means in terms of its structure and what it necessarily entails. I point out the conditions that must obtain if there is to be capitalism and I derive its necessary consequences in terms of competition, technical development, and conflict. Competition occurs between all commodity owners: between businesses, between workers, and between businesses and workers. The subsequent conflicts do not always mirror these competitions, however; the conflict between business and labor is the most significant of all of them. The second part of the chapter places these processes in geographic context. In particular it shows how and why space enters into the conflicts generated by capitalism; how, that is, both businesses and workers, individually and collectively, try to exploit the advantages of geography in order to achieve advantage in the struggle over profits and wages; and how at the same time this transforms the geography with respect to which future struggles must take place.

Understanding Capitalism

As I have noted, if we want to understand why things happen in the contemporary world a good place to start is with capitalism. It provides the incentives to developing people's productive capabilities but in a conflict-ridden manner. Its distinctive feature as a way of organizing production can be stated quite briefly: it is *the production of commodities with commodities.*

Under capitalism everything that enters into the production process is bought as a commodity: markets mediate production. This includes labor power or the ability to labor: this is bought and sold, or more accurately rented, in markets.[1] Capitalism cannot exist without labor markets and cannot, therefore, be reduced to the presence of exchange in a society. Only if exchange extends to labor power can one reasonably talk about capitalism. Moreover, once money has been laid out for raw materials, machinery and tools, and labor power the only way the owner of that money, the capitalist, can get it back is by selling the finished products. He or she, therefore, is caught up in an endless but necessary sequence of commodity exchanges: entering into exchange in order to obtain the necessary conditions of production – raw materials, tools, and labor power – and then exchanging the finished product in order to retrieve the sums originally laid out and so obtain the money with which to start production over again.

Preconditions

The preconditions for capitalism are historical. I have just talked about the necessity for a labor market. Labor markets are by no means given. They are historical creations and, when viewed against the total span of human history hitherto, prove not to be omnipresent features of social life. In order for labor markets to form the immediate producers have to want to work for a wage. If they have access to the land and tools with which they can produce their own means of subsistence it is unlikely that they will. This is because they have an alternative means of gaining access to food, drink, shelter, and the like. A first precondition for capitalism making its historic entry, therefore, is the separation of immediate producers from the means of production. As we will see, this separation is often of a forcible, violent kind.

---

1   Not "bought and sold" since that would be tantamount to slavery.

A second precondition comes from the fact that immediate producers must be *free* to work for a wage. Slaves are not free since they are owned in their persons by someone else. Serfs are not free since their lords have a claim on part of their labor in the form of labor services and possibly the product of their labor, and can restrict their movement away from one particular fiefdom to another. Capitalism, therefore, presupposes the abolition of slavery, serfdom, and other forms of servile labor.

Third and finally there have to be people with sums of money sufficiently large to: (a) employ workers who both want to and are able to work for a wage; and (b) purchase the means of production that have been separated from the immediate producers. This money may have been accumulated through trade or through lending. But henceforth it can be reproduced and expanded through the labor of others.

Competition

Capitalism has had important consequences. One of these is competition in all its varied forms. Under capitalism production decisions are private: it is capitalists that decide what to produce and they do so without consultation with each other and without any other sort of purposeful coordination. But this means that there is no guarantee that the product can be sold. It is taken to market with a view to sale but it may not be sold, or at least not at a price sufficient to cover the outlays made for its production. It is only after the fact of exchange that capitalists can adjust their production – what they produce, how they produce it, and how much they produce. Uncertainty reigns and it is an uncertainty the resolution of which is of a highly consequential kind. This is because it can mean the difference between staying in business and going bankrupt and hence being forced into the ranks of the wage workers.

To stay in business, to make sure that it is your product that is sold and you are not the one with irretrievable expenses, competition becomes a necessity. This assumes diverse forms but the most obvious are those of cheapening the product and developing it. Cheapening the product can involve capitalists in a search for more efficient technologies which can economize on raw materials or on workers: new machinery or new ways of organizing the labor process as in an intensification of the division of labor. Alternatively it may be that there are opportunities in the form of cheaper labor or cheaper raw materials elsewhere. Gaining access to these, however, may be conditional upon a development of the technology of transportation or of production itself, thus allowing unskilled workers to do the job where formerly it required the skilled. In these ways capitalism, through competition, develops the productivity of workers: their ability to produce.

A second competitive strategy is the development of the product. This may involve improvements in existing products as exemplified by the history of the automobile or the house. Or it may mean the identification, development, and bringing to market of entirely new products like the video player, artificial fibers such as nylon, or new drugs. Every product has a history and capitalism, through the spur of competition, has greatly expanded the variety of products and variations around a single product that are available to us. Capitalism stimulates, therefore, not just the development of technological capabilities but also the development of social needs. But these are unintended consequences. The goal is profit and as much profit as possible. Given the inherent uncertainty of markets, only through amassing profits can capitalists hope to survive: to put together the resources that will allow them to endure the vagaries of business and to invest in those new technologies and new products that will give them a competitive edge. This means pressures towards the investment of profits rather than their consumption.

Think and Learn

In popular discussions of the economy the greedy businessperson, out to drive a hard bargain in order to amass more wealth, often looms large. What do you think about this view? Does capitalist development occur because there are greedy people?

As suppliers of a commodity, their power to labor, workers are also subject to competition. The most fundamental way in which they compete with one another is through the wage they are willing to accept: all other things being equal you can be the one to be offered a job if you are willing to work for less than other candidates. But there are obviously limits to wage competition that stem from the need for a certain minimum standard of living. This is not to imply a physiological minimum of caloric intake or clothing since what is an acceptable standard of living is in part cultural. And cultural issues aside, the worker's minimum consumption needs have changed greatly over time. Today it is virtually impossible in the United States to be a wage worker without a car: it has become *the* necessary means of getting to work and for most people it is a case of "no car, no job."

Other strategies that wage workers resort to include making their labor power more scarce by developing it in the direction of skills in short supply. Much of this is anticipatory as young adults take night classes and university degrees in occupations or occupation-related skills that they believe are more in demand. But for many investment in training and retraining is ongoing.

Finally, and holding skills constant, workers shift from one geographic location to another or from one sector of the economy to another. Places and sectors develop unevenly so that labor scarcities and hence wages can show significant variation. In such a context workers can improve the wages they get by moving from lower wage places/sectors to ones where higher wages are available: geographic and inter-sectoral mobility.

The development of the forces of production

Production is a relationship to nature. In order to produce our material requirements – shelter, food, clothing, means of transportation, etc. – we work on naturally occurring substances or on ones that have been partially transformed by others: cotton thread instead of raw cotton, for instance. And in working we mobilize our own naturally given capacities for conceptual thought along with instruments of labor which are transformations of naturally occurring substances and forces.[2]

How effective a worker is in producing, how much he or she can produce in a given period of time, depends on many things. These include: the skill of the worker; the instrument of labor; the object of labor. Skill requires little elaboration. In talking about the instrument of labor there is clearly a difference from the standpoint of productivity between a bolt-tightener that is manually operated as opposed to one that is electrically driven. Likewise, workers producing pig iron in a blast furnace will be more productive if the iron ore has a higher iron content. These are what are known as productive forces. But these three productive forces – the skill of the worker, the instrument of labor, the object of labor – have to be brought together; productivity depends on how the worker employs his or her skill to use the instrument of labor in working on the object of labor. This process is what is known as the labor process. So the labor process also has to be included as a productive force.

Most labor processes are collective in character. The work needed in order to produce (e.g.) an electric cooker or a CD player is divided among several people or perhaps people in different firms that produce different components of the finished product. The object of labor proceeds through several work stations and at each one it is transformed in some way by the application of a worker's distinctive skill combined with an instrument of labor appropriate to the task: a hammer, screwdriver, lathe, die, etc. In other instances it is not individual labor that is applied to the object of labor in a succession of tasks but labor as a collective. In the hunting of large animals by indigenous peoples the cornering of the animal so as to drive it into a trap is done by several people who coordinate their work with each other simultaneously rather than consecutively. In these instances, while some of the coordination of the labor required is in the hands of the individual – even in the hunting case the individual has to know how and when to use his or her stick or stone – there also has to be coordination of the collective. In other words, into the division of the labor process – the division of labor necessary to produce a particular item – is inserted a new role: that of leader.

Over the span of world history, of course, the productive forces have undergone great transformation. People's ability to use nature in either its

2   For example, a worker may work with a machine that is driven by electricity and both of these represent "naturally occurring substances and forces." The machine is a set of steel parts and the steel is a transformation of iron ore and coal. Electricity comes from coal or from natural forces like falling water or, much more rarely, winds.

raw or transformed state in order to produce has consequently been greatly enhanced. People have become more productive. They have learnt how to produce more in a given period of time or the same in a shorter period. This development in turn depends on the way in which production is socially mediated. Production is a social process. It depends on the coordination of many different types of labor one with another, with many different instruments and objects of labor. This means it has to be regulated, coordinated. This is what is meant when it is said that "production is socially mediated." Essential elements of that social mediation are the rules governing access to property, and these are quite crucial to the development of the productive forces. Different property relations correspond to differences in the rate at which the productive forces develop because of the variable stimulus that they provide towards that development. And so far, of all the different property relations through which production has been organized by far the most successful in developing the forces of production has been capitalism. This is typically attributed to the way in which it enforces a regime of competition between firms that stimulates the development of worker productivity. In order to retain their position in the market for their product firms have to cut their costs and typically the way in which they do it is by reorganizing the division of labor, equipping their workers with improved machinery, so that worker productivity increases and, assuming that wages do not increase in tandem, the (labor) cost per unit product produced decreases.[3] Yet, and paradoxically given the centrality of competition to capitalism, this is dependent on what is known as "the socialization of production"; on developing the social character of production.

Think and Learn

The labor process is an omnipresent feature of human existence. Everything we do can be thought of as a labor process. Think about some of the labor processes you are involved in. How, for example, have the processes of studying been transformed over the past twenty years or so? To what do you attribute that transformation? Has the skill of the student increased, the instruments of labor changed, the object of labor? Or all three? What about some of the other labor processes you are habitually involved in, such as housework?

---

3 It might be objected at this point that firms compete not only through the adoption of techniques that facilitate worker productivity but also through the development of new products. That is true. But when capitalist development as a whole is considered the development of new products is subordinate to increasing worker productivity. This is because the drive to increase worker productivity is necessary if firms are to find markets for new products assuming the form of new machines and raw materials. And on the other hand, the addition of new products to the typical shopping basket of goods assumes that real incomes have risen: which they will to the extent that worker productivity increases.

Productivity develops through a deepening of the division of labor and hence the interdependence of one firm and one worker with another. Firms specialize in different things. Workers develop different skills. As specialization proceeds and with it the discovery of quicker ways of doing things, so too can the tools and machines with which people work be improved so that they facilitate worker productivity. But this deepening of the division of labor means increased social interdependence. Productivity advances through the enhancement of the social character of production. This specialization and the ensuing socialization of production extends beyond the immediate production of commodities to those activities that service the immediate producers. Production is separated from finance and transportation, for instance, and these become the responsibility of specialized firms with specialized knowledges and technologies. As in all branches of capitalist production, competition also develops the productive forces in *these* sectors – new and cheaper means of transportation or the ATM machine, for instance.

Think and Learn

The labor process is social in character. Does it have a geography? What are the implications of that geography for the productive forces? Is there a geography of the layout of the workplace that is important for worker productivity, that enhances worker productivity as compared with alternative layouts? How might it enhance worker productivity? Can you think of other "production geographies" at geographic scales larger than the workplace? Think of how different places specialize in different things; how might that facilitate productivity?

The development of the forces of production is not uniform. Workers develop their skills and their understandings of the labor process at different rates. Some are more productive than others, perhaps for reasons that are not completely evident, as any employer will affirm. To some degree this is a matter of basic literacy and numeracy: the ability to read instructions or to perform simple calculations on the job can make all the difference. So there are things that employers can expect of employees in more developed countries – for the most part at least! – that they could not possibly expect in countries where formal education is only available for a short number of years and where many if not most children will not go to school anyway. But the particular skills that workers develop, the specific technical capacities that they invest in themselves and which raise their productivity depend in considerable degree on the demands being made by employers.

Some firms develop their productivity at a faster pace than others. The number of person hours it takes to produce an automobile in the US varies considerably from one auto firm to another; so too does it when the figures are viewed internationally. This is not to say that a firm whose productivity lags behind that of others is necessarily teetering on the brink of bankruptcy.

Its situation may be one in which it can take advantage of compensating factors. A firm with relatively low productivity per worker may more than make up for this lag by paying unusually low wages. This means that firms in less developed countries may be able to survive in competition with those from more advanced countries even though their productivity is much lower. But having said that, the fact remains that the development of the productive forces is what "development" as we know it – the capacity to produce an ever growing stream of products without increasing the labor force – is all about. The workers employed by firms in more developed countries tend to be more productive than those employed by firms in less developed countries that produce the same product; which is why they can generally afford to pay their workers more.

Dilemmas of Capitalist Development

While stupendously productive, and while developing great social powers, the capitalist form of development is also one that is deeply problematic. It is, for a start, one characterized by social tensions of a remarkably intense sort, tensions that continually threaten to, and often do, break out into open conflict. Obvious here are the tensions and contestations between employers and employees around pay and work conditions. But equally there are those that have in the past led to imperialism and colonialism and, by so doing, generated still more tension and conflict and violence. There are also arguments that link nationalism and the rise of the nation state with capitalist development; and the nation state has, as we all know, been a vehicle for visiting immense oppression on others.

A second point here has to do with our relation under capitalism to the rest of nature: to other forms of life, to the earth, and, as intimations of global warming suggest, to the atmosphere as well. As indicated earlier in the chapter our relation to nature, including our own nature, is the most fundamental of all our relations. As those relations change, our very existence, certainly our health, our own natural forces, can be seriously compromised. Like any mode of production capitalism depends on natural forces and substances. It requires continual supplies of raw materials for the products it fabricates, continuing supplies of foodstuffs, shelter, and so forth for its workers. But some have suggested that the relation between capitalism and nature is a contradictory one; that its tendency is to undermine the ecological conditions on which it depends. It is argued that as a result of its tendency to the infinite development of the forces of production, the continual massing in ever greater quantities of various sorts of use value, the capitalist form of development imposes demands on nature that are just unsustainable. And while some forms of energy are, in effect, inexhaustible, there are many raw materials which are finite in their availability.

In short capitalist development throws up as by-products, and clearly unintended ones, both a social question and an environmental one. The response to these from advocates of the capitalist form of development as *the* form which development should assume has typically been by reference to some version of Adam Smith's hidden hand: that the only unintended consequences of capitalist development are wholly benign, indeed positive. This is the belief that by pursuing their own private interests people satisfy a wider social interest. Competition results in an efficient organization of production: it maximizes the product and minimizes the costs that have to be sustained in order to produce it. Similarly, in a more recent application of this doctrine, there is the view that market forces result in a harmony with nature in the same way as the interests of producers and consumers, capitalists and workers are harmonized; that as, for example, raw materials are exhausted so their increasing scarcity sends price signals to producers which result in the development of cheaper substitutes. As I now want to demonstrate, however, these defenses are not impregnable ones.

The social question

Not very far beneath the surface of societies in which commodities are produced with commodities lies a profound, pervasive fear. Capitalism is an angst-ridden, tension-generating, form of production. People can starve, they can go bankrupt, they can be the objects of the bailiff's daily round. The competition of firm with firm, of worker with worker, is no mere game. On it can ride the ability not just to live with oneself, to maintain one's self-respect, but for millions and millions of people in the world today to live at all.

Even so, and as I will argue at length later, while this is a problem for all commodity owners as they enter into exchange relations with one another, the most aggravated tension, the central one in capitalist societies, is that between the owners of productive capital, the capitalists, and those they employ, the workers. Social power is unequally distributed. And the biggest inequality is that between, on the one hand, those who own the money that is used to hire workers and the means of production, and on the other, the workers themselves.

Think and Learn

Bankruptcy is a common feature of life in the advanced capitalist societies. Think about someone you know who has gone through bankruptcy. What is your own explanation for it? Were they profligate in their spending? Did they drink or gamble away their money? Was the problem an illness that was extremely expensive to treat? Having thought about these issues, consider the following question. If there were no profligate spenders in the world, no gamblers, no alcoholics, no savings-threatening illnesses, no personal shortcomings or mishaps in other words, would there be no bankruptcy? What do you conclude from that regarding the relation between bankruptcy and the capitalist form of development? Is the bankruptcy of some an inevitable, necessary feature of such development?

Continued

Apply the same analysis to unemployment. Why do you think people are unemployed? Because they lack skills? Because they are poor timekeepers? Again, if we eliminate these as reasons for unemployment, and also allowing for temporary unemployment as people move between jobs, can we imagine a capitalist society in which there was no unemployment? Why do you think that?

The sources of tension are concretely quite various. But they all hinge on the way in which labor power becomes a commodity under capital and, as a result of competition, has to be treated as such by employers. The consequence for workers, variously experienced, is one of lack, or more accurately, "lacks": a lack of security, a lack of sufficiency, and a lack of significance.

(1) *Insecurity.* We have seen how under capitalism there is an intense impetus to the development of the productive forces. Typically this means that each worker, equipped with improved machinery, organized in more efficient work configurations, produces more and more. Production costs go down, which means that prices can be reduced. This puts pressure on other firms to respond in like manner. But as more and more is produced a ceiling will be reached in terms of the ability to sell the product; at which point some firms and their workers will be squeezed out. Or alternatively, in order to stay in the market they may fire their existing workforce and hire workers who are willing to work for less. Employment by any one employer is therefore an unstable experience. No worker is indispensable.

This might not be such a serious issue if workers could find employment with other firms. Indeed this happens. There is tremendous turnover in labor markets as workers seek better terms with other employers, as yet others are released, or leave of their own accord. But some of the turnover will inevitably be from the status of employed to unemployed. The unemployed, like the poor, are always with us. They may vary as a fraction of the workforce but capitalist labor markets and 100 percent employment cannot go together. This is because as labor markets become tighter wages will tend to increase and threaten profitability. Employers respond either by laying workers off and so increasing the supply of labor and lowering wages for others, relocating to where labor is more plentiful and therefore cheaper, or simply substituting machines for workers. And at times, unemployment has reached levels such as to pose quite serious challenges to social stability.

(2) *Insufficiency.* A common source of dispute between employers and their workers is, of course, wage and benefit levels. Workers want higher wages, improved pensions, and health care benefits. This is only partly related to the sense of insecurity, the desire to save for the proverbial rainy day, and for retirement. For while capital has a tendency to create a class of unemployed, it also has a tendency to increase the needs of workers and hence their need for income. Firms compete not just by improving levels of worker productiv-

ity and so being able to engage in price competition. They also compete through the introduction of new products and services. But new products, if they are to achieve the purpose firms had in producing them, then require that people have the means to buy them.

Often, though not always, they will indeed answer some vaguely felt need and demand will take off. Furthermore, what is initially seen as a discretionary item often tends to assume the status of a necessity: a first car, a second car, and, increasingly it would seem, a third car as well. For as more and more bought cars urban form changed in such a way, as the economics of mass transit became more and more adverse, that it became difficult to move around in any other way. So people feel the need for the money to buy these items.

Think and Learn

The automobile is a good example of how discretionary items often tend to get converted into necessities. Can you think of an example of your own? What is the mechanism, the logic, through which a discretionary good became a necessity? Is the residential suburb on the way to becoming a necessity?

But while firms are anxious to sell they are less anxious to provide their workers with the money with which to make their purchases. This is because each firm relates to its own workers as a production cost and not as a potential market; wages therefore have to be kept down. The potential market for the product is seen as consisting of the employees of all other firms. Since, however, every firm follows that logic the result is insufficiency relative to the consumption needs of workers and a feeling on their part that wages should be increased.

(3) *Insignificance.* People have material needs, therefore. But they also have needs for a sense of significance, respect, human dignity. Under conditions of capitalist production, providing for these needs is once again problematic. The employer purchases the worker's labor power and within the terms of the contract and labor law it is the employer's to use as she sees fit: in other words, in such a way as to maximize the firm's profitability. If in the employer's view workers need to be reassigned to new jobs in the plant, or new machines need to be introduced and workers retrained, then that is up to the employer. There is no need for the workers to be consulted on this. In fact consultation can pose problems since the workers may object to the changes planned, and for diverse reasons – health and safety, an intensified work pace.

The worker's power to labor, in other words, is treated as a thing no different from the coke and iron ore that go into the blast furnace or the blast furnace itself. If the worker's labor power could be separated from the worker herself this would be of no matter in other than material terms – the need, that is, for the worker to conserve her single commodity. The situation would be no different from that of someone hiring out a piece of equipment to someone else. But in fact the worker and her labor power are inseparable. The labor power can't be treated as a thing without treating the owner of it, the worker, as a thing. The failure to consult, the failure to take into consideration the worker's feelings as a human being, is therefore experienced as highly alienating.

There are variations in the degree to which these various "lacks" are experienced. Some feel more insecure than others, some more insignificant. Employees are treated differentially according to how much leverage they have over the employer. Those with skills that are difficult to replace will be favored. They will be retained even when business is lagging and others are laid off, simply because they will be hard to rehire once they are let go and possibly hired by someone else. The fact that they dispose of a unique knowledge and ability may also entail a degree of consultation, and therefore a sense of significance denied to the broad mass of workers. But the fact is, the tendencies are there. And as business constantly works to reduce its costs by making workers more and more substitutable one for another they are not going to go away.

Think and Learn

"... as business constantly works to reduce its costs by making workers more and more substitutable one for another." How *do* workers become more substitutable for one another? What is the role of the development of machinery in this process? Can you imagine a situation in which all workers were substitutable for each other? Given what you know so far of the logics of capitalist development, why might this be impossible?

This is not to say that the division between employer and employee is the only cleavage around which conflict can occur. There are also those that attach to the division of labor. This can be viewed from two angles: there is the social division of labor and there is the technical division of labor. In the former, labor is divided according to the nature of the final product or service: agriculture, chemicals, retailing, for example. In the latter labor is divided according to the function carried out in the particular production unit: the firm or the plant. These particular divisions might include, for example, managerial, technical, maintenance, and line workers. Clearly there are variations in the way both of these can be conceptualized. The division between white collar and blue collar labor falls under the technical division of labor, as do classifications according to skill. The social division of labor should also be viewed flexibly: broad rather than narrow definitions of products and services, for example, so as to include different branches of agriculture under the same heading, government services versus those provided privately, etc.

Both of these forms of the division of labor can be the source of social cleavage. A common categorization of the different branches of the social division of labor is as "sunrise" or "sunset" industries. "Sunset" industries are those which are facing static or declining markets and hence profitability; in the US today they would include much of the shoe and textile industries. "Sunrise" industries, on the other hand, are those whose markets are expanding and business prospects are buoyant: the various industries that are defined as "hi-tech" are obvious cases in point. Cleavage in this case is often apparent in disputes over trade policy: sunset industries want protection against imports from other countries, while sunrise industries are afraid that if that happens the excluded countries will lack the foreign exchange with which to buy their own particular products and services.

Think and Learn

What is a "sunset industry" in one country can be a "sunrise industry" in another. Think of an example. (Hint: where are most American shoes now produced?) Is there a relation between the fact that in some countries an industry is a sunrise one while in another country it is a sunset one? What is that connection?

The technical division of labor corresponds to variations in income and benefits. It is a hierarchical division with managerial elements at the top, unskilled workers at the bottom and technical, skilled, and semi-skilled strata assuming more intermediate positions. Much of the conflict here, latent or manifest, centers on the struggle for the more desirable positions: "desirable" in both material and status terms. This is a conflict, for example, that animates debates about equality of opportunity and affirmative action. It is also reflected outside the workplace. Parents seek out those school districts that they believe will give their children an advantage and then try to deny others the same advantage on the grounds that they will hold "their own children back." And once positions have been acquired there is further struggle over issues of taxation and public services. Those lower in the hierarchy seek more progressive forms of taxation which will shift the burden of financing public services away from them; while those onto whose shoulders the burdens would be placed protest at the injustice of it all, and how "they worked for what they have got."

To some degree cleavages around the social division of labor are reflected in party politics. At various times in the histories of the Western democracies parties have organized themselves around a division between agriculture and industry. In late nineteenth-century Britain this was true of the Conservatives, who were the party of the rural areas, and the Liberals who brought together both the so-called manufacturing classes *and* their workers.[4] In the United

---

4 Though the fact that a universal male franchise was not enacted until very late in the nineteenth century sheds some light on that. Prior to that time the franchise had been a property one. Significantly the introduction of a universal male franchise was quickly followed by the emergence of the precursors of the modern Labour Party and the ultimate demise of the Liberals as *the* opposition to the Conservatives.

States the resentments of family farmers and their subsequent hostility to the so-called urban monopolies of banking, food processing, and railroads have fed into party politics at various times. The Greenback movement which swept across the Midwest and South in the late nineteenth century is a case in point.

But the main party political cleavage is that between capital and labor. The typical way in which the parties are usually distinguished is in terms of where they lie on a so-called right–left spectrum. Right-wing parties, whatever else they may be in particular national contexts, tend to be the party of choice for business. They tend to stand, in other words, for such pro-business policies as limits on the power of labor unions, low taxation especially of capital gains, limits to the welfare state, freedom to set the prices of essential consumer goods like housing, transportation, foodstuffs, and health care, and privatization of everything, including education, health care, and housing. The support base of left-wing parties, on the other hand, will comprise the labor unions, the less well off, the wage earners. To the fore will be issues of union rights, improvements in health and safety in the workplace, limits to the length of the workday, more vacation time, expansion of the social safety net (unemployment compensation, health care, etc.), and public provision of essential goods and services like education, health care, and housing. It is precisely in these ways, with some national variations, that one can characterize the dominant cleavage in the democracies of Europe, Australasia, and North America: the British Conservative and Labour Parties, the German Christian Democrats and the Social Democrats, the Gaullist and the Socialist Parties in France, the American Republican and Democratic Parties, the Christian Democrats and Socialists in Italy, the Liberals and the Labor Party in Australia.[5]

This is not to say that there are neat splits in party support between capitalists and wage workers. Right-wing parties could never gain power if they relied only on the support of large stockholders, bondholders, and the owners of private firms. Rather, and of necessity, they appeal as well to the upper echelons of the working class: those located towards the upper end of the technical division of labor. For not only do these people earn more, and have anxieties about protecting their privileges from those lower down, they will also own items of personal property, particularly houses and some stock, and this makes them susceptible to arguments about the sanctity of private property rights. On the other hand, the support base of left-wing parties is not exclusively working class either. Some businesses may find it in their interest to support left-wing parties. The public spending that is often central to the policy plans of left-wing parties – new schools, new hospitals, more money for education and public health in general – has its own business constituency, including the construction companies, and the drug and textbook firms.

---

5   Though not so clearly in Canada where separatist (e.g. Quebec) and sectionalist (e.g. the Western provinces) feeling has worked counter to a clear polarization of the parties along class lines.

The environmental question

More recently we have become aware of problems of capitalist development that are less easily expressed in terms of tensions between capitalists and workers, employers and employees. Talk of global warming, the depletion of non-renewable resources, the loss of bio-diversity, over-fishing of the world's oceans, pollution, the so-called "throwaway society," and so forth has served to focus attention on what have come to be known as environmental issues. Since the 1960s the environment has become a major focus of debate and legislative redress, both within countries and at an international level. In some countries political parties departing from the old left–right spectrum have emerged around these issues: the so-called "Greens."

Capitalism is at issue here for a variety of related reasons. First is its huge propensity to the growth of production, a tendency fueled by the forces of competition. This in turn puts increasing demands on nature as a source of raw materials and energy and also as a resource for absorbing various wastes. Many of the resources drawn on are effectively non-renewable, like oil. Many of those that are renewable are consumed at a rate that exceeds their rate of replacement. The depletions of the world's oceanic fisheries and timber resources are cases in point. Second, there is capitalism's tendency to externalize its costs wherever and whenever it can. To the extent that its costs can be imposed on others elsewhere in the form (e.g.) of air pollution or on future generations through the costs of developing substitutes for those resources that have been depleted, they will be since that makes for enhanced profitability.

There are of course "answers" to these dilemmas. One is that the capitalist market, through its incentive framework, can take care of problems of depletion. As the ultimate reduction in the supply of oil and coal occurs, so their prices will increase. This will provide incentives for capitalist firms to develop the technologies through which non-exhaustible forms of energy can be harnessed: tidal power, solar power, wind power, for example. Similarly as gasoline prices increase so there will be incentives for commuters to shift to mass transit and for developers to shift to higher-density, energy-conserving forms of urban development. In other instances, it is argued, recycling will provide the answer. As the price of paper increases as the world's timber resources are exploited at a rate exceeding their rate of natural replacement, recycling of paper will become economically more attractive.

Think and Learn

Think of your own examples of how this market response to depletion and shortage might work. How might food scarcity result, through the price mechanism, in changes in the content of North American and Western European diets, for example? If newspapers became more and more expensive because of a shortage of newsprint, how might people's habits of keeping up with the news change? Given the increasing depletion of oceanic fisheries how is the market responding to keep people supplied with fish?

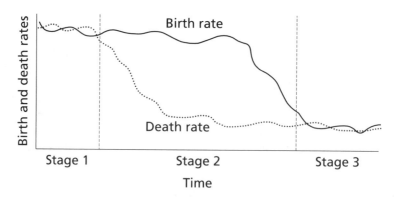

Figure 2.1   The demographic transition: in stage 1 high birth and death rates both result in a relatively static population; in stage 2 death rates decline but birth rates tend to continue at their previously high levels, resulting in high rates of population growth; in stage 3 birth rates also come down to match relatively low death rates; the result is a return to relatively static populations, albeit at much higher absolute levels.

A second answer is in terms of the better specification of property rights. Problems of air and water pollution and of the over-fishing of the world's oceans exist because of the status of the air and the oceans as common resources: resources over which there are no private property rights. Rather, since they are common resources they are treated as free goods and this means they are over-used. If, on the other hand, and so the argument goes, the state (e.g.) charged firms for the right to use the atmosphere as a dump for their effluents then they would adjust their technologies, the products they produce from pollution-intensive to pollution-extensive, accordingly. Likewise, as far as ocean fisheries are concerned the answer to over-fishing is a system of licenses to catch a certain number of fish and no more. The number of licenses sold and the catches they allowed would be set so as not to exceed the ability of the various fish species to reproduce themselves. But given the fact that both the air and the ocean are global in character, given that the movement of air and of oceanic waters is not controllable by any single national government, the authority charging for the use of the atmosphere or issuing licenses to catch fish would have to be international in character.

Even so, implementing this sort of "environmental" agenda makes heroic assumptions, not simply of achieving the sort of international cooperation that would make a reality of emission charges and oceanic fishing licenses but also with respect to the technological advances that would have to be made in order to (e.g.) mobilize forms of energy not subject to depletion. On top of that there is the whole question of population growth: the problem is not simply one of increasing consumption per person, it is also one of the increasing number of persons! The typical maneuver here is to appeal to the notion of demographic transition (figure 2.1). This is an empirical regularity discovered

by demographers based on the historical experience of the advanced capitalist societies. According to this pre-industrial societies are characterized by relatively high death rates and birth rates, with the end result that population
grows only slowly, if at all. With the beginnings of industrial development,
however, death rates are brought down through (e.g.) public health improvements while birth rates remain high: the results are the sorts of high rates
of population increase characteristic of contemporary India and, until very
recently, of China. But as societies move into phases of high consumption so
birth rates decrease and, as they approach death rates, total populations stabilize. But the problem is: at what levels? And given the assumption of transformation into a high consumption society, what does that imply for future
demands on the world's ecological base?

In class terms environmental issues have been hard to categorize. On the
one hand one can point to the tendency for the incidence of many environmental hazards – exposure to pollution, to toxic wastes – to exhibit a social
bias: typically those with money are able to buy into environmentally benign
neighborhoods while the most working class of neighborhoods often lack the
political clout that would allow them to successfully resist the location there
of polluting land uses. On the other hand, business has been effective in building coalitions with workers against environmental pressure groups on the
grounds that environmental legislation is a threat not just to profits but also
to jobs.

Significant in this regard is, indeed, the emergence of Green Parties in a
number of countries since this suggests that the environment cannot be a priority of parties either on the right or the left of the political spectrum. And
consistent with this is the social base of the Greens. This is because they tend
to draw on those social groups that are less directly dependent on the capitalist development process and who tend to be sheltered from its insecurities,
i.e. those in government employment of various sorts. Yet the more theoretical analysis above suggests that the environment matters to us all, and that
the logic of capitalist development is antithetical to the creation of a more
enduring, sustainable relation with it.

The Difference Geography Makes

We have seen that in terms of what is conventionally assumed to be politics
the central conflict of capitalist societies is expressed in terms of a right/
left party political spectrum: on the one hand those dependent on profits
and property income of various sorts, and the more affluent of the working
class voting for right-wing parties; and on the other hand, the less well-to-do
of the working class supporting parties of the left. Geographically this is manifest in highly predictable geographies of voting. In the British case the Labour
Party polls extraordinarily well in dominantly working class areas like the
former coalfields of South Wales, Northeastern England, and Central Scotland,
and also in inner city areas. The more "middle class" suburbs, on the other
hand, along with retirement centers for the affluent, like some of the seaside

resorts on the south coast, have tended to be Conservative Party strongholds. Similar patterns can be observed in the United States.

But there is also a more complex political geography in which class effects are refracted by issues of territory. This is a political geography which is only partially captured by geographies of voting. Two interrelated processes are at work here and both have their geographic expressions.

The first of these involves a dialectic of action and reaction between the opposing forces of the workers' movement on the one hand and business on the other. Workers seek to organize themselves collectively so as to enhance their bargaining power with employers. This is the rationale of the labor union. It is also the motivation for the formation of workers' political parties, like the British Labour Party, the French Socialist Party, and the German Social Democrats. The argument here is that such a party, once in power, would pass legislation that would put a floor under wages; legislation like a minimum wage or the introduction of a social safety net. But just as workers strive to structure their exchange relation with business in these ways, so business strives to find loopholes, strategies, escape hatches through which it can frustrate organized labor, and shift power in the bargaining relation in its own direction. The most obvious, to the extent that it is technically feasible, is to replace (expensive) labor with machinery: in short to develop the productive forces. The threat of (e.g.) replacing people as dishwashers with the mechanical variety is always part of the armory business draws on in its attempts to resist minimum wage legislation, for example. Alternatively a firm may hive off some of its operations, creating a new non-unionized offshoot to supply it with the parts it previously manufactured with organized labor. Or again it may bring its operations to a close, releasing its existing, unionized workforce; and then start up again with workers willing to work for a lower wage, like housewives, university students, or the elderly. This is emphatically not to argue, however, that the frustration of the labor unions in these ways is always intentional on the part of business. New firms are continually coming into being just as old ones are disappearing. Typically new firms start out with workers who are unorganized and for a while at least will be a challenge for the unions.

Once geography is brought into the picture the possibilities for business of eluding, deliberately or otherwise, the organizing efforts of workers broaden considerably. For the workers' movement, inevitably, tends to develop in a geographically very uneven way. There are *always* variations in the degree to which collective action occurs and, when it does, what it is able to accomplish in terms of labor law, income supplements, the level of the minimum wage, etc. In the United States union membership varies greatly from one State to another, and also within States: between major metropolitan areas and small towns, for instance.

So moving in the direction of lower wages and less militant, more pliable labor is often[6] a strategy for business as it attempts to elude the challenge

---

6 "Often" because some firms may find it difficult to relocate. Their needs for particular sorts of labor skills or access to markets may be an obstacle to this particular strategy, for instance.

of the labor movement. And to be sure from the seventies on there was a major hemorrhage of employment from the relatively unionized, high-wage American Midwest and Northeast in the direction of the weakly organized, low-wage South and small towns of the so-called Sunbelt. Unions have interpreted the investment of major multinationals in the Third World in similar terms. But equally, to the extent that firms are successful, deliberately or otherwise, in avoiding the protections workers have put in place in particular places the labor movement can be expected to take measures designed to protect its members. To the extent that workers elsewhere remain unorganized, unable to strike harder bargains with employers, then this puts the wages and benefits, not to say jobs, of unionized members at risk. Accordingly we can expect them to try to extend union protections to workers elsewhere so that the ability of firms to compete in terms of labor costs is voided and the security of workers' wages and jobs enhanced. To the degree that businesses have invested overseas in an attempt to avoid higher labor costs, and unionizing foreign workers is difficult for the labor movement, it may bring pressure to bear on the government to discourage that investment by (e.g.) imposing a stiff tax on repatriated profits. It needs to be added, however, that these efforts are by no means always successful.

Think and Learn

Think of the wide variety of manifestations of the dynamism of capitalist geographies of production. List some of them. Then consider the degree to which each of them can be explained in terms of the geography of labor cost. Can we always explain shifts in the geography of production in terms of the geographically uneven development of the labor movement?

The second process is that, if the labor movement is unevenly developed, and constantly shifts in its geography as it tries to cope with the changing geography of its adversary, so too is it the case with the productive forces. As a physical process capitalist development is geographically highly uneven. The skills necessary in particular labor processes are not found everywhere and neither are the raw materials. In nineteenth-century England and Wales the distribution of industry closely matched the outlines of the coalfields; this was because of dependence on coal-driven steam engines and the expense of transporting coal (figure 2.2). Similarly, despite the scare stories of the havoc that the establishment of branch plants by Western multinationals in less developed countries will wreak on employment in the more developed, the fact remains that it is extremely difficult to relocate those labor processes requiring high levels of skill since the skills are so geographically concentrated there.

The same goes for geography as a productive force. The productivity of labor depends on how it is combined with the instrument and object of labor.

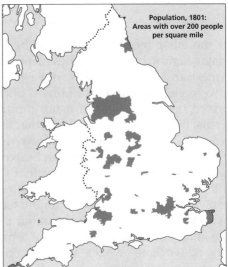

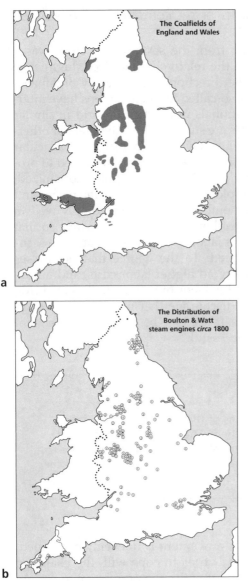

a

c

b

Figure 2.2 Coal, steam power and population distribution: England and Wales in 1800. There are clearly very close associations between these three maps. The invention of the steam engine revolutionized the location of industry in Britain and this had important effects on the distribution of population. Steam engines tended to be located on or close to the coalfields since the cost of transporting coal in a pre-railroad era was so high. Note, however, that ocean transport could make steam engines viable sources of power elsewhere; this is particularly evident in the use of steam engines in the Southwest, where they were used primarily in the tin and copper mining industry, and in London. London, for example, obtained most of its coal by sea from the Northeast coalfield.

Source: **b**, after Figure 94, p. 452 in H. C. Prince, "England *circa* 1800." In H. C. Darby (ed.), *A New Historical Geography of England.* Cambridge: Cambridge University Press, chapter 8. **c**, after Figure 83, p. 393, *op. cit.* Reprinted with the permission of Cambridge University Press.

Yet the labor process is social; and since it is social it necessarily has a geography. People work with each other. The labor process is elongated through successive stages in a division of labor. But elongation is obviously costly since it requires employing others for purposes of transportation. In terms of squeezing more out of fewer workers, therefore, geographic arrangement counts: which is one of the reasons why large metropolitan areas with their highly developed infrastructures and transportation systems are so attractive to many firms.

So the productive forces with respect to which capital locates, which it mobilizes to productive effect, are geographically uneven. But any particular pattern of unevenness is unlikely to be reproduced for long. Capitalist geographies turn out to be highly inconstant in their configurations. This is in part a result of the attempts of firms to elude the challenge of the labor movement, seeking out lower wages, a more favorable labor law regime, and lower taxes elsewhere. The exploitation of this uneven development may, however, be also due to the advantage that firms already located in low-wage areas can gain relative to firms elsewhere. To some degree this has been the story of the newly industrializing countries (NICs) of the Far East. But there is also a dynamism to capitalist geography that has sources less directly related to the uneven development of the labor movement.

For example, new growth areas emerge as new sectors of the economy are constructed around new products and services. As one looks at the changing geography of the British or the American economy the succession of sectors is clearly imprinted on it. The rise of the automobile industry led to growth in Southern Michigan and Northern Ohio in the American case and in the Midlands and to a lesser extent the Southeast parts of Britain. The emergence of hi-tech has brought in its wake new growth poles around Silicon Valley, Southern California, Seattle, and, in the British case, in an arc close to and stretching around London from the southwest to the northeast. These shifts occur at a wide diversity of geographic scales. Alongside the international and interregional changes illustrated here there is also a dynamism to the geography of metropolitan areas. This includes the burgeoning of suburban office employment as well as the closure of industrial plants in old urban cores (and often enough their relocation to the edge of wider metropolitan regions).

In their political implications, however, these shifts are far from benign. For as the geography of capital changes so the prospects of firms and workers in different places change, and not necessarily for the better. The establishment of more efficient firms using newer technology and possibly cheaper labor can be a challenge to older firms elsewhere. The emergence of new growth regions around sunrise industries threatens older ones that grew under the impetus of what are now seen as sunset industries. And as the productive forces change in their geography so too does the flow of value. Revenue flows to the more efficient firms in more efficient locations and away from those whose plant, worker skills (perhaps), and attendant physical infrastructures are becoming obsolete. Investment shifts in similar directions: away from the less profitable firms and localities towards the more profitable ones.

This is problematic for firms and workers in the areas which are being marginalized by these changes. For a start there are investments of long life embodied in physical facilities – factories, municipal infrastructures, docks, railroads, power stations, workers' housing – whose value may not have been completely amortized. Loans still have to be paid back, bonds retired, mortgages paid off, but the stream of revenue out of which to pay has now diminished. So while people and firms could relocate to where the new investments are occurring and take advantage of changes in the geography of the productive forces this would have to be traded off against these, often considerable, losses.

In addition to physical facilities, social infrastructures are also at risk. Firms develop understandings and relationships with other firms that are important to their profitability and which would be difficult to replicate elsewhere. There are place-specific knowledges that are productively important and which, again, depend on the continuing existence of a particular cluster of firms and workers and which would be lost with relocation elsewhere.

So with the changing geography of capitalism, profits and wages in particular places may be threatened and there are obstacles to making the sort of geographic adjustments that might restore them. Some firms, some workers at least, are likely to be trapped in space. The changing geography of value flow is a problem for them. This is the context, therefore, in which one can expect various forms of coalition, bringing firms and workers together, around policies designed to control those wider movements to local advantage: various forms of territorial coalition, in other words, of a cross-class nature which will lobby the state for remedial measures, or put in place policies designed to attract new investment into their particular regions and localities.

However, seemingly, at least, a problem with the discussion so far is that it refers almost entirely to what goes on in the sphere of production, the workplace. Yet obviously there is more to life and more to politics for that matter. There is, in particular, a very distinct politics of the living place involving residents' groups, school boards, city planning departments, zoning hearings, property values, rents, residential displacement, and the like. It is, of course, commonplace to see these two forms of politics as distinct and to be sure I am going to start out treating them that way. The next chapter focuses on the political geography of the workplace and the one after that on the political geography of the living place. This mirrors the separation between working and living that comes about with capitalist development. In previous modes of production the distinction had been more blurred. The home was the point from which the peasant organized production, the blacksmith worked at a forge attached to his house, and so on. With capitalist development, however, production is removed from the home and placed in factories. The farm worker lives in a house separate from the farm and so on. So it is not surprising that we have this sense of two different forms of politics and of political geography. Yet it is also necessary that we move to an understanding that breaks down this distinction, and this will be brought to the point towards the end of chapter 4.

For the time being, therefore, I want to stress that we can approach the political geography of the living place both as an aspect of that of the workplace and also as something which in its logic is analogous to that of the workplace. We can see very readily, for example, that with the construction of factories, the sinking of mines, a whole physical infrastructure for living was required – houses, schools, churches. The importance of this for production was signaled by the fact that in many cases, in order for production to go ahead, the captains of industry had to construct these infrastructures themselves, producing the so-called company town. Yet by the same token that infrastructure for living had to conform in its geography to the needs of production. It had to be in the right place, obviously, and it had to conform with the needs of the employer to economize so as to remain part of a competitive enterprise. The company towns are long gone but employers still worry about the cost of housing simply because it affects the wages they have to pay to attract workers. And if the geography of living gets in the way of the geography of production then it has to be transformed – houses have to be cleared to make way for freeways, airports expanded despite the protests of nearby residents, and so on.

The other point to be made here is that when we turn and examine the political geography of the living place apart from that of the workplace, what we find at work are very similar logics indeed. There is, for example, a similar antagonism to the forces of capitalist development, particularly the developers and the banks, which through their plans for changing urban geographies are seen as threatening neighborhood amenities and the home values of existing residents. There is also, however, a territorial competition among resident groups as they try to push off the "undesirable" onto each other – the shelters for the homeless, the school bus parks, and the like – and as they try to attract in what will enhance neighborhood amenities, including "desirable" residents. And just as employers and workers can join together around a common program of enhancing a locality's position in a wider geographical division of production, so we can find developers and existing residents making common cause around programs of improving a neighborhood's standing in a geography of consumption. Moreover, these practices can be understood in very similar terms to those we encounter in the political geography of the workplace. There is a similar tension between local dependence and the wider flux of the space economy. People become embedded in particular neighborhoods, and for diverse reasons. This means that as the geography of the city around them changes, as it invariably does, so their values in place will be threatened, generating the various forms of coalitional activity we have identified above.

Geographies of Alternatives

One of the problems of this analysis, however, is that it assumes that the way in which capitalism evaluates the world is the only way. The logic of capitalism is to commodify: to convert everything into something that can be bought and sold, including people's labor power. And to the extent that people "buy" into that view of themselves – and the metaphor is an apt one! – viewing themselves primarily as sellers of labor power and purchasers of their means of subsistence, then those values are clearly reproduced. Recent history is littered with attempts to live life according to different values, however, and geography has played a part in this. This is because this utopian impulse has typically taken the form of communities which try to separate themselves off from the world of the commodity. Separation has been seen as fundamental, in other words. Few of these communities survived for long, suggesting to some at least that as a value system capitalism *does* have something to recommend it. And of course, it does. Quite apart from its immense productivity it also emancipates. It instantiates a particular form of freedom, a freedom to sell to, or buy from, whomsoever one can strike a bargain with: a very different world, in other words, from slavery or feudalism. Likewise there is a certain sense of equality: one person's money is no better than anyone else's. Even so, the dilemmas and tensions persist, so it is unlikely the utopian spirit will die.

The attempt to realize a different set of values through separation is perhaps most clear today in the environmental movement. But an examination of that experience also underlines the problems. With capitalist development, nature, like everything else, tends to get valued purely in monetary terms. It becomes a commodity and is accordingly defined as "natural resources." One problem is that there are alternative systems of evaluation and these result in different views as to how nature should be used. Where mining companies see minerals and the money that can be made from their extraction others may see a wilderness that should be preserved. Where English farmers see hedgerows as obstructions to efficient farming and want to grub them up, others see them as a habitat for threatened wildlife or as important in the formation of a landscape – the patchwork of small fields separated by thick lines of vegetation that one sees from the air – that is seen as defining what is distinctively English. These are reasons that, seemingly at least, have nothing to do with the imperatives of a capitalist economy. Rather they may have more to do with the preservation of national symbols (e.g. the importance Americans ascribe to the wilderness) or the preservation of recreational space.

A problem then is to how to reconcile these quite different approaches. Perhaps the most common way of handling it has been through a spatial separation: the establishment of national parks, nature preserves, and game parks which will be off limits to mining, agriculture, the construction of housing, real estate dealing, and other forms of commodifying activity. This has a superficial appeal but there are also difficulties.

One is that, if experience is anything to go by, the tension between the commodifying and the urge to keep areas immune to those forces is never resolved. Some cases to ponder:

1   Alaska's Arctic Wildlife Refuge (figure 2.3). This seemed sacrosanct until oil was discovered adjacent to its western border around Point Barrow. Since then there has been continual agitation on the part of oil companies to open the area up to exploration. As Alaska's oil output has dwindled

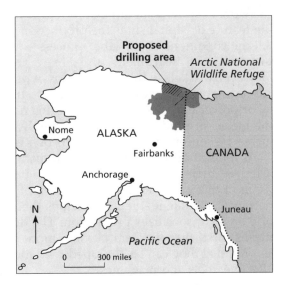

Figure 2.3 The Arctic National Wildlife Refuge in regional context.

and so its contribution to State revenue, so the Alaskan State government has joined in the chorus.

2   Kenya provides equally informative if more complex examples. There, game parks are big business because tourism is such an important element in the country's economy. It is, therefore, important not just to state revenues but also to the country's hotel and safari industry. On the other hand, Kenya is a poor country. The game animals in the parks are viewed as a commodifiable resource by those living adjacent. Poaching for the international ivory trade is a constant problem. And when parks are expanded people's livelihoods have to give way, so increasing the tensions between the commodifiable and what is not to be commodified. One can also ask the question: if, for some reason, Kenya was to fall into disfavor as a tourist destination, what then would become of the game parks?

The tension won't go away therefore. Furthermore, it is always one that threatens to be resolved in favor of the forces of commodification. Setting areas aside for the contemplation of nature is not, we will be reminded by business's spokespeople, cost-free; and those costs are always calibrated in money terms. Not that it is constructed as an issue of profit. Rather it is always defined as a matter of livelihoods, people's material well-being, a few elitists who want to prevent the broad mass of the population from enjoying the same material standard of living as they do.

Summary

The economic is a central focus of politics and hence of political geography. Central to economic life in the contemporary world is capitalism. We need therefore to know something of its logics. In the first place, note that the defining feature of capitalism is the production of commodities with commodities. All the conditions of the labor process – labor power, object of labor or raw material, and instrument of labor – are bought in respective markets. However, markets in these conditions of production are historically produced. Accordingly the entry of capitalism onto the world stage depends on making labor power into a commodity and this requires that workers be stripped of the means of production; that they be dispossessed and ownership of those means be concentrated in the hands of a few.

As a result of the mediation of production by exchange capitalism is characterized by intense competition on the part of both firms and workers. One result of this is that it has become *the* motor of development. There has been nothing quite like it – yet. It develops the productivity of workers as a result of the way it socializes production. The division of labor is deepened which means, first, that workers become more specialized at what they do; and, second, that they can be equipped with more specialized, and therefore more effective, tools and machines.

Yet for all this questions remain about capitalism as the mode of producing people's material needs. For a start there is a social question. Capitalist development is a source of great tension between business and labor. Among workers it generates chronic insecurity, as well as insufficiency and a sense of insignificance. The degree to which people feel this will vary from one person to another but that everyone will feel some aspect of this at some time in their lives is incontrovertible. And bearing further witness to the cleavage that results between employers and workers is the characteristic left–right form assumed by party politics in the advanced capitalist societies; though clearly, if right-wing parties had to depend purely on the votes of the owners of productive property they would never win an election. A coalition with the better-paid workers is therefore typical.

There is also an environmental question. The expansion of production that occurs with capitalist development imposes growing demands on nature as a source of raw materials and energy, and also as a waste absorbing agency, posing questions about the sustainability of this mode of production. Part of the problem is the common, as opposed to private, nature of many of the resources drawn upon, like the ocean. Given the competitive relation in which they exist with respect to each other, capitalist firms have an incentive to overuse any resource that is essentially free.

Capitalist development unfolds in a geographic context. Space is a weapon deployed by both business and labor as they seek to enhance their respective leverages. The labor movement, the ability of workers to impose their will on the exchange relation with employers, varies in its strength geographically. It is in part with respect to this geographically uneven development that businesses locate, grow, or stagnate. But to the extent that business evades the organizing drive of workers, they try to reassert their ability to exercise upward pressure on wages and work conditions by geographically extending their organization. This is one of the ways in which the competition and conflict between firms and employees is reflected geographically. As the map of labor organizing unfolds business seeks to colonize holes in it; and as it does so labor moves to stop them up, only to see new ones appear elsewhere.

But the relation between business and labor, employer and worker, is quite a bit more complex than this. The struggle between them can also dissolve and be replaced by coalitions that come together in different places to compete with similar coalitions elsewhere. This is because firms, like workers, exhibit a dependence on particular places, at particular geographic scales. If they are to grow it has to be in that particular place and for that they may need the cooperation of workers. But since they are competing in similar final markets this can be a source of tension with similar coalitions that have formed for similar reasons in other places. So for this reason as well as the geographically uneven development of the labor movement the capital–labor relationship is always characterized by a strong territorial element, which is why it is of such intense interest to the political geographer.

REFERENCE

Fukuyama, F. (1992) *The End of History and the Last Man*. New York: Free Press.

FURTHER READING

It is hard to know what to recommend for further reading on political economy since most of the readings are rather inaccessible to undergraduate audiences. On the other hand, dipping into Adam Smith is not difficult and certainly is worthwhile. I would suggest that students take a look at his writings on the division of labor and exchange (chapters 1–4 of Book 1 of *The Wealth of Nations*). The remainder of Book 1 could also be read with profit. Smith defines one strand in political economy and Marx another. Marx is especially inaccessible but he did write one popular piece, *Wage Labor and Capital*, which should be useful to the novice. This can be accompanied by a book written by John Eaton and entitled *Political Economy* (New York: International Publishers).

Among the geographers who have worked with political economy the written work of David Harvey and Doreen Massey is especially helpful and provocative. David Harvey's (1989) *The Urban Experience* (Baltimore: Johns Hopkins University Press) and Doreen Massey's (1995) *Spatial Divisions of Labour* (London: Macmillan) should be on the shelves of all budding political geographers. There are also some excellent collections. These include R. J. Johnston, P. J. Taylor, and M. J. Watts (eds) (1995) *Geographies of Global Change* (Oxford: Blackwell) and R. Lee and J. Wills (eds) (1997) *Geographies of Economies* (London: Arnold).

On some of the more specialized topics discussed in this chapter:

- on the demographic transition see the entry under the same name in R. J. Johnston et al. (eds) (2000) *The Dictionary of Human Geography* (Oxford: Blackwell);
- on utopia and its geography, although rather advanced there is a rewarding discussion with some excellent illustrations in chapter 8 of David Harvey's *Spaces of Hope* (Berkeley and Los Angeles: University of California Press).

Chapter 3

The Political Geography
of Capitalist Development I:
The Workplace

Context

As we saw in chapter 2, the capitalist development process is geographically
a highly uneven one. This is apparent in the case of the productive forces. The
development of labor skills, the provision of a dense network of transportation and communication media, and the emergence of agglomerations of
interrelated plants and offices are all distributed across space in a highly
differentiated manner. It is also the case with the labor movement. With capitalist development workers organize in order to secure improved wages and
work conditions, more job security and the like. But this occurs in a geographically uneven manner. There are areas that are intensely unionized,
exhibiting high levels of worker solidarity. And by the same token there
are ones where the level of militancy is sharply subdued and the degree of
collaboration with employers often heightened.

   At the same time we have seen how the geography of capitalist development is extraordinarily dynamic. New growth areas open up with rapidly
increasing employment and wages. Still others experience increased unemployment and wage stagnation. This is a process that occurs at all geographical scales. There are "newly industrializing regions" quite as much as the
celebrated newly industrializing countries. Along with this dynamism,
however, goes a good deal of social tension. To some degree this is associated
with its impact on wages and employment. In those cities and regions, even
countries, experiencing disinvestment and falling employment, the history of
the labor movement is one of attempts to limit capital's mobility: to force it
into channels that will minimize its impact on the income and wage prospects
of its members. And to the extent that local businesses try to compete with
more advantaged firms elsewhere by importing labor from other regions or
countries, that too will become anathema. These sorts of struggle are the focus
of the first section of this chapter.

But we would be wrong in concluding that workers and employers in particular places are invariably locked in an antagonistic embrace. Rather there are good reasons why in particular situations they may join together in common cause. And when they do, struggle between worker and employer can seem to give way to a competition and a conflict between one place and another. For it is not just workers that are adversely affected by capital's changing geography. Plant closures hurt not just those thrown out of work but also those firms which (e.g.) supplied the (closed) firm with services or even inputs, and those which serviced the workers in terms of retail goods, housing, banking services, and the like. At larger scales, to the extent that whole countries suffer from disinvestment then this is a matter of concern not just for workers and those firms that deal in so-called untradables[1] but also for the state, since its own revenues will show a sharp decline. Some workers, some firms will be able to adapt to these changes by moving. But many will be locally (or regionally, or nationally) embedded. For them the investment that will buoy the economy has to come to them; moving to it poses problems. And of course, states don't move:[2] they are dependent on growth occurring within respective boundaries.

It is in this context that alliances emerge, often led by business or state agencies, in order to recapitalize local/regional/national economies: to revitalize the local economy. But to the extent that they are not the only ones attempting this, they are brought into a competition with similar coalitions elsewhere: a competition, that is, to attract new investment, to persuade the government to put its infrastructure, its research facilities, its military bases, its offices, there so as to stimulate employment in the remainder of the economy. But at the same time this territorial competition can expose tears in the social fabric. Competition requires sacrifices and the distribution of the sacrifices can become an issue. To attract business organized labor may have to make concessions, for example. So a thin line has to be trod between the business–labor unity that will make the area attractive to new investment and the friction that will deter it.

Struggle around the Employment Relation: The Difference that Geography Makes

A major site of conflict between business and labor is the labor market. The labor market determines the degree to which labor power is a scarce commodity and this in turn conditions the bargaining power that they have with

1 Products and services which typically can't be, or are not, exported. Obvious examples include instant printing services, newspapers, retailing and wholesaling, the construction of housing, most repair services, mass transit, etc.
2 Though their boundaries may change and this can have important implications for their revenues. In the United States central city municipalities which can annex unincorporated territory are able to recapture for their tax bases suburbanizing businesses; a possibility which is denied those cities which are already surrounded by other municipalities and so are no longer contiguous with unincorporated land.

respect to each other. If labor power is scarce, if employers have trouble finding workers, then the advantage lies with labor; but if there are lots of workers for each job employers can pick and choose and drive down wages and benefit packages.

Here, as in so much else, geography matters. For a start labor markets have geographies. For many people the labor market is quite local. It corresponds to the urban region within which they happen to live. It is within this area that people can substitute one job for another and in which business can, conversely, substitute one worker for another with relative ease. It defines the area within which workers compete for jobs and employers compete for workers.[3]

The second, related, way in which geography matters is through the various comings and goings into and out of these geographically defined markets. This is because they affect the supply of labor relative to demand. Holding the availability of jobs in an area constant, then influxes of labor from elsewhere reduce labor's bargaining power in that particular market. On the other hand, increased investment in the area, the creation of new jobs, can, all other things being equal, result in a shortage of labor and upward pressure on wages. These comings and goings, therefore, are of intense interest to workers and employers in particular – geographically defined – labor markets and generate conflicts as both factions try to regulate them to their respective advantages.

Think and Learn

How might labor, through its representative organizations, seek to control the movement of labor from outside into local labor markets? How relevant would actions of the national government be in this regard? What sorts of actions do you think?

Third and finally there is the fact of the state. The state has the power to affect movements of workers and investment and hence labor scarcity and relative bargaining power in particular labor markets, or indeed across all labor markets in a particular country. This is one reason why the state is a focus for the lobbying of labor unions and employer organizations respectively. But states can also affect labor scarcity in other ways. This is because they are typically responsible for the provision of a variety of income supplements that can compete with wages. These include unemployment compensation, old age pensions, rent rebates, and so forth. From the employer standpoint these can affect labor supply in adverse directions. To bring workers to the

3 For a minority, however, labor markets are national or even *inter*national. This would be the case for higher level managerial grades, engineers, and university professors. A quick perusal of the job advertisements in *The Economist* or *The Wall Street Journal* would affirm this.

labor market it is important that their dependence for subsistence on the wage be maximized.[4] States, moreover, are territorially defined. Income supplement policies can vary from one to another and to the extent that employers take advantage of the geographically differentiated bargaining powers thus produced by moving their investments around this can also become an issue.

Conditions in local labor markets are continually changing. Inward investment brings new jobs. Plant closures or employment reductions as firms relocate elsewhere can reduce the jobs available. Workers arrive and some depart. And this, of course, is superimposed on the effects of reinvestment by existing firms and, over the longer term, the rate of reproduction and hence the supply of workers by existing households. But to the extent that workers in particular labor markets experience wide fluctuations in their fortunes, a sharp increase in their level of job insecurity, and requests from employers for renegotiated labor contracts on pain of their relocation elsewhere, then the movement of new workers into that labor market and the exodus of employers out of it will likely become issues. In particular they are likely to become the focus of strategies designed to return the balance of advantage to the workers. Of course, some workers are able and willing to move out of that labor market in search of improved conditions elsewhere. But others will be more dependent on it. They may be socially embedded through networks of friends and relatives. They may own houses which, given the presumably depressed state of the local labor market, they would have to sell at a loss. Or they may simply be too old to leave their current job and hope to be employed by someone else.

Think and Learn

Thinking of people you know in your home town – parents, friends of parents, peers, parents of friends, former schoolteachers – how easy or difficult would it be for them to relocate elsewhere if they were faced with difficulties of securing employment there? What is the source of these difficulties? Would it be easier for the younger people you know, or harder? Why might that be?

During the seventies and early eighties many local labor markets in the Northeast and Midwest of the United States experienced sharp downturns. This area, moreover, happened to be the stronghold of organized labor in

---

4   This is by no means to argue that employers would prefer no income supplements at all. As business conditions fluctuate over time they want to be in a position where they can release labor but rehire as business prospects improve. But this means that workers have to be persuaded to stay around until that happens. Without some sort of income this is unlikely to happen. So employers do have stakes in the provision of some sort of social safety net. Where they differ from labor is over its magnitude.

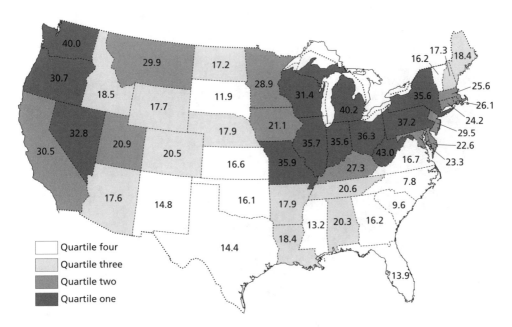

Figure 3.1 Rates of unionization of non-agricultural workers, by State, in the US, 1970.

Source: US Statistical Abstract. Washington, DC: US Government Printing Office, 1971, Table 390, p. 242.

the US (figure 3.1). These declines were linked to the changing geography of American corporations. A common view was that in terms of their production locations[5] they were fleeing the higher wages and more militant labor of what came to be known as the Coldbelt, for the lower wages and less organized labor in the Sunbelt, particularly small towns there and the States of the old South. There was certainly anecdotal evidence for this and the net drift of industrial employment was clearly in the direction of the Sunbelt (figure 3.2). This was the context for a drive to unionize the Sunbelt. In this way, it was believed, interregional differentials in wages and organization would be evened out and this would deprive corporations of an incentive for relocating out of Coldbelt locations. At the same time their ability to strike and so exert upward pressure on wages would be enhanced since employers would not be able to switch production to non-unionized sites (presumably in the Sunbelt).

An example of this organizing strategy comes from the experience of the United Mine Workers' Union (UMW) in the US. For many years coal production in the US was concentrated in Eastern States like West Virginia, Kentucky,

5 Not necessarily the locations of their headquarters or research facilities.

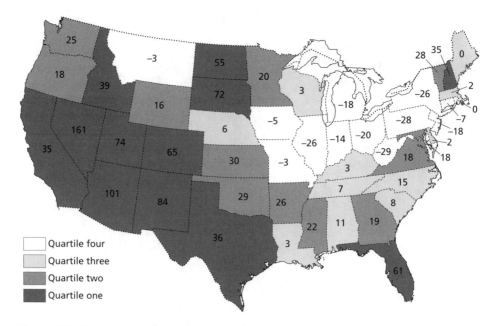

Figure 3.2 Percentage change in manufacturing employment by State, in the US, 1970–1985.

Source: US Statistical Abstract, 1971 and *US Statistical Abstract, 1986*. Washington, DC: US Government Printing Office.

and Pennsylvania. Over the past thirty years, however, coal production has expanded greatly in such Western States as Colorado, Utah, and Wyoming. This coal has a number of advantages over Eastern coal. These include its relative cheapness and also the fact that – unlike a good deal of Eastern coal – it is low in sulfur, which makes it attractive to power companies anxious to come into compliance with the federal Clean Air Act of 1993. But to make matters more difficult for the UMW with its membership base in the East, most of the miners in the West are unorganized and to the extent that they *are* organized, they are organized by a rival union. It is in this context that the UMW has tried to organize mines in the West, but, as box 3.1 indicates, this has not been easy.

A more successful expression of attempts to control local labor markets to the advantage of the workers there, or at least the organized workers, was, and remains, prevailing wage legislation. In the US *prevailing wage laws* (see box 3.2) came into existence in the 1930s in the form of the federal legislation known as the Davis–Bacon Act. This requires federal contractors to pay the local "prevailing" wage. This wage is determined by the Department of Labor and is typically consistent with the wages paid locally by unionized contractors. The effect is to limit the competition from non-unionized firms bringing in workers from outside. The States have similar laws for construction work involving State funds.

Box 3.1 *Organizing Coal Miners in the West*

The problem of organized labor's leverage when business expands into areas where labor is not organized, or is organized by a competing union, is vividly illustrated by the problem of the United Mine Workers' Union (UMW) in Western States like Wyoming, Colorado, and Utah. The UMW has not been very successful in this regard because of its refusal to recognize the peculiarities of coal production there.

A major stumbling block has been in the UMW insistence that all operators it contracts with make the same contributions per hour worked to the union health and retirement fund. For reasons intrinsic to coal mining in the area, however, Western operators can provide the same benefits to their workers for less simply because they have younger workforces (hence, less retirement) and – due to the surface nature of production there – no black lung disease (hence, lower health costs). Consequently they can offer a higher wage to their miners than they would otherwise be able to. They have been willing to do this to keep out the UMW.

This ability to pay a higher wage, and so provide incentives to workers to reject the UMW, is further enhanced by the high productivity of the Western coal mining industry. In contrast to the East where shaft mining predominates, mining in the West simply involves the use of power shovels to strip away a relatively thin surface layer of rock and so expose and excavate the thick seams of coal lying directly underneath. Even after the high costs of transportation involved in getting the coal to Eastern markets costs of production are so cheap that Eastern coal can be undercut.

Conditions of work in the Western coal mining industry also make it difficult to organize workers there. Sitting in the air conditioned cab of a bulldozer or shovel is very different from lying on one's side at a coal face. This different work experience, moreover, has provided an opportunity for a rival union in the West, the International Union of Operating Engineers. This is the union which organizes the workers who build highways. Many of the workers, to the extent that they belong to a union, belong to the Operating Engineers. This also suits the operators since that particular union is much less militant than the UMW and, in accord with the youth and limited health problems of its members, has a much cheaper pension and medical plan.

On the other hand, to the extent that firms seek to lower their labor costs by relocating parts of the labor process elsewhere, this relocation may meet with stiff opposition. Unions may threaten to "black"[6] parts coming from the branch plants in question should they be established (see box 3.3 for an example of this from Britain). Likewise, at the national level the overseas investment of multinational corporations has been seen as a threat to employment and elicited calls from the labor movement that would discourage it: in particular the imposition of stiff taxes on repatriated profits.

---

6 "Refuse to handle."

Box 3.2   *The Political Geography of Prevailing Wage Laws*

Prevailing wage laws exist at both federal and State levels in the US. They serve to protect local construction workers who are union members from the competition of non-union workers from outside the locality. The crucial provision of prevailing wage laws is that on projects involving public funds above a certain dollar amount so-called prevailing wages must be paid. These wages are determined by union scale wages in specifically delineated – usually multi-county – areas.

The event which originally stimulated the federal legislation – the Davis–Bacon Act of 1931 – is indicative since it was drafted to keep out low-wage black workers from the South. The bill was introduced by New York Representative Robert Bacon after an Alabama contractor won the bid to build a federal hospital in Bacon's district and brought in large numbers of non-union workers from Alabama.

Effectively, prevailing wage laws discriminate against construction workers who are black and/or from low wage areas of the country like the South. This is for two reasons. On the one hand such workers are less likely to belong to a construction union and so will not be protected by prevailing wage laws; and on the other hand their generally lower level of skill makes them not so productive and therefore less able to make paying the prevailing wage worthwhile for their employers.

Prevailing wage laws continue to be hotly contested. Minority contractors and black construction tradesmen generally are part of the opposition; but any construction firm that would have difficulty, perhaps on account of the productivity of its employees, from paying union scale wages and therefore bidding on a project involving public money will be opposed.

Local governments also tend to be opposed, particularly in outlying areas. Obviously they will be affected by prevailing wage laws since their construction work inevitably involves public money. But in small towns and rural areas local governments often find themselves paying much higher bills than what the job could be done for by local non-union labor simply because the prevailing wage will be based on that in a distant metropolitan center where overall wage levels are much higher. For local governments, therefore, battling prevailing wage laws is one more strategy in their ongoing attempt to live within their budgets.

However, much of the focus of union strategies is likely to be at the national level: attempts, in other words, to control the movement of workers into the country and the foreign investment of domestic corporations. As far as the latter is concerned, a widely expressed concern is that firms are not only relocating some of their operations out of union heartlands into Sunbelt locations, but they also relocate overseas. This is a problem for highly unionized workers everywhere in the world. Its salience has been greatly heightened by the publicity given to what has become known as "globalization." Common strategies have included the following:

Box 3.3 *Labor Unions and Low-wage Branch Plants*

Since the Second World War the town of Dundee in Scotland has been both the site of persistent unemployment and the focus of government efforts to entice new industry to locate there. In 1987 it seemed that these efforts had been crowned with success when Ford announced it would establish a plant in Dundee producing computerized engine controllers and employing over 400 workers. Unlike other workers at Ford factories in Britain the workers were to be represented by the Amalgamated Engineering Union (AEU) and working conditions, including wages, were *not* to be on a par with those of Ford workers elsewhere in the country. This was acceptable to the AEU on the grounds that some jobs were better than no jobs. The plant, therefore, would not be subject to the Ford National Joint Negotiating Committee (FNJNC) that bargained on wages and work conditions with management for the other 22 Ford plants and it would not be subject to the agreements already made by the FNJNC. This led in turn, however, to a threat on the part of the FNJNC to refuse to handle any parts coming from the Dundee plant. This was followed by an announcement from Ford that they were not going to continue with the plant, though other factors like the increasing value of the pound at that time may also have played a part in the decision. The plant was eventually located in Spain.

The danger from the standpoint of the FNJNC was of a slow attrition of its position and of the wages of the workers it represented as Ford established other plants like the one it planned for Dundee. But the fact that the plant was eventually located outside the country points to the continuing weakness of labor unions in regulating to their own benefit multinational corporations that can move from one country to another.

Source: Foster and Wolfson (1988).

1 *The enactment of legislation to deter capital export.* A common suggestion here is that the US government should impose heavy taxes on profits repatriated to the US; the assumption is that if American corporations could be dissuaded from investing overseas they would be more likely to invest in the US and so create more employment there. In its anticipated effects this is analogous to the idea of runaway shop legislation.

2 *Enact "fair labor" standards.* To the extent that the products of high-wage labor come into competition with the products of low-wage labor elsewhere it will always be the aim of that high-wage labor to extend the same standards to those low-wage workers. By so doing the low-wage advantage will be taken away and businesses will be less likely to relocate, so keeping jobs in the high-wage area. A good example of this type of approach comes from Germany, where the labor unions are pushing for uniform labor legislation within the European Union. The wages of German workers are, on average, higher than anywhere else in the EU. German workers tend to be more unionized though their 50 percent rate does not compare to the unionization rate of 75 percent in Belgium. In

addition they have negotiated with employers conditions of health and safety and hours of work that workers elsewhere in the EU have yet to approach. In 1992 the remaining barriers to trade between the member countries of the EU came down and at the same time there were a number of highly publicized location decisions in which less developed countries like Spain were chosen over Germany. Consequently the German labor unions have led the fight in the EU to standardize across countries, at high levels, social benefits, safety regulations, and work hours.

3   *Form international labor unions.* The same logic that applied to the attempts of US labor unions to organize workers in Sunbelt locations also applies at the international level and has led to attempts to form international labor unions with locals in different countries. Hitherto this has not been very successful. More has been achieved, on the other hand, in the construction of links between the labor movements of different countries, especially where there is a common employer: thus labor unions in one country may offer financial support to striking workers in another, come out on sympathy strikes, or refuse to handle extra work diverted from the striking plant. National federations of unions have also worked to make it easier for workers in other countries to organize. One point of leverage has been an international preferential trade system designed to help less developed countries. This system is called the General System of Preferences. According to its rules a developing country's exports to an industrialized nation enjoy reduced tariffs or are duty free; but to qualify a nation has to allow workers to form labor unions. This proviso was used by the AFL-CIO to bring pressure to bear on the Malaysian government to lift the ban on unions there.

But if firms may choose to solve what they regard as their labor problems by relocating, by the same token they can solve them by importing workers who are willing to work for less and, perhaps, less susceptible to organization by labor unions. For the American worker and for organized labor in the US, Mexican immigration, whether legal or illegal, has long been a problem. They have consequently been a major pressure group for restricting immigration and for more vigorous enforcement of the law by the Immigration and Naturalization Service (INS). It is not just a matter of labor market competition. The immigrant is a threat to that union organization which has allowed American workers to bargain for higher wages. Hispanic immigrants and illegal aliens in particular are a major pool of strike breakers. This is because they are willing to work for a non-union – i.e. lower – wage. In addition the illegal alien is extremely difficult to organize into a labor union. For if they join the union the employer may fire them, and while this is illegal they are unlikely to protest for fear of being identified by the state as illegal and so subject to deportation.[7]

---

7   Significantly hispanic-dominated labor unions feel as strongly about the matter as the rest of the labor movement in the US. A case in point is the primarily hispanic United Farm Workers Union, which feels the competition of the illegal alien in California and has subsequently called for a much more vigorous pursuit of illegal aliens by the INS.

On the other hand, it seems unlikely that American labor unions and their members will find an easy salvation in the form of more vigorous policing of the borders and greater restrictions on immigration. This is because the Mexican immigrant, particularly the illegal alien, has become so important to American businesses, especially in the Southwest of the country. The cheapness of the immigrant has allowed numerous American businesses – like the garment industry of Los Angeles – to compete with producers in Third World countries that pay much lower wages. Businesses, moreover, have been set up in the expectation that this cheap labor supply will continue to exist. As a result restricting Mexican immigration is extremely difficult.

Think and Learn

It was pointed out earlier in the chapter that for most workers, labor markets are very local and correspond to the particular urban region they happen to live in. This means that concerns about labor market conditions are likely to be local concerns. Why, therefore, do you think that when it comes to mobilizing the government to mitigate those local conditions it is to the central government that workers look rather than to local government?

Transforming Class Conflict into Territorial Competition:
From Profits versus Wages to the Competition for "Development"

In their struggle with each other over respective abilities to determine wages and work conditions we have seen how both labor and capital try to mobilize geography to their respective advantages. In its relocations, business is often trying to take advantage of the uneven geographic development of the labor movement. This is in order to secure for itself lower costs of operation and facilitate its profitability and hence competitive survival. The same goes for its strategy of importing cheaper, more easily exploitable, labor from other countries. But just as business seeks to circumvent the controls put in place by labor so, in its turn, labor, particularly through labor unions and its political parties, seeks to frustrate business by putting barriers in its way: by organizing workers elsewhere, for example, by putting limits on the repatriation of profits or on immigration. By the same token capital is driven once again to find ways round these new barriers. If investment overseas becomes more difficult, for example, then an alternative may be altering immigration laws. And so it goes.

But this direct confrontation is only a part of the political geography of the workplace under capitalism. For despite their obvious conflicts of interest we also find situations in which the businesses in particular places at diverse geographic scales – cities, regions, countries – form coalitions with workers there

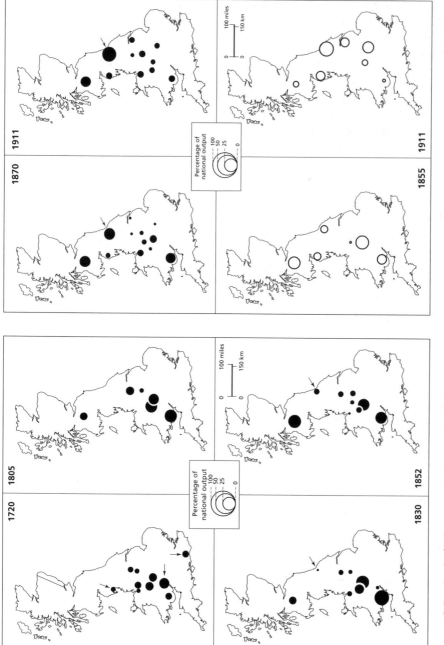

Figure 3.3 Capital's inconstant geography I: the British pig iron industry, 1720–1911. Over this period there was continual change in the industry's geography. In the map for 1720 the arrows indicate areas of the country where iron ore was still smelted with charcoal. These areas no longer produced pig iron by 1805, but had been eclipsed by areas where coke was readily available, like South Wales and the Birmingham area in the Midlands. Note also the rise of Middlesborough from 1852 on, again indicated by arrows. This was associated with the discovery of iron ore deposits in the vicinity (see maps in the lower panel). Further south these same iron ore-bearing strata led to some shift of the center of gravity of British pig iron production eastwards.

Source: J. Langton and R. J. Morris (eds) (1986) *Atlas of Industrializing Britain 1780–1914.* London and New York: Methuen, pp. 129, 131.

around policies that are touted, at least, as providing benefits for all, increased wages as well as profits, and often under the label of "development." These are policies in which the opposition between business and labor is somehow converted into an opposition between places: a competition between places and their respective coalitions of forces. And this in turn is a competition which can easily give way to a situation of conflict in which instead of labor claiming that business is exploiting it, regions and countries claim that they are being exploited by still other regions and countries. So the question that we confront in this section is: how is it that a social class cleavage can somehow be transformed into a cleavage of a territorial sort, a cleavage, moreover, from which the traces of class conflict are somehow expunged?

The roots of this transformation can be traced to a contradiction between the needs of firms and workers for some sort of geographic fixity on the one hand and the geographic turbulence that, as we have seen, capitalist development exhibits. Interests in profits, wages, tax base, are always interests in *particular* places. A mom-and-pop store depends upon the patronage of the immediate neighborhood: relocation into another neighborhood in search of a better market is often very difficult for small businesses. Likewise, and despite the much vaunted multinational corporation, most American businesses would have difficulty operating profitably overseas. And to all intents and purposes (e.g.) French workers depend upon the demand for labor in France. They are quite indifferent to the demand for labor in Germany or Italy since a variety of considerations – language, institutions, family – make relocation difficult. Similarly the State of Ohio must, by and large, depend for its revenues on the level of business activity in Ohio. In other words, the typical situation is one of *spatial entrapment* or *local dependence*. For wages, profits, state revenues, specifically local conditions – resources, social relationships that either cannot be moved in principle or which would be very hard to move elsewhere – are crucial. Substituting alternative locations – labor markets, tax bases – is difficult, though as we will see the geographic scale at which that difficulty is experienced can vary a great deal.

Now from the standpoint of the particular agents involved, this local dependence is a problem. No locality, no region or nation state is an island, at least not under capitalism and its tendencies to unite markets over seemingly ever-increasing geographical scales. Rather they are all connected one to another by various flows: flows of investment, of labor, of finished or semi-finished commodities, tourists, shoppers, etc. These flows in turn reflect the abilities of firms – manufacturers, retailers, hotels, etc. – in different places to compete in wider market places; to compete with firms in other localities or countries. The map of competitiveness is constantly changing. Capital's geography is *inconstant* to a very high degree (figure 3.3). New firms in new localities emerge to challenge old firms in old localities. As those older firms are pressed to the wall, so the effects are transmitted to other firms, their suppliers in the same "old" places. Alternatively the old firm may be able to compete but only by closing its old plant and opening a new one somewhere else where wages are lower. The result is that the space economy undergoes further

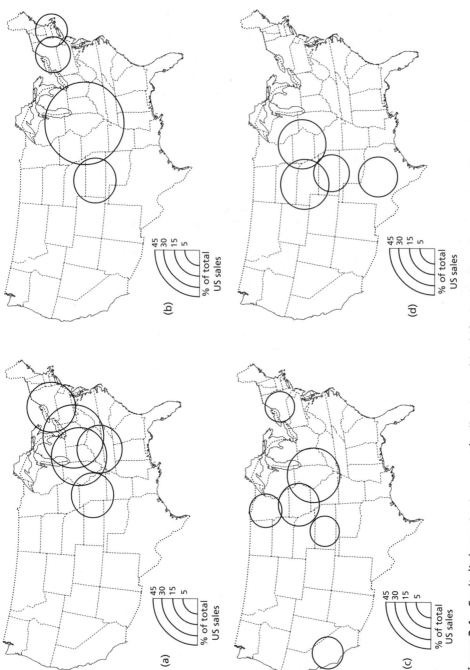

Figure 3.4 Capital's inconstant geography II: geographic shifts in the American meatpacking industry 1850–1982 (**a**, 1850; **b**, 1890; **c**, 1940; **d**, 1982). Note how the early focus of the industry on the Ohio Valley (Ohio and Kentucky) is displaced by 1890 by Illinois and the rise of Chicago. By 1940 Chicago had lost some of its pre-eminence and states like Iowa and Minnesota had increased in relative significance. By 1982 meatpacking had shifted even further west, becoming centered, along with Iowa, in states like Texas, Kansas, and Nebraska. Figures are percentages of US sales by State; only States with more than 5 percent of the industry are indicated.

Source: M. Storper and R. Walker (1989) *The Capitalist Imperative.* Oxford: Blackwell, pp. 94–5.

change setting off shifts in wages and in tax base in the locality the firm left behind.

This is not to say that there are no elements of relative constancy. There *is* considerable continuity in the map of capitalist development from one point in time to the next. But there are also dramatic changes affecting particular places; the dramatic changes are highly localized in other words.[8] New growth areas in particular products emerge, threatening existing places specializing in that product. It is a long time since Chicago was hog butcher to the world and since Lancashire was a major exporter of cotton goods. On the other hand, the emergence of major meat processing centers in the Great Plains and the development of a garment industry in the Far East organized from Hong Kong are part of the story of the changing roles that Chicago and Lancashire perform in wider geographic divisions of labor (figure 3.4).

In addition changes in methods of production may widen the window of locational discretion and allow businesses, in the interests of competitive effectiveness, to relocate at least part of their production. This can result in plant closures. Where the plants in question – rubber tire factories in Akron, a steel plant in Consett, County Durham, a truck plant in Allentown, Pennsylvania perhaps – have been important to local employment the result can be devastating, not only for the workers themselves, but for all those businesses and local governments that cater to their needs. If they were not spatially entrapped they could follow the relocating plants. And again it bears emphasis that these are problems which are experienced at a variety of geographical scales. Whole nation states can suffer eclipse: the case of the United Kingdom, at least until the recovery of the past decade, is exemplary. At the same time other national economies may boom, as in the case of the Far East's newly industrializing countries or NICs. Regions experience different economic fates: California versus the US Farm Belt, for instance, Britain's "London and the Southeast" versus the North; while every country has its rustbelts – its Youngstowns, Consetts, and the like.

These changes create dilemmas for those who, by virtue of their local dependence, cannot adjust to them through their own relocation. And to some degree this dilemma may be shared by business and labor. This creates a space within which a coalition can be constructed around a plan to, or so it will be argued, "raise all boats." But this involves counterposing "local fortunes" to those elsewhere in a competition, for example, for inward investment, or to displace plant closures on to other localities. Typically, though not always, these coalitions are business led, but in order to push through their plans they need the support of a wider constituency, and the local dependence of others – workers, local government, even retirees concerned about their property taxes – will make them susceptible to their overtures.

Part of the offensive is invariably discursive in character: arguing for a unity of interest where everybody thought there was nothing but conflict. How this can work is illustrated by the cartoon shown in figure 3.5. The cartoon is in

8 "Localized," that is, relative to a particular geographic scale: a metro area, a region, a nation state.

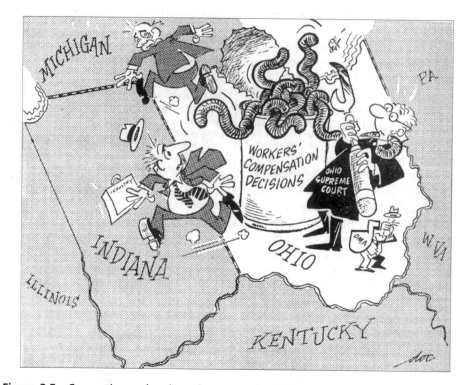

Figure 3.5   Converting a class issue into a territorial issue: a cartoonist's view.
Source: Columbus Dispatch, February 21, 1985, p. 7C. Reprinted, with permission, from *The Columbus (Ohio) Dispatch*.

reference to a debate that was going on in the State of Ohio at that time about workers' compensation policies. Workers' compensation for injury on the job is something for which labor unions have always fought vigorously. For employers, on the other hand, it is an expense since they have to pay premiums to cover the insurance of their workers against that sort of injury. Workers' compensation, in other words, has always been fought over between workers and employers. One of the ways that employers have learnt as a strategy to combat worker militance on this issue is to identify workers' compensation levels as a problem for the state, i.e. the territory, in competing for inward investment. It is identified, therefore, as something that not only is not in the interests of employers; it is not in the interests of workers either since they would otherwise benefit from the jobs brought by new investment coming into the state from outside. But obviously this means situating the problem with respect to the idea of a territorial competition.

The cartoon makes this argument very clear. At the time that it appeared the view was being promulgated by the Ohio Manufacturers Association (shown in the cartoon as the OMA), among others, that workers' compensation payments were driving businesses into other States (as represented by the business person fleeing in the direction of Indiana, for example). Evidently this territorial interpretation, although counter to the view that workers' compensation is a question of equity between employer and employee, was one endorsed by the newspaper in question.

Another crucial aspect of this discursive "softening up" is the deployment of the term "development." The talk, for example, is of "local economic development," promoting "national economic development" and the "developed versus less developed countries." "Development" has a wonderfully positive ring to it: how could anyone possibly be opposed to it? And since it applies to places rather than to people it avoids awkward questions about who is really going to gain from the projects coalition leaders have in mind. The distributional question can therefore be marginalized, even though a moment's thought will quickly convince that the idea of places developing is a nonsensical one and that it is the people in places who develop by virtue of (e.g.) their ability to capture and divert to their own purposes the flows of value moving through the hands of those located in a particular place. But this is perhaps to anticipate a more concrete knowledge of these coalitions, territorial competitions, and what is being competed for. It is to some particular instances, therefore, that we now turn.

American local growth coalitions

In urban areas of the US it is common to find a set of businesses coming together with local government around policies designed to expand the local economy, i.e. to increase the amount of money circulating through it in the form of payments for their services and products or in the form of tax revenue for local government. The businesses typically involved are the utilities and the developers but they are often joined by others like the banks, the local newspaper, and some retailers. The attribute they all share is a dependence on the health of the local economy. In turn, this health depends on the magnitude of the locality's export sector, or what is sometimes called the basic sector.

For example, it is as a result of their local dependence that the gas and the electric companies often take the lead in encouraging new investment in the area from outside. They are franchised to serve very limited service areas and if demand does not materialize there or, worse yet, declines, then their bottom lines will suffer. To make matters worse they make huge investments of long life that they need to recoup – investments in power stations, pumping stations, grids, etc. These are investments which are inevitably of a speculative kind since no one can predict with accuracy what the local market will be like fifteen or twenty years hence. Not surprisingly every gas or electric utility has an economic development department assigned purely to expanding business demand for their services.[9]

---

9 As far as the American electric utilities are concerned this may be changing. Deregulation calls for a separation of the two functions of power generation and distribution. The old electric utilities will continue to exist but will gain their revenues from selling distribution to the power

Developers, on the other hand, are dependent on a local knowledge that takes time to build up and cannot be readily transferred to some other urban area. Developers need to know their market and also to be known. If they are to make money they need to understand the specificities of a particular local market: the sorts of real estate products that will sell there, in terms of design, price range, and location; they need inside knowledge of where the city is going to extend its water- and its sewer-lines so that they can buy up land cheap and then sell later at a premium. All this requires integration into a network of informants, both fellow developers and those in the know at city hall. At the same time they need to build up a local reputation: a reputation with builders who will purchase the lots in their subdivisions, and with the banks that will extend the loans for development of the land. All this knowledge – being known as well as knowing – takes a long time to acquire, and relocation to some other real estate market means starting all over again: hardly an attractive prospect. As a result, if developers are to expand their businesses it usually has to be locally. And the growth of the local market is one of the conditions for that expansion.

Along with the utilities and the developers, the other major element of local growth coalitions in the US is local government. Local governments want to see their local economies expand since that increases the values that can be taxed: the sales in local stores, real estate, and incomes. This is because local governments in the United States still depend quite heavily for their revenues on a local tax base. It is not as if, if that tax base fails, they can move someplace else in the US where the economy is expanding rather than going into a retreat. They depend on a local tax base and that makes them dependent on the health of the local economy.[10]

companies. No longer do they have to worry about recouping the costs of investments in power stations, therefore. But they do have to worry about obtaining sufficient revenue from the power generators. At present it is unclear whether they will continue to have a serious interest in local economic development and maintain their economic development departments: departments whose brief historically was attracting in new industrial firms. New industrial firms would be nice, presumably, but only if they locate on parts of the distribution network that are currently underused. The fact that their prices will be wholly determined by the market and they cannot appeal to the state for price increases to cover their expenditures may also temper any speculative activities they might indulge in in extending their networks.

10   Banks and newspapers also deserve mention. Newspapers, to the extent that they are locally owned and operated, have very local markets. If they are to expand their revenues then the local economy has to expand for it is this which determines the amount of advertising they carry. Many newspapers, of course, are part of a chain and this means that they are *not* locally dependent; they are able to spread their risks through their multilocationality. So if indeed revenues deteriorate in one location that is likely to be offset by increases elsewhere. The same logic applies to the banks. Forty years ago banks in the United States were extraordinarily locally dependent. In most cases they were not allowed to branch beyond county lines. And if they could they were not allowed to branch into other States. All this has changed and many banks have responded by acquiring a regional and in some instances a national presence. Even so, there are still cases where a bank is a county bank and has no branches elsewhere. In such instances they are likely to align with those other forces that wish to expand the local economy, for in this way they can hope to increase both their depositor base and the demand for loans. In small towns reliant on one major employer they are especially vulnerable. For if that employer should fail the banks'

The major focus of the local economic development policies promulgated by these interests has been attracting in "export" activities in the form of new plants or offices which serve wider markets and which will therefore stimulate other forms of local growth – retail activity, housing, the expansion of residential markets for the utilities – as well as, of course, increasing the local tax base; for as the local economy expands so too do sales taxes, local income taxes and property tax revenues.

Think and Learn

Local growth coalitions concentrate on attracting in "export" activities. This raises several questions. (1) Which of the following do you think would likely qualify as an "export" activity? A state office? An automobile plant? A distribution center? A shopping center? A custom machining shop? Why do you think they would qualify or not qualify? (2) Why do you think local growth coalitions are not interested in attracting in activities serving the local market rather than export markets, e.g. an automobile dealership? (3) What is an "export" activity depends on scale. What might be an "export" activity for a local government but not for the whole metropolitan area? How do you account for this?

The policy tools used to stimulate investment are diverse. A common inducement is the tax abatement.[11] City government may also provide public facilities – new highways, sewer and water line extensions – to help bring a large employer within its boundaries. Convention centers may be built to stimulate investment in hotels; industrial parks built to provide a congenial environment for longed-for investments by hi-tech firms; or airports expanded to make them attractive to airlines considering the city for a hub operation. In turn the establishment of such a hub may make the city more attractive to firms considering it as a location for their corporate headquarters.

Once the inward investments are made, of course, the industrial plants and office employment, the establishment of an airline hub, then the problem becomes one of retaining them. The threatened closure of a major plant, the rumor that it will be relocated elsewhere, commands as much attention, logically enough, as the prospect of attracting in a major Japanese automobile

---

debtors might have a hard time paying back their loans: loans taken out to buy houses or to expand retail businesses, for instance. Diversifying the local economy is, in consequence, often a priority for them.

11 This is an agreement to forgo property taxes on the investment for a period of time that is variable but which can be as long as twenty years. The property tax is levied on the land and the improvements to the land (i.e. buildings and physical infrastructure). When an abatement is granted it is confined to the improvements. The owner of the land will still pay property taxes on the assessed value of the land.

transplant or a corporate headquarters. Rescue packages will be put together involving a mix of worker concessions on wages and benefits, tax breaks for the firm, the creation of a freeway link, perhaps a purchase–leaseback arrangement in which the city will purchase the plant with public tax money and lease it back to the company.

So whether it is making the city more attractive for investments yet to be made, or attempting to retain existing major employers, money will have to be spent, land use changes will have to be made, and, in the case of the threat of plant closures, wage concessions offered. In all cases this risks courting public opposition since almost invariably they will be asked to foot the bill, either in the form of wage concessions or the property or sales tax hikes required to pay off the bonds sold to build (e.g.) a convention center, construct a major new airport, etc.; or in the form of the various sacrifices entailed by land use change. These latter can vary from declines in home values in the vicinity of major industrial/commercial projects, through congestion effects, to the residential displacement often accompanying downtown revival and concomitant gentrification. People can protest, resist, and they will. But the answer almost inevitably will be that in the first place the development of the city is at stake: jobs, an increased property tax, improved local amenities, such as a major hub airport; and that in order to attract in the export activities that will produce these benefits, or indeed to retain these activities, money has to be laid out, and some people in some neighborhoods have to be adversely affected since the investors coming in from outside have lots of choices elsewhere and where people are more than willing to foot the bills. In other words, as in the Ohio workers' compensation case "our" city/locality is in a competition with *other* cities/localities for the inward investment that will lay the golden eggs. In order to compete "we" have to make our city or locality at least as attractive to inward investors as others.

Among many, though not all, these arguments are likely to resonate positively. For they too, like the utilities, banks, developers, and local governments pushing the plans, have their own forms of local dependence. Family relations embed people in particular localities. There are children in school, spouses that have good jobs, aging parents to care for.[12] Many people own a house and given that it will likely be one of their major assets they are quite reasonably concerned about it retaining, or even accruing, value. Moving elsewhere also entails major risks, particularly for those in jobs that require only modest qualifications. The attraction of moving is to obtain employment that is otherwise not available but for most jobs recruitment is very local: through local newspapers or by word of mouth. So you have to be living in a town in order to get a job.[13] Moving elsewhere without a job can be quite foolhardy. In conse-

---

12   It is significant that at the upper echelons of the job market "pots" have to be "sweetened" to entice executive level people to move: in particular employers are finding increasingly that in order to hire they have to agree to help the spouse obtain employment in the new location and sometimes pay for the removal costs of aging parents.

13   Only in the case of the highly qualified will a potential employer foot your travel expenses in order to interview you, and then help with removal costs (Gordon, 1995).

quence, for many people the idea of bringing in more employment, forestalling plant closures, and also therefore protecting local home values is an attractive one.

This is not a blanket characterization, of course. Younger people, less likely to be married, less likely to be homeowners, are less geographically constrained. And not all those who are locally dependent will buy growth coalition rhetoric. The retired on fixed incomes are more likely to be affected by the prospect of increased property taxes than by the possibility of increased employment in the city. And people in particular neighborhoods affected by highly dislocating land use change also face a different sort of tradeoff from those who have to concede a mild increase in local tax rates.

On the other hand, some inward investment and therefore some local growth is more attractive than others: sunrise versus sunset industries; corporate headquarters and research establishments as opposed to branch plants; the environmentally benign versus the pollution-intensive. As a result the competition of local governments and their growth coalition supporters for inward investment often occurs through attempts to develop and exploit a more desirable niche or position in a wider spatial division of labor.[14] Some cities are better able to do this than others since they have more to offer businesses seeking new locations.

Nevertheless at the other end of the spectrum some places have trouble competing for *any* position in the spatial division of labor. Having lost one position they find they are attractive neither to branch plants, offices, research and development nor to any of the relatively more desirable forms of economic development. They are left, therefore, to scramble for either the undesirable or at least the highly contentious. Particularly vulnerable in this regard are small towns in rural areas that are losing their functions as service centers for immediate hinterlands, mining towns where the mine closed down, and many of the Rustbelt towns, small and not so small, affected by deindustrialization and plant closure. For communities like these a hazardous waste dump, a State or federal prison, or a casino may appear as their only salvation. The feature that all these activities have, of course, is that environmentally they are not particularly desirable. Hazardous waste conjures up the image of public health problems while prisons evoke concerns about escaped prisoners at large in the community. Casinos have associations, imagined if not real, with the threat of organized crime as well as the stigma of something

14   On the other hand relying on one highly specific position in the spatial division of labor can be dangerous since no one can anticipate when its market might collapse. Accordingly in some cities a major concern has been injecting some diversity into the local economy. Risks need to be spread across more than one position in the spatial division of labor, therefore. A case in point is Las Vegas. For many years the mainstay of the local economy has been the gambling industry. This, however, was hit hard in the early 1980s by national recession. As part of a diversification strategy local growth interests have tried to attract other sorts of business, especially in light manufacturing and high technology. In particular the city has been vigorously marketed to Japanese corporations as a low-cost alternative to Southern California. The construction of a high-speed rail line to Los Angeles will further increase its attractiveness by reducing the city's relative inaccessibility.

that many still regard as not quite within the pale. This serves to draw attention to the fact that advantage in the spatial division of labor is calculated not just in terms of whether or not the industry is a growth industry and what sorts of wages it pays but also how environmentally benign the activities are.

Finally, we should note that in their pursuit of local economic growth there are many things local governments are *not* able to do. This may be because it's not in their power to do them; or, if it is, it's too expensive. For example, employment legislation – workers' compensation, unemployment compensation, union law, tax law, business regulation law – can have important effects on whether a city is an attractive place for doing business. But it is not in a city's power to change these things since control is vested substantially at the level of the States, along with the assignment of some functions to the federal level. Likewise the sorts of sums of money that cities look for in order to attract new businesses through various financial incentives are often beyond their own financial capacity. States, however, are much more effective at raising money through taxation than local governments. In consequence they have become important sources of low interest loans, and of grants for the incentive packages local governments put together to attract particular businesses. The States are therefore very substantially involved indeed in the competition for inward investment. A good deal of their policy orientation revolves around the problems of maintaining a competitive business climate: low labor costs, weak or non-existent labor unions, low State taxes.

Excursus: the American model versus the Western European model

What we have been describing is the politics of local economic development found in the United States. It bears emphasis, however, that this is not the only form that this politics can assume. In Western Europe there is a quite different one. This is not to say that we are talking about a black and white contrast. Elements of the American model can be found in Western Europe and vice versa. But there are also quite clear differences.

In the American case the politics of local economic development is energized by strong local interests. In the case of the local growth coalitions we have attached these to the utilities, developers, local governments, and sometimes the banks and local newspapers. These all have a strong interest in the expansion of particular local economies by virtue of their dependence on them. It is these interests which push for policies aimed at attracting inward investment and which bring the different localities into competition one with another.

In Western Europe this sort of locally driven politics is not nearly so apparent. This is in significant part because major agents dependent on the expansion of local economies, and which might take the lead, are less common. Historically the utilities have not been locally dependent because they have been nationalized industries. The result has been that their markets have been national rather than local. Local governments have not been so dependent on a local tax base since they have received the bulk of their revenues from respective central governments and inequalities resulting from deficiencies in local tax base have been made up by additional contributions from the center. Local newspapers have not been so important because, and unlike in the United States, there is usually a national press. More local newspapers are often taken as supplements to a national paper and many are published weekly rather than daily. Similarly bank ownership has been extraordinarily centralized so that small banks dependent on local economies are a rare species indeed. In short the conditions that would have made possible the sort of politics of local economic development found in the US just have not existed. Rather it is *national* as opposed to local interests in local economic development that have tended to prevail. Local economies have been of interest as objects of intervention more as a means of achieving goals that are less local and more national in character.

This is not just for the negative reason that strong local interests in local economic development tend to be lacking. There are also more positive reasons having to do with the strength of national level interests. Not the least of these has been that of the labor movement. Organized labor has tended to be stronger in Western Europe. Its geographic development has also been much less uneven. Unions have a strong presence in most parts of Britain, France, and Germany, for instance, in contrast to the US where the differentials in rates of unionization between, say, Ohio and North Carolina are extreme. In the US it has been easy for States to try to lure new investment by advertising their propitious labor climates. In Western Europe this has been more difficult and for the most part the unions would not have allowed it anyway (recall box 3.3 earlier in this chapter). Rather they have been extraordinarily vigilant with respect to the possibility of their wages and hence jobs being undercut by cheaper labor elsewhere.

Nevertheless, there is a politics of local economic development in Western Europe. It just happens to be different. By way of example, consider the following policy instruments that have been employed there at various times.

(1) *Depressed area policies.* Unemployment is a fact of life in capitalist economies, though its severity tends to vary over time. It also varies considerably over space. In Britain areas characterized by relatively high unemployment are often referred to as "depressed areas." Every Western European country has its depressed areas. These include such well known examples as Southern Italy and some of the old coalfield areas like the area around Lille and Roubaix in Northeastern France, South Wales, and Northeastern England. Before the reunification of the two Germanies that portion of West Germany adjacent to the border with East Germany had especially elevated unemployment rates.

Every Western European government has presided over policies designed to channel investment, public and private, into these areas. The nature of these programs has varied considerably. There have, for example, been payroll subsidies to firms investing in these areas (see figure 3.6). In some cases these "carrots" were complemented by "sticks." For a long time during the post-

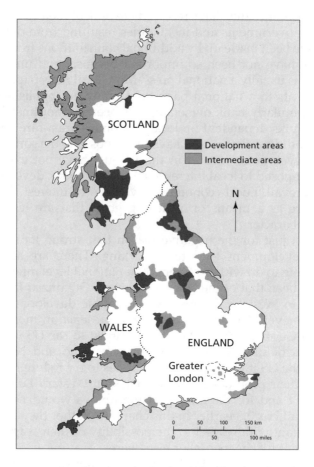

Figure 3.6 Government assisted regions in England, Scotland and Wales, 1993.
Source: P. Balchin and L. Sykora, with G. Bull (1999) *Regional Policy and Planning in Europe.* London: Routledge, p. 110.

war period in Britain any firm planning to expand in the low unemployment areas of Central and Southeastern England had to obtain permission from a central government department, the Board of Trade. In a number of cases the government used its own control of investment in nationalized industries to boost employment in the depressed areas. The Italian government invested in steel mills in Southern Italy, for instance. The British government moved a number of government offices out of London to provincial centers located in depressed areas: the department in charge of drivers' licenses moved to Swansea adjacent to the depressed coalfield area of South Wales; the Department for Social Security moved to Newcastle in Northeastern England.

(2) *Urban growth control policies.* A number of countries have tried to limit the growth of employment in very large cities. These have typically been accom-

Figure 3.7 Britain's new towns.

Source: After M. Clawson and P. Hall (1973) *Planning and Urban Growth*. Baltimore: The Johns Hopkins University Press, p. 203.

panied by attempts to create new growth poles outside of the major centers and to which new office employment – and manufacturing too – could be directed. In Britain after the Second World War successive governments sought to reduce the populations of the larger urban centers by the establishment of so-called new towns (figure 3.7). Sites were selected by a government agency, land assembled, and the necessary physical infrastructure of water, sewer lines, and highways implanted. Much of the housing was built by the state and rented out as public housing. In line with the idea of decentralization and reducing the pressures on the largest cities, particular efforts were made to ensure that these would not be simply dormitory suburbs. This partly explains their location, for the most part at least, at some remove from these larger cities. In addition, in order to provide a local employment base, firms were recruited, often on favorable terms, to locate there. People on public

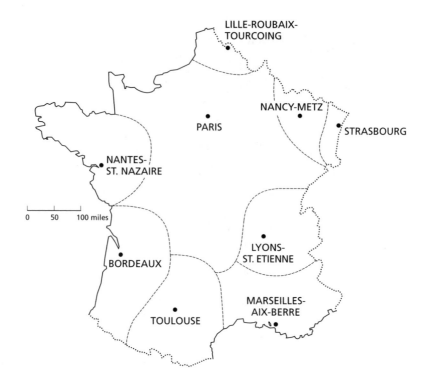

Figure 3.8   France's eight provincial *métropoles d'équilibre*.
Source: K. R. Cox (1979) *Location and Public Problems.* Chicago: Maaroufa Press, p. 154.

housing waiting lists in the major cities were given the option of housing in a new town and firms often brought workers with them.

France has also had decentralization policies but, and given the tremendous preponderance that Paris has in the country as a whole, these have focused primarily on reducing the concentration of the population in that city. In part they worked though restrictions on office development there along with attempts to divert growth to eight major provincial cities defined as *métropoles d'équilibre* (figure 3.8). In addition five new towns were planned and implemented at a distance of just less than twenty miles from Paris though again, as in the British case, with attempts to give them independent employment bases so as to reduce commuting (Goursolas, 1980) (figure 3.9).

The motivations behind these different policies have been diverse. Originally the aim of depressed area policies was almost entirely one of inter-regional equity: trying to ensure that development in the country as a whole would not favor people in some areas relative to people elsewhere. But from the sixties on, as the West European economies boomed and inflation began to emerge as a major policy issue, a new purpose was injected into regional policy. The view became that, to the extent that employment could be shifted from areas where labor was in short supply to those where it was in surplus

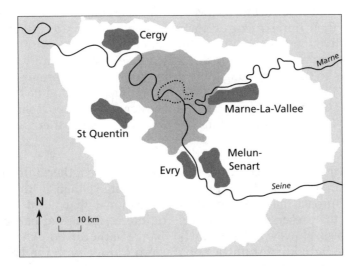

Figure 3.9 The five new towns of the Paris region. Paris is the shaded area in the center.

Source: after Georg Falkenberg (1987) "Die 5 Villes Nouvelles im Raum Paris." *Geographische Rundschau*, December 12, p. 684.

– i.e. the depressed areas – then this would reduce inflationary pressures. Typically towards the end of the growth cycle demand expands ahead of supply. This forces up prices and workers respond to the subsequent increase in the cost of living by demanding higher wages. This further feeds the inflation so that it assumes spiral-like qualities: price increases provoking wage increases resulting in further price increases, etc. Inflation is problematic for a variety of reasons, not the least being that it can price a country's products out of export markets. The medicine for the condition is typically an increase in interest rates. This reduces the ability of firms to invest and that of consumers to buy. It was believed, however, that if growth could be shifted to the depressed areas where the surplus of labor would tend to suppress wage demands then inflationary tendencies could be at least postponed and ideally nipped in the bud.

Controlling urban growth had similar motivations in mind: moving employment out of the overheated labor and housing markets of the big cities. The housing market was an important consideration because as prices there increased so workers would demand increased wages. But there were also efficiency arguments. In order to attract workers, firms in the major urban agglomerations had to pay higher wages so that the higher housing prices could be afforded. Inflation aside this could limit the competitive ability of firms relative to firms in other countries. On the other hand, it was recognized that there were good reasons why firms stayed in the big cities. The major centers offered infrastructural advantages, for example, like high-speed rail transport, airports with good connections, a large pool of labor on which to draw. So in order to attract firms away from these congested areas other cities had to have their infrastructures, their attractiveness to executives as places in which to live, perhaps, built up. In Britain this took the form of the new towns, though note how these are all located in relative proximity to the major urban centers so that by moving there firms would not compromise their need for access to the facilities of those centers too much. London has a constellation of new towns within fifty miles. Manchester and Liverpool share five new towns, as do Glasgow and Edinburgh. There is a similar pattern around Paris (see figure 3.9).

So within the different countries of Western Europe rather different approaches to local economic development have emerged than in the American case. The approach has been much more national than local and this in turn reflects the presence – or absence – of interest groups at particular geographical scales. However, there may be change in the air. For when one compares not the individual countries of Western Europe with the USA but the European Union, a rather different conclusion emerges. This is because there are now clear tendencies for the member countries of the EU to function with respect to each other much as do the States of the United States. In short, the same competition for investment seems to be emerging. The point of the EU is to create the same sort of internal market for goods, capital, and labor as exists in the United States. The move to a single currency is only the most recent step in this direction. As this has happened so the horizons of investors have become community-wide. A firm like Hoover now locates so as to serve the market for the EU as a whole rather than to serve a particular country. The old pattern of foreign investment was to serve country-specific markets since each of the Western European countries was surrounded by tariff barriers. If you located in France then France would probably have to be your market owing to restrictions on getting the products into Germany, Italy, etc. Indeed, simply by erecting tariff barriers a country might be able to attract inward investment since that would be the only way in which access to its market could be obtained.

All that has now changed. Firms like Toyota or Hoover now build to serve the market of the EU as a whole. If they locate in France they know that they will be competing with firms that have decided to locate in the Netherlands instead. Accordingly their field of location choice has become the EU as a whole. If countries are to get the investment, whereas previously they could just erect tariff barriers, now they have to compete for it by offering various incentives. This is for the most part quite new and it is contentious. Countries with relatively low wages and more liberal labor law are attractive to firms that are wage-sensitive. So when Hoover decides to close its plant in Dijon and expand capacity at its plant in Glasgow there are loud protests in France at what has become known as "social dumping." The same concerns have been expressed about differences in national taxation. One of the reasons for Ireland's recent quite extraordinary success in attracting new investment from North America and the rest of Western Europe is supposedly its generous taxation provisions. In the United States, States have learnt to be resigned to that sort of competition. In the EU, as yet, they haven't.[15]

Think and Learn

We have seen that there are important differences between the United States and Western Europe in the politics of local economic development. Think now about how changes in the United States might influence the formation of local growth coalitions, or rather their failure to form. In this instance assume away the local dependence of the banks and utilities and concentrate on the implications of revenue equalization. What would be the effect on the formation of local growth coalitions if: (1) local governments were funded by a mix of local taxes and subsidies from respective States, and the latter payments were scaled so as to increase with local needs and decrease with local tax base; (2) instead of local government funding being shared with the States in this way, the States were by-passed and local governments were subsidized by the federal government?

National growth coalitions

The sorts of local growth coalitions encountered in the US, the interest that local government there shows in enhancing local economic growth, clearly seem a little odd when contrasted with the experience of other countries. This is much less so, however, when we turn and examine other geographical scales: in particular, the space shared by the different national governments. Invariably one of their major concerns is enhancing national economic growth or, as it is sometimes referred to, particularly in less developed countries, development; and this brings them into a relation one with another as they compete to channel and capture the worldwide flux of values.

There are good reasons for this contrast between the local and the national governments. Not the least is that while in most countries local governments get most of their revenue from the national or central government, and so can afford to be indifferent to local rates of economic growth, that logic can in no way apply to national governments: in order to spend they need revenue, and revenue depends on a healthy economy. There are also reasons of an electoral nature. How the economy is doing affects how good people feel about their material prospects and this has been shown to be an important determinant of how they vote. Incumbent parties can expect to be voted back into office if unemployment is low, and wages are increasing; but otherwise, they need to be concerned. The rate of economic growth, moreover, affects the ability of the

---

15 And indeed the idea of enhancing the economies of the more backward regions of the EU continues. Although there are no direct incentives to firms, less developed regions have benefited from community funds granted with a view to improving their infrastructures and so making them more attractive to investment.

government to spend and bolster its support among marginal constituencies, particularly in the runup to elections. As economic growth increases so do government revenues, and government can spend without the unpopular measure of raising taxes or going into budgetary deficit and risking an increase in the rate of interest charged to business.[16]

Finally, there is the fact that governments need money to pay for defense. Defense is a major part of government expenditure. As hereditary enemies arm so there has to be a response. Greece keeps a careful eye on Turkey, Argentina on Chile, Iraq on Iran, Pakistan on India, Israel on most of its neighbors, and so on. Not coincidentally, in countries where defense needs have been felt as highly pressing but development has been lagging, national governments have often taken the lead in organizing for economic development, creating capitalist classes almost from nothing. South Korea and Taiwan, significantly two of the Far Eastern NICs, are cases in point.

As far as specific policies are concerned, in discussing local growth coalitions in the US I pointed out the emphasis on attracting inward investment. Similar initiatives can be observed at the national level, though they are by no means so predominant and we will have to qualify their significance. For many less developed countries attracting in some multinational investment has been regarded as one of the few options they have for economic development. In that instance multinational corporations (MNCs) are seen as bringing employment, as providing some training in industrial processes and, perhaps, in stimulating the demand for locally produced inputs and services.

The sorts of developmental policies which many less developed countries introduced after the Second World War were actually not intended to attract investment from outside. But this in fact turned out to be their effect. These were the so-called import substitution policies. Their aim was to stimulate the production for the domestic market of consumer products which had hitherto been imported. The state facilitated this through protectionist measures against competing imports. But often the investment that took place was by foreign companies. Many of these had formerly serviced markets in less developed parts of the world through exports from their production bases in North America and Western Europe. The imposition of tariffs on their exports altered their calculus and instead in many instances they shifted production for those markets behind the tariff barriers, i.e. into the less developed country. From the standpoint of the foreign company this is a strategy of establishing *clones* around the world: plants that do exactly the same thing, are organized around exactly the same labor process, and do not relate one to another except through a common ownership relationship.

More recently, however, the pursuit of investments by MNCs has become an explicit part of the portfolio of developmental strategies in the Third World. For many such corporations Third World sites are attractive because of their low labor costs. But given the repertoire of labor skills available in less developed countries this sort of decentralization of production is only

16   And so in effect undermining rather than stimulating economic growth.

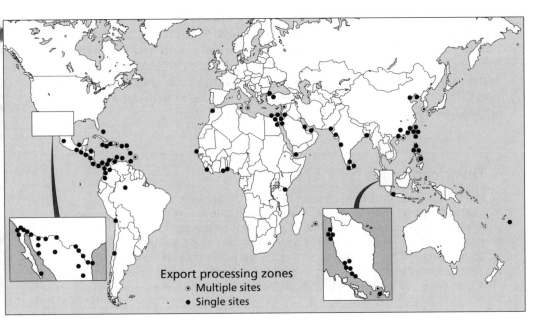

Figure 3.10 Export processing zones.

Source: P. Dicken (1998) *Global Shift*. London: Paul Chapman Publishing, p. 131.

possible for those parts of the labor process which can be carried out with little or no previous skill. Not all parts of the labor process are amenable in this regard so a common strategy for many MNCs has been to separate the different phases of the labor process geographically, assigning low-skill phases to plants in less developed countries (LDCs) and keeping the higher-skill aspects in the more developed ones (DCs). In the garment industry, for example, the skilled design work and pattern cutting may be retained in the DCs while the less skilled and labor-intensive sewing and making up is either done at a branch in an LDC or subcontracted out to companies there. In contrast to a clone division of labor this is referred to as a *parts-process* division of labor. More to the immediate point, however, this geographical subdivision of the MNCs' operations has consequently created new industrializing opportunities for LDCs. Accordingly, many of them now compete for the branch plants of MNCs.

But in contrast to the clone form, this assumes some dismantling of the trade barriers of less developed countries so that raw materials and parts can be imported for their ongoing fabrication. A favorite competitive strategy for LDCs in this regard is the free trade zone (FTZ). FTZs have been established in numerous LDCs or what used to be LDCs, including Sri Lanka, Bangladesh, Indonesia, the Phillipines, Taiwan, South Korea, Chile, South Africa, and some of the Caribbean countries (see figure 3.10). Firms establishing plants in the FTZ are able to import raw materials and export finished products duty free, so long as none of the products are destined for the domestic market. Other attractions include tax incentives and no limits on equity holdings by foreigners. This has been a very common approach to attracting foreign investors into the LDCs. This is partly because the concessions offered fit in with what MNCs want and what they are indifferent to: they're interested neither in using local raw materials and parts, nor in selling the finished product locally; what they're most interested in is importing the raw materials and parts without trade restrictions, and employing cheap labor.

Think and Learn

The establishment of parts-process divisions of labor that straddle international boundaries has its own conditions in terms of trade restrictions, or rather their absence. In the US a good deal of the production of semiconductors for incorporation into computers to be produced back in the US has been shifted to Far Eastern locations. In the case of the European Union this shift has not taken place. What does this suggest about the respective trade policies of the US and the EU towards semiconductors?

One of the more well known FTZs has been established in Mexico just over the border from the US and is now in process of being vigorously exploited by US corporations (see figure 3.11). This particular instance has come in for heavy criticism from the American federation of labor unions, the AFL-CIO, since the closure of some industrial operation in the US is often subsequent to investment in Mexico. But *all* the FTZs are matters of concern for the trade union federations of the DCs since they are seen as taking away their employment base.

Relying on inward investment, however, has its problems. Not the least of these is the intense competition it induces among nation states. As corporations shop around for sites there is a view that this induces what has been called "a race to the bottom": the promulgation of equally low environmental standards, taxation levels, and labor protection standards with a view to providing what is called "an attractive business climate." This clearly does not sit well with labor or environmental lobbies, so these policies often elicit staunch resistance. Moreover, for a good deal of low-skill employment the more developed countries just *cannot* offer the same business climate and hope to keep wages at an acceptable level.

The other problem is that by making the growth of government revenues, incomes, and the profitability of existing industry dependent on drafts of inward investment the government forfeits its control of the sort of development that will occur. An important consideration for LDCs, for example, is whether or not it will facilitate processes of industrial learning. MNCs, on the other hand, often look at LDCs primarily as reserves of cheap, unskilled labor and are quite happy for them to remain so.

Think and Learn

The creation of parts-process divisions of labor – so important for the possibility of competition between countries for inward investment – is a relatively recent development. If it has been a condition for a form of territorial competition at the global level – competing for the different parts of the process – what have been the conditions for parts-process divisions of labor that span international boundaries? How might changing skill levels in different parts of the labor process be important? And what about changing transportation and communication technologies?

Think and Learn (Again)

The distinction between parts-process and clone divisions of labor allows us to link up with an earlier discussion in this chapter. Consider once again box 3.3 on the Ford proposal to establish a parts plant at Dundee which would not be covered by existing agreements with the unions to which Ford workers belonged, and the threat to "black" parts coming from the Dundee plant. What does this suggest about the relative leverage that labor enjoys where the geographic division of labor is a parts-process one rather than clones?

These considerations have been the context for a different policy emphasis: one on enhancing the productive capabilities of workers and of existing firms, and where those firms don't exist, as is often the case in LDCs, forming them. Less developed countries want to control the developmental process so that there is some upward movement in wage levels. In more developed countries the emphasis has been on developing capabilities that will make them immune to competition from the less developed.

In the more developed countries the continual drumbeat of the importance of education is inescapable. All the governments of the developed countries, moreover, invest heavily in underwriting research and development costs. A problem with the development of any new knowledge by a private firm is that its advantages tend to be appropriated to some degree by other firms. They introduce imitations of new products, lure away key scientists and technologists who bring their basic knowledge with them, and so forth. So the payoff to the individual firm from investing in research and development may be diminished. There is, in other words, a case for states providing incentives, perhaps in the form of tax breaks, for firms engaging in such activities, or simply for the states funding research establishments themselves. The goal is to create new niches in the international division of labor in which a country's firms will enjoy a quasi-monopolistic advantage.

Box 3.4   *Upgrading Technical Abilities: South Korea and Taiwan*

The cases of South Korea and Taiwan give some indication of what these polices have entailed:

1   *A very high level of state intervention.* In both South Korea and Taiwan state ownership of the banking system has been highly important. By restricting alternative sources of capital this has given the state the ability to allocate credit. Credit has been rationed to firms, often at low rates of interest, so as to build up those industries which promise not only profitability but also important learning experiences in the path to industrialization. State ownership of the banks has also facilitated control of foreign exchange; this in turn has meant that speculative overseas investments or simply capital flight could be prevented. Apart from state ownership of banks control has also been exercised through state-owned firms. Foreign investment has been monitored so as to favor linkage with domestic firms and the transfer of technology. All this has been carried out against a background of protection designed to facilitate the learning process in classic infant-industries fashion.

2   *Labor productivity.* A noted student of the industrialization process in Taiwan and South Korea, Alice Amsden (1990), has referred to what she calls "strategic shop floor focus." The fact that DC firms rely on product and technology innovation for their competitive advantage means a lot of corporate emphasis on R&D and marketing functions. In newly developing countries, however, which are borrowing the technology of others, the emphasis has to be more on the on the shop floor and on ways of increasing productivity through (e.g.) changes in the organization of labor there.

There has been a similar focus on transforming production capabilities in some of the (NICs) of the Far East, particularly South Korea and Taiwan. But in those instances the focus has been less on the development of new products. Rather the competitive advantage of the NICs in international markets has come from combining more advanced technologies with lower labor costs. In other words there has been a conscious attempt to upgrade technical proficiencies while at the same time maintaining an advantage over existing producers of those products through keeping labor costs down (see box 3.4).

On the other hand, it should be noted that the idea of using technologies developed elsewhere in order to develop is not a strategy confined to the NICs. In Britain successive Conservative governments during the eighties were aggressive in seeking out Japanese investment, particularly in automobiles and consumer electronics. Domestic production in both areas had been decimated. The Japanese were looked to as the vectors for the introduction of new approaches to labor relations, to quality control, and to efficient production. As far as the British automobile industry is concerned, this seems to have been an effective policy (see box 3.5). This suggests that it is not always easy

Box 3.5 *Japanese Investment and the Revival of the British Automobile Industry*

According to a report in the *Financial Times*: "car output [in Britain] is expected to exceed 2 m units a year by 2000. That compares with 1.69 m units last year and is double the 1.02 m cars made in 1986. . . . Such strong growth, which started with the arrival of Japanese car makers in the late 1980s, has transformed the UK's steady decline in the international car-making league. . . . Not since the 'golden era' between 1963 and 1974, when production exceeded 1.5 m units, has the UK built so many cars" ("Lessons from the East." *The Financial Times*, May 21, 1997, p. 29). The article goes on to list five crucial lessons that the Japanese auto companies brought with them and which have transformed the competitive capabilities of the British motor industry, or, as the article puts it, and more accurately, "the motor industry in Britain." These five lessons include: (a) the need for strong exports if investment is to be justified; (b) consistent quality of product; (c) an improving cost base, primarily through the adoption of the Japanese principle of *kaizen*, or continuous improvement in production technique; (d) the ability to transfer *kaizen* to suppliers, along with improvements in quality control; and (e) good labor relations: "The Japanese pioneered the non-hierarchical teamwork approach now prevalent in the UK motor industry. By stressing co-operation between workers and management, meticulous training and greater responsibility for line employees, the Japanese transformed the confrontational labor relations typical of the past," says Professor Garel Rhys, a motor industry expert at Cardiff University Business School (ibid.).

to separate out policies of transforming productive capabilities and strategies of market intervention; since in this instance the concessions that the British government made to Japanese auto companies, the financial incentives they provided, were with a view to more than simply boosting employment and pay rolls. Rather the idea was to enhance the competitiveness of the British auto industry through technical and organizational changes.

Summary

The politics of the workplace, of production, centers on the employment relation between business and labor. Accordingly the geography of labor markets is an essential ingredient of any understanding of the political geography of production. Labor markets are geographically dynamic. There are comings and goings, of business and labor, which affect the leverage that workers in particular localities have with respect to employers. Wage levels and work conditions can be at risk, and even if workers are not locally embedded by considerations of family, friends, home ownership and the like, movement elsewhere may not yield a compensation package comparable to what they have lost. Accordingly they seek, through their representative organizations and political parties, to intervene in and mould those dynamics to their own ends.

There is a standard repertoire of practices. To the degree that the geography of employment in their particular sector of the economy changes, for instance, to the extent that firms relocate to take advantage of cheaper, more pliable labor elsewhere, then one approach is to try to organize those workers. In this way they can put upward pressure on wage levels and so reduce the incentive of firms to relocate. It also allows workers in that sector as a whole to exercise a common upward pressure from which they can all gain. There are other tactics which can protect workers in local labor markets. These include prevailing wage laws, as in the Davis–Bacon Act, or refusing to handle parts coming from non-union plants elsewhere. This is not to assume, however, that the challenges to local labor markets come simply from the to-and-fro of workers and firms within particular national spaces. The economy is increasingly a global one. To the extent, therefore, that firms try to reduce their labor costs by relocating to another country, or alternatively importing cheaper labor from overseas, then we can expect additional strategies from the workers' movement: pushing the national government to exact higher taxes on repatriated profits so as to discourage investment overseas, or imposing severe limits on immigration.

These examples suggest that the idea of the brotherhood of labor is a rather tattered one. And to be sure workers often find themselves split territorially, protecting their own particular labor markets from newcomers, trying to limit the loss of firms elsewhere: an investment that would produce jobs in other places, perhaps in less developed countries. This makes them vulnerable to the appeals of business groups to join with them in place-based coalitions – local, regional, or national – in a struggle to channel value through respective places, and capture and share it in some way. Not that there is anything Machiavellian in this. While business does indeed divide and rule this is often simply a by-product of other strategies. For like workers, firms often have their own stakes in particular places: their own local dependences resulting from needs for particular labor skills difficult to find elsewhere, relations with other firms difficult to reconstitute in other places, and so on. And as with labor again, the continual geographic flux of value, the diversion of what they thought were their revenues to competitors elsewhere, is an enduring challenge. In these endeavors they are often joined by state agencies. Their revenues depend on the buoyancy of business and their jurisdictional boundaries can tie them down to the prosperity of particular places, and at the national level certainly will.

But as these different social forces come together in particular places around a program designed to channel value through it and to the advantage of their respective interests so they come into competition with similar place-based coalitions elsewhere struggling to do precisely the same thing. There is, as a result, a territorial competition, though the scales at which this occurs tend to vary. In the United States local growth coalitions are much more in evidence than in Western Europe. At the national level, however, territorial competition, either for inward investment or through the enhancement of local productive capacities, is the rule.

Furthermore, if the competition between workers and employers in labor markets often gives way to conflict, so too is it with territorial competition. As some place-based coalitions lose in their attempts to secure more desirable positions in wider geographic divisions of labor so the role of the state in structuring outcomes, intentionally or inadvertently, will come under scrutiny. Concepts of territorial justice will be dusted off, and those who are by those terms benefiting unjustly will find their own concepts of fairness which can work to *their* benefit in the ensuing debate. But this is something we will treat later on in chapter 9.

REFERENCES

Amsden, A. (1990) "Third World Industrialization: 'Global Fordism' or a New Model?" *New Left Review*, 182, 5–31.

Foster, J. and Wolfson, C. (1988) "Corporate Reconstruction and Business Unionism: The Lessons of Caterpillar and Ford." *New Left Review*, 174, 51–66.

Gordon, I. (1995) "Migration in a Segmented Labour Market." *Transactions of the Institute of British Geographers* NS, 20(2), 139–55.

Goursolas, J. M. (1980) "New Towns in the Paris Metropolitan Area: An Analytic Survey of the Experience 1965–79." *International Journal of Urban and Regional Research*, 4(3), 405–21.

FURTHER READING

On the politics of regional development a provocative discussion is provided by A. Markusen (1989) "Industrial Restructuring and Regional Politics," chapter 4 in R. Beauregard (ed.), *Economic Restructuring and Political Response* (Beverly Hills, CA: Sage Publications). A useful case study of the competition between localities to shift plant closures on to each other is J. P. Blair and B. Wechsler (1984) "A Tale of Two Cities: A Case Study of Urban Competition for Jobs," chapter 14 in R. D. Bingham and J. P. Blair (eds), *Urban Economic Development* (Beverly Hills, CA: Sage Publications). Competition can also extend to retail facilities. A good study of that process in a particular urban area is J. Paul Herr's (1982) "Metropolitan Political Fragmentation and Conflict in the Location of Commercial Facilities," in K. R. Cox and R. J. Johnston (eds), *Conflict, Politics and the Urban Scene* (London: Longman). On labor unions and their geographies see Andrew Herod (1997) "Labor's Spatial Praxis and the Geography of Contract Bargaining in the US East Coast Longshore Industry, 1953–89." *Political Geography*, 16(2), 145–69. On urban growth coalitions the essential reference is J. Logan and H. Molotch (1987) *Urban Fortunes* (Berkeley and Los Angeles: University of California Press).

On the distinctive approach to economic development adopted by a number of the Far Eastern NICs see A. Amsden (1990) "Third World Industrialization: 'Global Fordism' or a New Model?" *New Left Review*, 182, 5–31. In the Blackwell *Dictionary of Human Geography* (2000) see the entries for "growth coalitions," "new town."

Chapter 4

The Political Geography of Capitalist Development II: The Living Place

Introduction

A significant division in the politics of the advanced capitalist societies is that between the politics of *the workplace* and the politics of *the living place*. This is a distinction that comes about with the development of capitalist society and factory production. As a result of the fact that, with capitalism, people no longer work for themselves but for others, they no longer live, in the sense of residing, at their place of work. Rather the geography of production gets separated from the geography of *re*production. People restore themselves, get rest, enjoy recreation, spend time with their family, at some place other than where they work. The result is a politics *not* of the factory, office or call center: a politics, that is, of work conditions, wages, labor law, and production. Rather it is one of housing costs and conditions, the neighborhood and the amenities it provides, recreation and the "outdoors."

A common tendency, if not *the* tendency, is to treat the politics of the living place as somehow separate from, unrelated to, that of the workplace. This is a view that will be challenged in this chapter, and from a number of different angles. For a start, under capitalism the politics of the living place can be understood in very similar terms to that of the workplace. There is, for example, the same tension between fixity and mobility that we found underpinning the politics of production, and the same subsequent tendency on the part of the locally dependent to form coalitions, usually in the form of neighborhood organizations, to channel the wider flows of value to local advantage. There is also the fact that a good deal of that flux of value is owing to the efforts of local growth coalitions to reorganize the land use geography of the city so as to make it more attractive to inward investment. The resistance to the projects of local growth coalitions that we referred to in chapter 3 is, accordingly, often the resistance of people in their living places to the concomitant devaluation of their properties implied by those plans.

The approach in the chapter, however, is to start out in a way that does not challenge conventional preconceptions about the separateness of the political geography of the living place under capitalism. Rather I want to progressively assimilate the examples discussed here to the conceptual framework employed in chapter 3 and then to a more integrated conception of capitalist political geographies in which the separation of workplace and living place and their respective politics is seriously called into question. Accordingly I start out by describing a number of examples of the politics of the living place as it unfolds in the advanced capitalist societies of the contemporary world. These include the classic housing and rezoning issues, NIMBYs, gentrification, and schools issues. I also make reference to some international contrasts, for the way the politics of the living place is expressed varies from one country to another. I will pay particular attention here to some Anglo-American contrasts.

The second section looks at the broader social context within which these conflicts occur. Crucial here are the property market and the land use planning system. For the people who do the living in the living place are by no means the only ones with interests in it. Developers and local governments are also to the fore and their logics do not necessarily correspond to those of the people resident in the neighborhoods affected by their activities. This is not to say that all the animus of residents is directed in this way. This is because resident groups also compete one with another. They try to push the obnoxious land uses off on to each other just as they also try to attract in what they regard as more desirable.

In the final section of the chapter I go back to the distinction with which we started: that between the politics of the workplace and that of the living place. This is because in many respects this is a false dichotomy. The world of production intersects at many points with the world of reproduction. The land use changes that developers and local governments seek to bring about are often designed to make the built environment of the city more attractive to inward investors, for example. But finding places for an expanded airport, new freeways, new sports arenas, new housing and the like often creates conflicts with the people who live in the areas affected. Likewise the particular form of residential living can be of vital interest to industrial firms. In the US, suburban residential development is extremely low density which means that it is very difficult to live there without an automobile. So the auto industry and the big oil firms have major stakes in seeing that this particular form continue to be produced and reproduced. In short: although we have a separate chapter devoted to the politics of the living place we need to be constantly alert to the connections with production and the sorts of issues discussed in chapter 3.

Some Examples

The availability of housing

In the nineteenth and early twentieth centuries landlords dominated the provision of housing. Most people lived in housing that they rented. Earlier on the landlord might well have been their employer since in order to attract workers many manufacturing companies had to build their own housing. For mining companies this was more typically the case since the location of mining enterprises was determined by the geography of the mineral in question rather than by any pre-existing urban infrastructure. This was the origin of the company town. Later on in the nineteenth century, however, small business-people, professionals like doctors and lawyers, and other people of some means came to dominate the landlord stratum.

Most of the issues that emerged between landlord and tenant had to do with rents and the condition of the housing. The rapid growth of many cities meant that the demand for living space commonly outpaced its supply. This allowed landlords to increase their rents, often to the dismay of tenants who were being asked to pay more for the same amount of space. At the same time, weak building standards, poor sewage and drainage, tiny housing units, frequently in a context of increasing rents, led to growing public concern about housing conditions and the various pathologies of disease, not to say other forms of material deprivation and crime, with which they were frequently associated. The housing question, as it was known, was a major issue in the nineteenth-century city.

Of course, to the extent that new housing units were constructed, to the extent that street car lines were built out from the center of the city to open up new areas for residential development on the urban periphery, so the supply of housing could expand and afford some relief from increasing rents and deteriorating housing conditions. Indeed changes in personal transportation in the city, especially the street car and later the bus and the automobile, were important preconditions for that increased supply of housing that would stabilize rental levels for the majority of the population. So an understanding of the housing cost issue is inextricably intertwined with questions of accessibility and therefore of geography. But more typical in the nineteenth century was housing shortage, with rents and housing conditions as major foci of political concern.

Today, and for the most part, divisions over housing in cities are much less about the availability and price of housing, and much more about the neighborhood advantages (or disadvantages) with which particular units come.[1] The subsequent tensions have been greatly enhanced by the fact of widespread home ownership. This is because home ownership gives people a stake in the value of their property and that value is greatly affected by what goes on in the immediate, and sometimes not-so-immediate, neighborhood. Even so the issue of availability and cost is not dead, nor is it likely to be. Consider just two examples of this:

---

1 This can be very easily grasped by a quick perusal of the real estate adverts in the local newspaper. When selling a house it is clearly a matter of listing not just the number of bedrooms and bathrooms and whether or not the living room has a wood burning fireplace, but also the neighborhood (why otherwise indicate the neighborhood rather than simply the street address?) and school district in which it is located.

1   *Gentrification.* The growth of office employment in many American down-towns has stimulated interest in older, adjacent residential areas typically inhabited until recently by people of lesser means. But as the professionals who work downtown show interest in buying into these neighborhoods the existing residents are no match for them simply by virtue of the way the property market functions: i.e. he or she who makes the highest bid gets the property. Increasing rents in the area mean that existing tenants can no longer afford to pay the rents landlords demand. Even those people who own their homes in the area may be vulnerable. This is because as home values increase so too do assessed values for property tax purposes. And if the owners can't afford the property taxes they have to sell out. True, they will be better off than the renters since they will take with them a nice capital gain. But even so the displacement can be problematic. This is because these neighborhoods often provide important qualities of access to the people living there. There may be some ethnic character to the area, for instance, along with ethnic-specific institutions. For renters it may be one of the very few areas of cheap housing accessible to the low-wage jobs they earn downtown as janitors, night watchmen, short order waitresses, parking lot attendants, and the like. In consequence this sort of displacement is often resisted. This is likely to be especially so where a landlord owning large amounts of property in the area undertakes a large-scale renovation of residential property and this results not so much in a slow, intermittent process of displacement but one of mass proportions.

Think and Learn

The idea of gentrification conjures up a particular set of associations: large cities, housing close to the downtown, residential displacement. But just how particular is the process? Can we think of cases in other contexts – British or French villages, small villages or farming communities on the edge of metropolitan areas that are overtaken by suburbanization – where similar processes of displacement might occur? What do you think are the forces producing displacement in these cases?

---

2   *Low-income populations.* There may be a more general problem of housing availability and cost that is not confined to those adversely affected by gentrification. Some clue to this is given by the prices that blacks pay for housing. It is widely known that for the same quality of housing they tend to pay more than white buyers. The reason for this is a shortage in the housing on offer to them *or* which they are willing to consider. Civil rights legislation notwithstanding there is still discrimination in housing markets and many real estate agents remain unwilling to show houses in "white" neighborhoods to

black buyers. This means that housing is more scarce for black buyers which tends to increase the prices they must pay. To some extent higher prices may also reflect recent separatist tendencies among some blacks: many prefer living in areas of black concentration since this means they can avoid the problems of living among whites. But the result is that they have to pay a premium since the supply of vacant housing that will thereby be available to them will be much lower than that available in the metropolitan area as a whole.

So with this as background, consider the claim often made by developers when they seek a rezoning in order to build housing for a low- to moderate-income clientele. More likely than not this request will be opposed by residents' organizations whose members live adjacent to the site in question. The retort of the developer is then likely to be that if the rezoning is turned down this will limit the supply of housing for the low- to moderate-income and force up the price they have to pay. Of course there is no organization of poor people to support the developer and lend credence to his claims. But it may well be that there *is* an effect on housing prices, unless, that is, the developer simply moves upmarket and the poor people move into the houses vacated by those moving into the upmarket homes.

But even setting aside these exclusions one can make out a case that the way the housing market functions there will *always* be a scarcity and therefore a cost problem which will hit lower-income people hardest simply because they are the ones who have least money to spend on housing. The price of housing depends on supply and demand. To the extent that supply races ahead of demand then price in a particular metropolitan housing market will fall. But as it falls countervailing processes will come into play such that the "surplus" housing is removed from the market. After all, and as large areas of St Louis, Chicago, Detroit, and Bedford-Stuyvesant in New York bear witness, landlords abandon housing. They abandon it because they can no longer make what is to them an acceptable rent, even though it may be structurally quite sound or at least was until they started disinvesting by forgoing maintenance expenditures. The property is then taken over by the municipality in lieu of property taxes that have gone unpaid by the landlord,[2] it is boarded up, and eventually demolished.

The politics of neighborhood

As the briefest perusal of a local newspaper will affirm, what goes on in people's neighborhoods, what sort of development is likely to occur there, is a lively political issue. Along with this goes a characteristic feature of local politics in the advanced industrial societies: the residents' or neighborhood association. These organizations may come into existence around a particular

---

2 Forgoing the payment of property taxes is one of the ways in which landlords "milk" value from properties which only bring in limited revenue because of falling rents. The other way they "milk" value is by studiously avoiding maintenance expenditures.

issue or in some cases may have a more enduring quality about them, retaining lawyers, levying dues, and having regular meetings. But however one looks at it neighborhood activism is a major element of local politics. Just about everybody, it would seem, has concerns about "the neighborhood."

Think and Learn

"Just about everybody . . . has concerns about 'the neighborhood'." How true do you think that is? Is it true of university students who rent private housing? Is it true of the residents of large retirement complexes? What sorts of neighborhood issues might they be concerned with, do you think? And are they likely to form a residents' association to deal with it? If so, why? And if not, why not?

The issues around which they organize are diverse. A concern for property values is often to the forefront and frequently embraced as such. But what is seen as a threat to these values, as well as to other interests, can vary a great deal and from country to country. The following, therefore, represents only the most partial of lists:

* *Schools.* In the advanced capitalist societies schools have become major arbiters of life chances. In job markets formal qualifications, acquired through the schools or through those institutions of higher learning to which schools give access, are now crucial determinants of the sort of job and the sort of income a child can expect in later life. Parents are alive to this. Schools differ in their reputations and this affects residential choices. Moreover the reputations of schools are now diffused in ways that they never used to be. Neighborhood schools and local school districts are listed in local newspapers according to the achievements of their students in state-wide examinations or certifications. A separate section is devoted to schools issues below.

* *"The Environment."* This covers a very large spectrum of issues. People don't want congestion: congestion of the schools their children attend, congestion of the highways that connect their neighborhood with the freeway. They don't want any trace of business activity in their neighborhoods: no gas stations on the corner, no neighborhood shopping centers, etc. The facts of production have to be purged from where they live. Relatedly pollution of all sorts will be resisted: not just the obvious ones like the smoke and noise from a factory but the less obvious ones. Visual pollution is an issue of increasing importance. It overlaps with what I said about purging the facts of production from the neighborhood but includes other things like houses that are "too large" for their lots, as in the case of the "monster houses" of Vancouver, and the increasingly ubiquitous relay towers of mobile phone firms.

- *Open space.* The overall thrust of new residential development has been on the periphery of the city. This owes something to the ease of assembling land for building purposes in such locations. But it is also a matter of popular preferences for views of distant fields and woods. In England possibly the most favored residential sites are in small villages in the so-called Greenbelts that surround most major towns and in which new development is strictly limited. But even the open space implied by large lot zoning is seen as preferable to high density townscapes.
- *Public safety.* Crime is a major public issue in the advanced capitalist societies, though more so in some like the US and to a lesser degree in Britain, than in others like Germany. Crimes on people and on private property are widely publicized. There is also a high degree of awareness that certain types of neighborhood have higher crime rates than others. This is a central part of the image conveyed by the term "inner city." Again, by their residential choices people, to the extent that they are able to, seek to minimize the risk of being a victim.
- *Local taxes.* In the United States, though much less in other of the more developed countries, local taxation is an important component of a local government's revenue stream. This is especially the case with American school districts which are highly reliant on local property taxes. American metropolitan areas, moreover, tend to be fragmented into numerous local government jurisdictions and school districts, each with their own property tax rate (figure 4.1). These tax rates can vary a great deal. In school districts where industrial and commercial property is a large component of the local tax base tax rates can be kept low even while spending significant amounts per pupil in the schools. But in other school districts where most of the tax base is residential and properties are not especially high value, then tax rates may have to be very high in order to generate the same sort of spending per pupil. These differences are reflected in people's residential preferences. On the other hand land use change can threaten to alter, say, the balance between residential and non-residential components with consequent implications for the property tax rates of current residents.

Given geographies of environmental amenity – or disamenity – tax rates, public safety, school quality, congestion, and so forth, some residents will be more advantaged and others less so. The challenge for the more advantaged is to defend their neighborhoods against those changes that might undermine what makes them so. The challenge for those who are less favored is to intervene in the flux of land use change, public spending decisions, and residential mobility so as to include land uses, publicly funded infrastructure, and households which will give them advantages that they do not presently enjoy.

Foremost among the weapons in this struggle to defend and enhance neighborhoods is the *zoning ordinance* and the power that is given to a local government to enforce it. In most cities zoning ordinances go back to the mid-twenties. They were introduced in order to separate what were believed to be incompatible land uses like residential and industrial: few residents

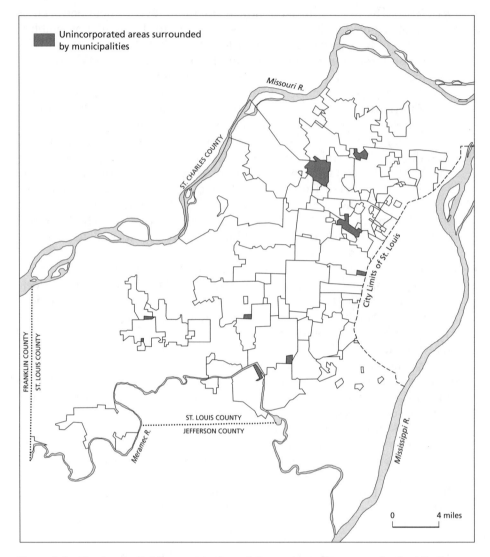

Figure 4.1 The territorial fragmentation of the metropolitan areas in the US: the case of St Louis.

Source: After Figure 4.2 in D. Phares and C. Louishomme, "St Louis: A Politically Fragmented Area." In H. V. Savitch and R. K. Vogel (eds), *Regional Politics*. Thousand Oaks, CA: Sage.

wanted belching chimneys overlooking their backyard. The essential characteristics of zoning ordinances are twofold:

1   They divide the land within a local government's jurisdiction into various land use categories. These might be "residential," "commercial," and "industrial." But in practice there are numerous subdivisions to these categories. "Residential" is often further divided into single-family housing

(or housing for owner occupancy) and rental; each of these categories may be further subdivided into low- and high-density.

2   The relations between the different land use categories are conceived as hierarchical: "residential" is "superior" to "commercial" and "commercial" is superior to "industrial"; "low-density" is superior to "high-density."

These hierarchical relations are essential to the enforcement of zoning ordinances. For while there can be no commercial or industrial uses in an area zoned "residential," residential uses can go into areas zoned "commercial" or "industrial," and commercial development can occur in areas zoned for industrial uses. Clearly, however, there are few residential developers who would want to place housing in areas zoned for commercial or industrial since there would be few buyers for such properties.

It is the rules governing zoning that are typically the focus of residential groups protecting their neighborhoods. Zoning designations can be changed. A developer can purchase land in an area zoned for single-family, low-density housing and then request a rezoning to high-density apartments. For existing residents in the area this is likely to be seen as undesirable. The bases for this judgment may be several. A common perception is that the residents of apartments are typically of lesser means than those in low-density single-family. This can carry over into concerns about how that will affect pupil compositions in local schools; the view, in other words, that the most effective education for the children of the affluent is one that segregates them from children from poorer backgrounds. Moreover, it may not only be the immediate residents who see a problem or just those with children in the local schools. Apartments are commonly regarded as "not paying their way": as adding less to tax base per resident than what the various agencies of local government, including the school board, have to pay on public services, and generally lowering the "social tone" of the area. What these instances suggest, of course, is that the fact of the stratification of the population into households with very different incomes is an important condition for the politics of neighborhood (see figure 4.2). Furthermore, to the extent that higher-income groups are able to prevail, an important consequence of that politics is residential segregation by income.

Think and Learn

Given that the stratification of the population by income gets reflected in the geography of the living place, how do you think that that segregation, in turn, reproduces that stratification? Do you think that it makes it more difficult for lower-income families to increase their wealth? And what about the children? Is the sort of neighborhood you grow up in important for your future life chances? Why might or might not that be? And what does that suggest about the rightness or wrongness of the use of zoning for exclusionary purposes?

Assume that $4,000 must be raised for each household in a school district in order to pay for public schools at the level people want them. Assume further that the only source of money for the schools is the tax on residential property.

RESIDENTIALLY SEGREGATED CASE

	"RICH" SCHOOL DISTRICT	"POOR" SCHOOL DISTRICT
VALUE OF EACH PROPERTY	$200,000	$100,000
TAX RATE	2%	4%
TOTAL PROPERTY TAXES PER HOUSEHOLD	$4,000	$4,000

Note: Since the rich and poor live in separate municipalities in this instance, and average property values differ between the two municipalities ($200,000 versus $100,000), in order to raise $4,000 for schools the tax rate in the "poor" school district will have to be twice as high as in the "rich" school district ($200,000 × 0.02 = $4,000 in the case of the "rich" school district; and $100,000 × 0.04 in the case of the "poor" school district).

RESIDENTIALLY INTEGRATED CASE

	THE RICH	THE POOR
VALUE OF EACH PROPERTY	$200,000	$100,000
TAX RATE	2.67%	2.67%
TOTAL PROPERTY TAXES PER HOUSEHOLD	$5,340	$2,670

Note: In this instance rich and poor households live in the same school district and will pay taxes at the same rate. However, in order to raise $4,000 per household where the *average* property value is $150,000 [($200,000 + $100,000)/2], then the tax has to be set at 2.67%. The result is that the wealthy pay more taxes than they did when they lived in their rich-only school district. On the other hand, the poor pay less than they did before. So the wealthy are, in effect, subsidizing the education received by children from poor families. This in turn is one of the bases for the residentially exclusionary policies of the more affluent.

Figure 4.2 Taxes and tax rates in separate and integrated school districts.

There are, in other words, numerous issues around which opposition can congeal. And the historical record in American metropolitan areas is that it indeed *will* congeal. That residents, in other words, will form a neighborhood organization, hire a lawyer, and make their presence felt at the rezoning hearings and council meetings which are part and parcel of the procedure for achieving a rezoning. The developer will also be represented, but it is rare that he or she will find it plain sailing. Of course, the rezoning may still be granted. But in most States residents have the recourse of putting the issue to a popular vote at the next election. And it is by no means uncommon for rezonings to be rejected when they are in fact voted on.[3]

3   Usually this requires a certain number of names on a petition. In Ohio this number is equal to 10 percent of the voters at the last gubernatorial election. This means that rezonings are much easier to get on the ballot in smaller local government jurisdictions than in larger ones.

There are other instances in which it is the residents that are instrumental in requesting rezonings. A common occurrence in older suburbs, and closer into the central city, is one of residential areas, consisting largely of single-family housing, which are zoned for apartments. The apartment designation is likely to go back to well before the Second World War and reflect a time in which the purchasers of a lot wanted the option of renting out their house if they should decide not to live in it. But today, to the extent that property is indeed converted to apartments or to the extent that the residents see advantages in selling their properties from having the area zoned single-family, then the respective neighborhood organization may indeed make such a request for a blanket rezoning. This is common in areas undergoing gentrification where a zoning designation that excludes apartments is seen as contributing to a more favorable investment climate.

In Britain there are similar mechanisms of land use control that can be harnessed by neighborhood groups. There, any proposal to develop or to redevelop has to run the gauntlet of "planning permission." This is a considerably stiffer test than the American "rezoning permission." Part of the reason for this is that it applies to all and every development whereas in the American case permission is only needed where there is in fact a desire to develop outside of the constraints of the current zoning. On the other hand, there are some zonations in the British case which act as guides to determining the outcome of requests for planning permission. Notable are the *greenbelts* which surround most major British cities and in which permission to develop is extremely difficult to obtain, unless, that is, it is defensibly what is called "infill" housing, filling in spaces in already existing villages; or development that adds to, transforms while retaining a part of, some existing structure. Opposition to development in greenbelts is obviously predicated on concerns about the retention of open space, though just as obviously existing residents gain from that in other ways. This is because the values of properties in the greenbelt greatly benefit from its preservation and from maintaining the scarcity of houses in it.

But to talk about land use control policy and the ability of residents to intervene, and possibly advantageously, is to seriously underestimate the diversity of strategies of which resident groups avail themselves. A recent and increasingly popular addition to this repertoire in the American case is the *impact fee*. The principle here is that when development occurs in a local government jurisdiction it also brings with it some congestion: congestion of local highways, of course, but also congestion of local schools. If this is to be alleviated then public expenditures must be made and in the American instance, given the reliance on local sources of revenue, that means raising local property taxes. The impact fee is a fee charged to the developer or builder for each house built. It goes towards defraying various public expenses like those incurred in the form of new schools, new highways, even the purchase of land for new public parks. This makes development more expensive for the developer and may deflect her elsewhere, thus preserving open space. But even if development does occur existing residents are insulated to some degree from its implications for their taxes. And given that the fee is an absolute amount levied on each unit it also exerts pressure on the developer in favor of higher priced housing where it will add less proportionally to the total price.

But the range of policies available to stem the intrusion of what are seen as undesirable land uses or even undesirable residents – though no one will publicly define them as such – knows seemingly few boundaries. It is an area where it is difficult to be definitive since new approaches are constantly being invented. A recent innovation in land use control has been to ask the developer to negotiate with local residents. Out of this may come some redesign of the development, the positioning of green space between the new development and existing housing, the redesign of highway patterns to take traffic away from existing residential areas, and so on. There is also growing use of the public purchase of development rights from existing landowners. Landowners sell their right to develop the land for housing or some other urban use in exchange for a payment that comes out of local taxes and has been approved by local residents.[4]

Where there is no land use change but simply social change, in the already developed parts of the city, for example, different strategies have to be employed. To make the area residentially more attractive, to retain the middle class or stimulate a gentrification process, neighborhood organizations may put requests to the city Traffic Department to convert some streets into cul-de-sacs. Another area of intervention is schools: alter the catchment boundaries of the local school, so as to make it more socially exclusive and improve the marketability of housing in the neighborhood, for instance.

Schools

As an element of the living place schools are clearly very important. One need look no further than the real estate adverts in the newspapers to gauge this: a set of specifications regarding numbers of rooms, garages, the size of the lot, followed by "Happy Valley Schools." The information about academic success rates in different schools that also appears in local newspapers, both in the United States and in Britain, has facilitated this process of "shopping around." Parents are interested in this because of the way in which formal educational certifications have become so important in the job market: not so much a high school diploma but the grades that will facilitate entry into university and later professional school. But it is not just parents. To the degree that "good schools" are in demand then the value of houses in school districts that have them will be bid up. So every homeowner, parent or not, can acquire a stake in the merits of local schools.

Quite how real the differences are between schools as opposed to being a matter of image would be hard to say. But people *think* there are differences, and act on them in their residential choices. They have also become a political issue. For if school does indeed bestow advantages then most parents will want those advantages for their children. And the fact of differentiation in

4   Typically it is farmers who sell their development rights while retaining the right to farm the land.

school quality suggests that not everybody does in fact enjoy them and this does not necessarily mean that they could if they wanted to.

School quality has become an issue in two, often related, ways: school funding and pupil composition. To take the funding issue first: in the US educational provision is delegated to local school boards which provide education to children living in local school districts. In any American metropolitan area there will be many of these, though typically the pattern will be of one large central city school district surrounded by many smaller suburban ones. School boards are given the responsibility to raise most of their revenue needs through local taxes. The local tax most commonly used is the tax on property, or property tax. This is levied on the value of all private property, residential and commercial. But these values tend to vary a great deal geographically. The consequence is that in some school districts, those with lots of people living in upmarket housing and with commercial real estate in the form of shopping centers and office parks, a very modest tax rate will raise large amounts of money per pupil. While in school districts that are less well endowed in real estate values the same tax rate will raise much less per pupil. Poorer school districts *could* generate the same amount per pupil but this would mean sharply raising the tax rate.

The implications for per pupil spending are highly consequential. In the first place the differences in many States between the poorest and the richest school districts are truly huge (see box 4.1 for an example). Second, money counts. Wealthier school districts can afford to attract better teachers, keep down class sizes, and invest in a broader range of physical facilities and activities (language laboratories, computers, field trips, providing classes in more specialized areas). And the third thing is that the distributional consequences of this arrangement are thoroughly perverse. If it was the children of poorer families who benefited from this arrangement then one might applaud it. The children of the wealthy have enough advantages already so that better, more experienced teachers and smaller class sizes might compensate to some degree for these handicaps. But in fact, and for the most part, the reverse applies.[5] It is the children of wealthier families who tend to live in the school districts with large amounts of assessed real estate value per pupil. In part this is because more expensive houses tend to be valued at higher levels and only the wealthy live in those houses.

At this point one might well ask: "why don't the people living in poorer school districts move into the better endowed ones?" The answer in brief is

5   Again, it is important to point out the exceptions. Some thoroughly working class school districts can afford to spend large amounts of money per pupil without stretching themselves financially. The explanation usually lies in a tax base which includes non-residential land uses of high value: shopping centers, or industrial parks, for instance. This is, incidentally, the reason why the central city school district, while poor in its residential population, is rarely bottom of the spending-per-pupil league in its respective metropolitan area: a relatively large proportion of its land area is under non-residential uses which, while producing no children to be educated, produce large amounts of property tax dollars per unit area. Nevertheless, the important point is that there are significant disparities in public provision, particularly education, which tend to be correlated with the relative affluence or poverty of the population.

that there is very little housing in them that they could afford and the existing residents typically ensure that there won't be much in the future either.
The reason for this has to do with the fact that, as we have seen, local governments have powers of land use regulation. These powers largely revolve
around the power to zone land and to change the zoning. Zoning, however,
is a tool that can be used to exclude, for it has effects on the subsequent cost
of housing and hence on the ability of poorer people to afford it. Zoning,
where it is for housing, can be for "single-family" or "multiple-family."
"Multiple-family" means apartments which usually opens up the housing to
a wider spectrum of qualifying incomes. "Single-family" housing, on the other
hand, can be high-density or low-density. Low-density residential zoning
forces up the price of land for each housing unit and so has greater exclusionary potential.[6] Adding to the shortage of low-income housing in the
wealthier suburbs is the fact that they typically distance themselves from
public housing projects. In the US public housing can only be located in a
municipality with its cooperation, a cooperation which is rarely forthcoming.

If things were different and these exclusionary barriers did not exist there
would be several effects, quite undesirable from the standpoint of the residents of the school districts affected. The first is, of course, that pupil composition would be affected. Their children would have to go to school with
children from less advantaged backgrounds. This might mean that the academic progress of their own children in school would not be as rapid as it might
otherwise be. There would also be concerns about undesirable moral influences. But the second is what we have primarily been talking about: school
taxes, the ability of a school district to generate for a given tax rate large
amounts of revenue per pupil. Lower-income residents mean lower-value residential properties which means that they don't contribute the same to the
local tax base as those living in high-value properties. In order to make up the
difference and maintain per pupil spending at its previous level tax rates have
to go up (figure 4.2).

This inequality, because of the way in which it discriminates against the
children of the poor – all the more embarrassing in a country which prides
itself on the ideal of equality of opportunity – has become a potent political
issue in many States.[7] Typically the challenge has come from poorer school
districts looking for a way to increase their revenues. The ameliorative actions
called for have varied. In some cases the request has been for redistribution
from the wealthy school districts to the poorer ones (see box 4.1 for an
example). In other instances the answer has been seen to lie in dispensing with
local revenue raising altogether and to have the State fund education, albeit

---

6   This is not to say that the exclusionary character of land use zoning is necessarily apparent.
On the edge of the city into which the built-up area is expanding zoning may have been for low-
density for a long time for reasons that have nothing to do with exclusion. Rather it may have
been imposed for public health reasons at a time when the only form of sewerage available was
the septic tank or the leach bed. Yet when developers attempt to obtain rezonings in such areas
in order to build lower-cost housing the rezonings are usually opposed by existing residents who
complain of the effects on congestion and property values.
7   But not at the federal level since it is the States and not the federal government which have
constitutional responsibility for education.

on an equal per pupil basis. But the existing system of school funding clearly has its defenders since some – the children in the wealthier school districts – so obviously benefit from it. Any struggle around inequalities in school funding is therefore likely to be bitterly fought.

Box 4.1   *Reforming School Finance in Vermont*

One of the few States to grasp the nettle of reforming schools finance has been Vermont. In that State the variations in per pupil spending were quite staggering. Manchester spent $5,844 per pupil in 1997 while Whitting, near Middlebury, spent less than 40 percent of that: $2,300. And in order to do that Whitting's tax rate had to be 35 percent higher than that for Manchester. But in 1997, and in response to a successful law suit challenging the (State) constitutionality of these differentials the State enacted a radical program of legislation. In the first place the power to spend for education has been taken away from the school districts and returned to the State. The State has fixed higher property tax rates in those school districts with strong tax bases per pupil and lower ones for those in school districts with less valuable real estate per pupil. Second, the State has equalized the amount of spending per pupil across all the school districts so that they each now get $5,010 per pupil. Third, local school boards can choose to tax themselves at a still higher rate in order to spend more than their allowance from the State but they will be penalized for it. A share of any increase in tax revenue must be forfeited to the State to be shared out among poorer school districts. *The Economist* does not comment on the improvements in the poorer schools, only on the dismay of those in the wealthier ones; but one can imagine that this is a highly controversial piece of legislation with strong advocates and equally strong, not to say vehement, opponents.

This discussion draws on "Education in Vermont: Robin Hood Rides Again." *The Economist*, December 12, 1998, p. 28.

Think and Learn

One of the more outspoken critics of the Vermont legislation has been the novelist John Irving, whose son attends school in a (formerly) richly endowed school district. According to *Time* magazine (June 15, 1998, pp. 34–5) this is what he had to say: "This is Marxism. It's leveling everything by decimating what works. . . . It's that vindictive. 'We've suffered, and now we're going to take money from your kid and watch you squirm.' . . . I'm not putting my child in an underfunded public school system . . . (and, if he can't set up his own private school) I'm moving out of here." And he's avoiding the local press "because I don't want to make my child a target of trailer-park envy." What do you make of this reaction? Do you think Irving is justified in his complaints? Is this "Marxism"? What about "trailer-park envy" and the idea that the reform is "vindictive"?

Now I remarked earlier that there were two major issues in the political geography of education in the US. We have talked about the funding issue. We now need to say something about pupil composition, to which there is more than the issue of "trailer-park envy." The argument here is that this is an important element of educational advantage. The children from more afflu-ent backgrounds often enjoy important advantages. They will likely have been brought up with more books in the house, a culture of reading, TV watching may have been carefully monitored, there is likely to have been more travel, and there will be expectations and encouragement regarding academic achievement and going on to university. The belief among some educational psychologists is that the ambition and motivation of these children can rub off on the less advantaged, often in rather roundabout ways, as in the sorts of assumptions the teacher can make as to what some of his or her pupils want in life. Another consideration has been that to the extent that pupil composi-tions vary, to the degree that some schools acquire a reputation for more ambi-tious, academically turned on students, this will affect the sorts of teachers schools get. All school principals want the more qualified, the more experi-enced teachers. If they can offer them students who are easier to teach, more willing, less difficult to deal with, they'll get them. Accordingly, the students in schools where most of the children are harder to teach for various reasons will, irony of ironies, get the least qualified, least experienced of the teachers. And as soon as those teachers have acquired the experience or an MA degree that makes them attractive to more school principals they will move on.

As a political issue this has lacked the potency of the school funding issue, but with one exception: where pupil composition has been a matter of race. From the sixties on this became a big issue. Blacks were residentially segre-gated from whites; and given the fact that pupils were assigned to schools on the basis of the so-called "neighborhood school concept" – going to the nearest school to where they lived – the result was bound to be racially segregated schools: black schools and white schools. Of course racial geographies had changed. Black areas had expanded in some directions and not others so that some black children might find themselves going to white schools. But the result, more often than not, had been for white-dominated school boards to put a stop to it by simply redrawing the boundaries of the "neighborhoods" served by particular schools.

For some, desegregation was seen as a matter of principle. The neighbor-hood school criterion for allocating children to schools had served as a ruse for keeping white schools white. For others desegregation was regarded as a way of improving the educational performance of blacks. In other words, race was seen as a shorthand for parental income; so bringing blacks, who were assumed to be from less affluent backgrounds, into contact with whites, equally assumed to be from more affluent families, would mean that the cul-tural capital that white children had acquired would spread, contagion-like, to the black children.

The result in short was busing for racial balance. Where it could be shown that school boards had intentionally influenced the racial segregation of schools federal courts ordered them to undo this disadvantage by allocating children to schools in such a way that every school would be racially balanced in its pupil composition. And since the school districts in question were almost always central city ones where children were able to walk to school this meant the introduction of busing, often over considerable distances, to get black children into white schools and white children into what had hitherto been entirely black schools: hence the term "busing for racial balance."

In almost every case busing for racial balance was vigorously opposed, largely by white groups with names like Alliance for Neighborhood Schools. The opposition never came to much. Even so, many avoided what they saw as the negative implications for their children of busing for racial balance by simply moving to suburban school districts. This was the so-called "white flight," though in point of fact it was not so much white as it was middle class since only the more affluent could afford suburban housing.[8] Blacks therefore participated as well as whites but because the proportion of the black population that is middle class is so much smaller the movement was predominantly white in its racial composition.

The result was a substantial compromising of the goal of desegregating schools. While busing for racial balance within central city school districts accomplished, at least in its early stages, a considerable desegregation across neighborhood schools, the effect of suburban white flight was to re-create segregation, but this time between rather than within school districts. It was this outcome that led to calls from civil rights groups, and also from whites who, for a variety of reasons, want to remain in the central city, for the institution of busing for racial balance on a *metropolitan* scale, i.e. the suburban school boards would no longer be providing education exclusively for the children of residents but for a pupil composition quite indeterminate in its geographical origins. The result, of course, was widespread opposition from suburban parents.

Think and Learn

Flight to the suburbs was only one, though perhaps numerically the most important, strategy that people employed in order to avoid busing for racial balance. But what does that mean? Does it mean that people sold their houses in the central city school district and bought new ones in the suburbs? Would there be others who shared the residential rejection of central city school districts but did not share the residential history of those moving from the central city? Other questions that "white flight" raises include: (a) was choosing to live in the suburbs the only strategy of avoiding busing for racial balance; (b) what do you think the implications of busing for racial balance would be on the demographic composition of gentrifying neighborhoods?

8   The "flight" element, in the sense of a movement from central city to suburban school districts, has also been exaggerated. Some of the growth in suburban school enrollments must have come from people moving into the city from outside and choosing a residential location in the context of busing for racial balance in the central city school district. Likewise, the childless who had plans for children almost certainly engaged in similar strategies.

"Redlining"

In many American cities a major neighborhood issue has been "redlining." This is a common practice of those financial agencies like banks and savings loans which lend money for the purchase of residential property – and bear in mind that when people buy houses they almost invariably take out a loan from a financial agency for that purpose. But this means that people in neighborhoods become vulnerable to the logics of those agencies and these are not necessarily consistent with their interests.

A major problem for financial agencies when lending money is to minimize their risk. In the housing market part of the risk attaches to the borrower: is this a person, knowing what one does about his or her sources of income, credit history, and so forth, who will be prompt in paying what is due on a monthly basis, maintain that promptness over a long period of time, not fall into arrears, etc.? But part also attaches to the area in which the housing to be purchased is located. Some areas/neighborhoods are seen as more risky than others. This is due to a variety of factors but two stand out: (a) the trajectory of housing values; (b) the availability of property insurance.

In some parts of the city housing values will be increasing while in others they will be decreasing. Increasing values are what mortgage lenders like because in the event of foreclosure they will definitely get back all the money they loaned. But in areas where values are decreasing they may not, depending on the rate at which values are declining and on the outstanding balance of the principal of the loan. So the risk is higher. Likewise property insurance is not available on a geographically even basis. Housing in some parts of the city has a higher fire risk than in others so that insurance is difficult to obtain. But this refusal by insurance companies is almost certain to be followed by refusal by the banks, since the house is collateral for the mortgage loan: what the bank can seize in the event of foreclosure. In the case of fire and failure to keep up the mortgage payments there may be nothing for the bank to seize.

So in their mortgage lending practices banks in most cities will declare certain areas out-of-bounds. These are the areas that have become known as *redlined*: an area on the map surrounded by a red line. Usually these are inner city areas. Suburban areas, on the other hand, are those in which banks and savings and loans feel that their investments will be secure and mortgage finance is usually readily available, so long, that is, as the borrower is judged to be minimal risk.

Redlining has received a very bad press. In part this is because it has been seen as limiting the housing choices of poorer people. Housing in areas that are redlined is typically cheaper housing. But the major reason for the conflicts it has engendered has to do with the interests of owners of domestic property in those areas. If you own a house in an area that has been redlined, then the market for it is greatly diminished. It is very unlikely that someone will buy it as a place in which to live since mortgage finance has dried up. Accordingly home values plummet and existing owners find it very difficult to move elsewhere. And bear in mind that regardless of the problems in any particular neighborhood people will want to move out for reasons that have nothing to do with those problems: a desire for more or less living space, a change of job necessitating moving to another city, etc.

This is not to say that there is *no* market for these properties. It is, however, a market towards which the existing residents are likely to be ambivalent. For while banks may be unwilling to make mortgage loans to individual buyers, landlords who want to buy the property for rental purposes are likely to be evaluated differently. This is because many landlords, when applying to a bank for mortgage finance, will have other properties elsewhere in the city on which they owe no outstanding mortgage balance. As a result they can offer them to the bank as collateral for the housing that they want to buy in red-lined areas and the bank will be very happy to accept. On top of that, some of the existing owners, despairing of ever selling their properties in order to move somewhere else, may also start renting out their housing.

For those homeowners who want to continue living in the area this can create problems. This is because the slow conversion to rental housing will likely be accompanied by some change in the social composition of the residents. People who might not have been able to afford to buy property in the area, assuming it had not been redlined, can afford to rent it. This will generate all manner of apprehensions, valid or not, among the existing population: concerns about personal safety, local schools, standards of tidiness, etc. The fact that landlords may have purchased with a view to milking the property of its value – continuing to collect rent but with no maintenance expenditures – to be followed by abandonment does nothing to alleviate resident anxiety. In short: residential exclusion raises its head once more.

So redlining has become a social issue. But it is difficult to legislate against. This is because it is so difficult to prove that financial agencies are in fact engaging in it. One can certainly point to the fact that banks are making few loans for purchase in certain areas of the city but they are likely to retort that they have received few requests. This may well be true. But is it that there are few requests since would-be buyers recognize that for property in certain areas of the city it will be difficult to secure loan finance?

Conflicts and Coalitions

How are we to understand these conflicts? In the first place a major conflict in the politics of the living place is clearly that between those who make their money from constructing, renting out, and financing the purchase of residential property – developers, financial agencies, and landlords – and on the other hand residents. It is, for example, the developers who request the rezonings that so often meet with opposition from neighborhood associations. Not surprisingly it is the developers who rail against the evils of exclusionary zoning, who shed crocodile tears for the people "deprived" of housing as a result, and who oppose the imposition of impact fees. Landlords, on the other hand, are the ones who more often than not are at the center of gentrification processes, raising rents which existing, less affluent residents can't afford in order to rent to someone who is so able – but often creating housing difficulties for the people displaced. Landlords also purchase property in redlined areas, converting houses into multi-occupancy apartments so that people of lesser means can move in. And while this serves to alleviate the housing shortage they create elsewhere through gentrification, it makes the long-term residents unhappy neighbors. And finally, of course, there are the financial agencies. To be a homeowner in a redlined area is not fun since the consequence is that you can't sell a house you want to get rid of; and it is the banks and savings and loans that are doing the redlining.

This serves to remind us, of course, and this is the second major point, that it is homeowners that are typically at the heart of the opposition in neighborhood issues.[9] They are more often than not the ones who organize resident associations to oppose rezonings and redlining. This returns us to arguments first introduced in chapter 3 having to do with local dependence and the geographic flux of the space economy. For the people who resist neighborhood change are the ones who can't get out of its way. They are embedded for various reasons, the most obvious of which is home ownership. For most households the home that they own is an important component of their wealth. Diminution in its value can be a serious threat to that wealth. It can, for example, reduce the amount of money that can be raised with a second mortgage in order to pay for the college education of children. Neighborhood change, new developments in the vicinity, can pose a threat to that value. But given the immobility of the home there is no way in which it can be moved and transported into some other location where values are increasing. The problem for the homeowner is to defend this important component of his or her wealth in a particular location. But given the relationship between home ownership and wealth this clearly has a lot to do with social stratification and defending the privileges that accrue to the higher strata.

Intensifying these problems stemming from the fixity of real estate are the costs of relocation. Moving house is not something that people do very often. Selling a house is a time-consuming process and in addition it involves a commission to the realtor.[10] Buying a house is also time-consuming and various fees will have to be paid in addition to the cost of the house. Residents may also be embedded in particular configurations of local social relations that could only be reconstituted elsewhere with difficulty. Children develop friendships through the schools they attend, they feel secure in a particular school, and parents hesitate to disturb them. School considerations are probably the most important of these "social relations" factors but there can be others, like an aging parent on a nearby street who needs fairly constant attention in the form of cooking meals, doing laundry, and so on.

---

9 Though clearly not necessarily in the case of gentrification controversies, where it is lower-income tenants who are much more likely to take the lead.

10 In Britain this may not be a major consideration since realtor (or estate agent, as they are called) fees are lower. But in the US the size of the fee is considerable: typically $7\frac{1}{2}$ percent of the price at which the house changes hands.

It may also be that there is just no substitute elsewhere that is reasonably attainable. The intricacies of mortgages can be important here. Inflation in housing prices means that, even for housing of similar quality, moving will result in a much higher monthly mortgage payment since a new mortgage agreement will have to be entered into. Accordingly, and assuming a constant rate of house price inflation, the longer one has lived in a particular house the more costly it is to move.[11] For those in the retirement age, on the other hand, the problem may not be so much one of the magnitude of the mortgage payments as finding a financial agency willing to grant them one at all. This is because in granting a mortgage banks and savings and loans will evaluate applicants according to whether or not they have an adequate stream of revenue out of which to pay back the mortgage; a regular salary is the most obvious source of such a revenue.

The third consideration in understanding these conflicts is the metropolitan property market. Metropolitan property markets are invariably highly dynamic and while those dynamics can pose problems for people in neighborhoods they can also, as we will see, provide opportunities. From the standpoint of developers, financial agencies, and landlords the metropolitan area is a field of investment opportunities, some attractive, some less so: investment not just in housing but in other land uses which can also affect the residential quality of neighborhoods – office parks, shopping centers, warehouses, and the like. As the investments occur or fail to, so people move in and out of them; into the suburbs and out of the inner city, for example, or away from areas that are undergoing conversion from residential to more commercial uses. These people in turn are not all regarded as equal from the standpoint of those in the neighborhoods receiving them. Some are seen as more desirable neighbors than others: people living in single-family housing rather than in apartments, for example, generally the more affluent as opposed to the less so. Once again, we cannot avoid the fact of social stratification and its implications for the politics of neighborhood.

So another way of looking at the conflicts that ensue in the living place is how different neighborhoods and their residents are situated with respect to this – highly dynamic and ever changing – metropolitan real estate market. For some, and as I pointed out above, it provides opportunities. In the case of gentrification it is not just landlords versus tenants. Those who purchase property from landlords and convert it for their own occupancy are clearly on the side of those promoting the gentrification process. They are the ones who often take the lead in the activities of neighborhood organizations lobbying the city for neighborhood improvements – tree plantings, an area-wide rezoning to single-family occupancy, the refurbishment of cobbled streets, traffic tranquilizers or speed bumps, altering neighborhood traffic patterns – which will make the area still more attractive to people with money. They are interested because they want to see their own property values appreciate and also because they want to see the social composition of the area shift in a direction

---

11   On the other hand, this may be no great problem if the buyer is willing to hold his or her wealth in the form of real estate rather than invest in other financial assets.

that they are more comfortable with: more people like them who, to use the standard stereotypes, don't repair their cars in the street while the portable radio blares, sit on the porch beer can in hand, paint their houses "hideous" colors, or keep pit bulls on the property.

In the suburbs major commercial real estate investments can also be attractive, so long as they aren't too close to the neighborhood. This is because of the fiscal advantages they bring to area taxpayers. A shopping center, depending on its magnitude, can be an attractive fillip to local government and school revenues and so help keep down the tax payments of residents. Indeed, a shopping center, if it brings in people from outside the local government jurisdiction, can be of major significance because it means additional sales tax revenues that are not contributed by residents. And a final advantage of commercial developments is that unlike residential they don't contribute children to the local schools: which means fewer children to educate and so lower taxes.[12]

For others, however, the dynamism of the metropolitan property market is just as clearly problematic: it is their neighborhoods that get the land uses nobody else wants – the rendering plant, the light industrial facility, perhaps the parking lot for the school district's buses. Instead of the boost to home values that would come from the location nearby of an upmarket residential development, home values may decline. And depending on the rapidity of the decline the neighborhood may become a candidate for redlining and a further, even more precipitous drop in home values. Likewise, as property values rise some will be displaced as we saw in the gentrification example.

So it would seem that, depending on the changing geography of metropolitan real estate markets, people in some neighborhoods will gain and others elsewhere will lose. This means that neighborhood organizations have stakes in structuring that property market to their advantage: by opposing rezonings that might have adverse effects, by supporting rezonings that would bring in a major shopping center (so long as it is not too close to their neighborhoods), by supporting measures that will promote gentrification (or vice versa if you stand to be displaced by the process). So people in different neighborhoods, existing residents and new residents, can find themselves opposed to each other and not just to the landlords, developers, and financial agencies that mediate the development process. They all try to attract in, what is for them, the desirable, the property-value enhancing, and push the less desirable off onto others. Accordingly every neighborhood organization can find a reason why the shelter for the homeless should be located elsewhere; why the new freeway should cut through some other neighborhood; or why some other part of the city should get the public housing.

Now what is being described here is, in more abstract terms, not that different from what was discussed in chapter 3, where we focused on political geographies of economic development. Recall the essential ingredients of that

---

12   The irony is, therefore, that American school districts look on children to be educated as a decidedly mixed blessing.

discussion: (a) locally dependent agents with strong stakes in particular places – in profits, wages, rents, and (government) revenues; (b) an inconstant, quite dynamic, space economy which placed the wages and profits on which people depended in doubt; (c) conflicts between workers and employers in particular places as firms sought to compete with firms elsewhere through lowering wages; (d) a competition between coalitions of workers and firms in particular places with similar coalitions elsewhere as firms or coalitions of firms – growth coalitions – sought to persuade their employees of the necessity of belt tightening if values were to flow through the area to the benefit of all.

In the case we are presently discussing we have similar configurations of forces and outcomes. There are certainly locally dependent agents. We have seen how homeowners depend on what happens in respective neighborhoods for their quality of life and for their property values. They are locally dependent because they can neither move their property to avoid its devaluation nor easily move elsewhere. Developers and builders too are locally dependent. They sink money into particular pieces of land and if the major department stores can't be persuaded to "anchor" the shopping centers, or the rezonings aren't given, then they will be in trouble. They also depend on particular, metropolitan, housing markets. They find it difficult to move and operate elsewhere because their business depends on knowing and being known: on knowing the highly specific characteristics of particular real estate markets, characteristics that will be found nowhere else, and on having a reputation with the banks that will lend the money and with the builders that will buy lots in a developer's subdivisions. This information is built up slowly over time so that moving to another property market puts the developer or the builder back to square one, and involves a lot of effort as they try to establish a network and build up a track record with the banks before they can make a profit.

Not only that, there is a dynamism about metropolitan housing markets that recalls the inconstant character of the wider space economy. There are growth areas and there are areas of decline and it is hard to anticipate exactly where they are. This dynamism can be threatening to homeowner and developer alike. They have to defend their values in particular places against the implications of the surrounding flux and this can bring them into the same sort of conflict as that between employer and employee. In order to save their investment in a large commercial development, an investment that looks compromised because the anchor stores aren't biting at the prospect of a shopping center, developers may embark on courses of action that bring them into direct confrontation with residents and their neighborhood associations: putting out the welcome mat for an outdoor amphitheater that will generate lots of noise and raucous crowds. Alternatively they may go cap in hand to the local government and request tax concessions that will mean local residential taxpayers having to pick up the bill, and so on. Or a developer, having bought some land, may find that his original plans for an upmarket development that were welcomed by the neighbors just aren't going to work; that in order to cope with changes in the local housing market she is going to have to ask for a rezoning to higher density or even to – perish the thought – apartments.

But equally there can be coalitions of developers and residents around the particular plans that the developers have for new development or the rehabbing of old. We have seen that in the case of gentrification, the "improving" landlords will have their supporters as they compete with landlords elsewhere for the infrastructural improvements that will help "turn the neighborhood around." These supporters will be homeowners who relish the prospect of the increasing home values that gentrification will bring. They will be more than willing partners as landlords try to alter the geographic flux of the broader property market in favor of their particular neighborhood. And the same goes further out in the suburbs. Residents may welcome the plans of developers for new shopping centers because of the contribution they will make to local tax revenues – money that will, consequently, not have to come out of residents' pockets – and be willing when asked by the developer to make concessions like funding highway improvements that will ensure that the shopping center locates there rather than in some other local government jurisdiction.

Think and Learn

Think more about the parallels between conflict around living place issues and conflict around workplace issues. There is the same combination of local dependence and geographic flux, the same tensions between capitalists on the one hand and those they employ (in the workplace) and affect (in the living place) on the other. There are also place-based coalitions around workplace and around living place issues. What about divisions among the workers and those in the living place, however? Workers in different places, we have seen, come into competition with one another. What is the analogue in the politics of the living place?

These remarks apply regardless of the particular advanced capitalist society we are talking about. Similar tendencies can be observed in Western Europe as in North America. But there are also differences between countries. The incentive framework remains that of a capitalist society – homeowners are anxious about their property values, developers about their profits, and banks about the money they extend for mortgages. But the particular concrete effects of acting within that incentive framework vary somewhat. I want to end this section, therefore, by giving some consideration to differences between the US and the United Kingdom.

(1)   For a start, education is organized differently and this makes an immense amount of difference to resident calculations and hence to those of developers as well. There are several aspects to this. The first is that the sorts of stark differences in per pupil expenditure and therefore in teacher quality and experience and school facilities found in the US just do not exist in the UK. This is

because of the way the schools are funded. Instead of calling on the local school districts ("educational authorities") to raise most of the revenue, the larger part of the money is provided by the central government and in accordance with such criteria as the number of children to be served by a local educational authority, the local tax base and so on.

Second, new schools are funded in the same sort of way. In the US the expense of constructing new schools has been a major reason for the agitation for impact fees: to take the burden of funding new schools away from existing residents and displace it to the developer or the people who will live in his or her houses.[13] Builders typically oppose impact fees since they believe it makes it harder for them to compete with the owners of existing houses who are looking to sell. Conflicts like that just do not exist in the United Kingdom because the existing residents do not pay for the new schools.

Third and finally with respect to education, there is the fact of private schools. These are considerably more important than in the American case. In the United States the children who would go to private schools in Britain go to the public schools in some of the more exclusive suburbs. In consequence, while in the British case exclusion is enforced through the ability to pay fees, in the American case it is through opposing rezonings. So in addition to a variation in the amount spent per pupil, school districts in the US case also vary a great deal in terms of their pupil composition – much more so than in the British instance. This variation means that in the United States schools can become a much more salient element in the calculations of home buyers and therefore of developers there and so energize housing markets and their politics.

Think and Learn

How do you think schools issues and concerns in the United States "energize housing markets and their politics"? Do you think that this would have been true, say, in the earlier part of the twentieth century? If so, why? And if not, why not?

---

(2)   A second difference is that the balance between new development and the old is different. In the US there is always vigorous development of new housing, new shopping centers on the edge of the city. This has effects on the demand for property in more central locations with corresponding tendencies to redlining and abandonment. The United Kingdom is different. Redlining and abandonment exist in some British cities too but not to the same degree. Likewise there is suburbanization but it is slower, and altogether less dramatic

---

13   Who gets to pay the impact fee depends on local housing market conditions. Although nominally it is the builder who pays it, whether or not it can be passed on to the buyer depends on how tight or slack the local housing market is.

in its scope. In fact housing in Britain is just more scarce than it is in the United States and this means that inner city housing markets tend to be more lively and that there are also stronger impulses towards densification: the replacement of older houses on relatively large lots by more houses or even by apartment blocks. Likewise the suburban shopping center – the "out-of-town shopping center" as it is known there – is not in evidence to the same degree.

Part of the reason for this different geographic balance is the fact of the so-called greenbelt. Every British city has one of these. As I mentioned earlier, it is an area surrounding the city in which permission for new development is very, very difficult to obtain. Any American visitor will be impressed by how stark the separation of urban from rural is in Britain. Unlike in the American case, when you get to the edge of a British city you know it. There are villages within the greenbelt but these are also protected from new development except what is called "infill" development or conversions of cottages. These villages are *the* desirable locations and demand for housing there has resulted in a sort of rural gentrification: the displacement of lower-income families who can no longer afford to live there by those who can. But beyond that, the beginning of the greenbelt serves at any one time as a limit to new development and deflects development interest more to the center of urban areas and plans for densification than would be true in the US (figure 4.3).

If, however, the incentives were different in the British case the greenbelt might be more vulnerable. But there are two important differences from the US:

• In Britain owners of farm land do not pay taxes on it. In the United States, as development moves further out, so the value of farmland rises and with it the taxes farmers have to pay. Eventually it reaches the point where it is very difficult to make money farming and the balance swings in favor of selling out for a use – residential or commercial – that *can* afford the taxes.
• In the United States the revenue needs of local governments make them very favorable to certain sorts of development: in particular shopping centers and office parks that can bring in hefty sums of tax revenue. Many residents are in favor of this because it relieves them of some of the tax burden. There is no such incentive in the British case. The major element of local government revenue is the grant from the central government, and to the degree that a local government succeeds in expanding its tax base that grant will go down.

There are other differences worth noting. There is much more central control of local development processes in the British case. The Department of the Environment, Transport and the Regions periodically sends out circulars to local planning departments indicating what its priorities are and these do not necessarily work to the advantage of local development interests. In addition the Department can "call in" what appear to be particularly controversial decisions on the part of local planning authorities and arbitrate them, again, in accordance with the national priorities of the day. One result of this

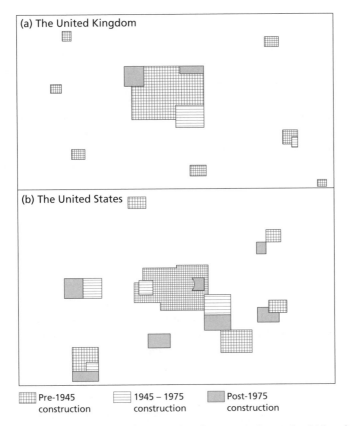

Figure 4.3 Contrasting patterns of urban development: the United Kingdom and the United States. In the United Kingdom cities have tended to be contained by strict planning legislation which has mandated a so-called greenbelt around the city. Within the greenbelt it is very difficult to obtain the necessary planning permission. Expansion to meet the city's needs for additional housing typically occurs either on the edge of the city, as demonstrated in the schematic figure, through densification of already developed areas and, at one time, through the construction of high rise, publicly owned, blocks of flats. In surrounding villages (indicated by places that have clearly not grown since 1945) development is limited by the fact that they are "in the greenbelt." What development occurs is largely what is called "infill" or through the conversion of existing structures from lower-income to higher-income use. The typical pattern in the US is very different. There is no *cordon sanitaire* in the form of a greenbelt, and in consequence housing densities can be much lower as the city expands at its edge. Nearby villages also add new residential development until it joins up the developing edge of the nearby city into a low-density pattern that has become known as "urban sprawl."

has been the difficulty developers have experienced in obtaining planning permission for suburban shopping centers. National governments have tended to be acutely sensitive to the view that to allow more of them would doom downtowns, or what the British call "High Streets," to decay. And knowledge of what has happened in the US has only served to strengthen their resolve.

A False Dichotomy

We have been talking about living place issues: issues that people confront in their neighborhoods. The implicit contrast has been with the workplace and *its* politics. This distinction, however, is in a number of respects a false one. We have had some intimation of that in the previous section of the chapter: the notion that there are important parallels between the processes generating the political geography of the workplace and those producing the political geography of the living place. There is also, however, a more substantive unity between working and living; one can't do one without the other and we should expect that unity to be reflected in the concrete issues that emerge in the living place.

The meaning of the living place

For a start the very concept of the living place in opposition to the workplace does not reflect anything inevitable about social life. The opposition, rather, has been socially constructed. This is not just a matter of the physical separation that came about with the industrial revolution and the factory system: the separation of where one worked from where one lived. It also has to do with meanings. For the meaning of the living place has been constructed – in the media, in politics, in the arts and literature, and with our complicity – in opposition to the meaning of the workplace. What we value in the living place are those things that are denied us in the workplace. In many respects the living place is constructed as a retreat from the workplace.

Consider some of the oppositions involved here and why we might cherish the living place in opposition to the workplace:

- The workplace is where our time is not our own; the living place is where it is. In the workplace all is hustle and bustle. At home we can relax and take our own [sic] sweet time.
- The workplace is the sphere of instrumentality where we do things that we don't necessarily like for some other purpose – to make a living [sic, again]. The living place is where we do things for their own sake – the pleasure of work around the house, pottering around in the garden, playing with the children, watching a favorite TV program, planning holidays, taking days out in the country.
- The workplace is where nature appears as an, often unpleasant, condition of production: the coal face, the raw material, the incessant noise and smell, the heat of the blast furnace, the horror of the slaughterhouse, the danger of the fishing boat. On the other hand, the living place is where nature is experienced as something pleasurable: as winding streets through a park-like setting, distant views of the ocean, mountains or simply of fields of wheat, ravines and babbling brooks, forests that are not obviously cultivated for some commercial purpose.

- The workplace is where our lives are organized according to the principles of the market, where the relation between one person and another is determined by the cold logic of the cash nexus. The living place, on the other hand, is where life is organized according to another logic: that of the close relations of the family, acquaintance with neighbors, volunteering as "Big Brother or Sister," to coach soccer, and the like.

In short, the workplace has a set of undesirable associations and we try to construct the living place as what the workplace is not. We don't want to be reminded of those associations so we strive to purge the facts of production from where we live, and this is one of the effects that land use zoning secures.

Think and Learn

What do you make of the idea that the facts of production are purged from the living place? Is there no production in the living place? Is it more devoid of the facts of production for some than for others?

This pattern of trying to escape the facts of production is as old as the industrial revolution. The industrial masters led the way by establishing homes in the country, often very big homes like William Randolph Hearst's Hearst Castle in California, or else, as in England, they simply purchased a castle or a country house. But the masses followed as soon as they could. As the workday diminished in length and mass transportation improved so began the trek to the suburbs where they hoped to re-create, if on a more modest scale, the Arcadian idyll that their employers had already established as worthy of emulation.

On the other hand, it is also clearly a dualism for some rather than others. Women, employed or not, would protest at the idea of the home as a place where their time is their own and where they don't have to do things they don't want to and they can take their time doing it. Cooking, cleaning, making beds are for most chores, and are as much work as working on the assembly line. So this suggests that the living place as the sphere of consumption, of leisure, is more a male fantasy which women may find difficult to share. For many women the living place *is* the workplace. Likewise there is even an element of coercion for men. People may enjoy mowing the lawn, but there again, they may not. Yet it is not a good idea to refrain, for apart from the moral pressure of neighbors many cities have ordinances that can compel keeping the lawn tidy. And how can one say that the living place is a way of getting away from the pressures of the market when every homeowner is concerned about the value of his or her home?

The living place as complement to the workplace

For business the city is first and foremost a workplace and must therefore conform to the needs of profitability. But since the city also includes where workers live, it necessarily follows that living place conditions must also be consistent with those ends; and where they are not, then they must be reorganized.

Business needs workers who have the correct skills, are not too expensive, and are compliant. This reflects on living place conditions in a number of different ways. Different sorts of business require different sorts of labor competencies. Blue collar households will produce one sort and white collar households another. White collar households tend to produce white collar children. This means that to the extent that employment in a city has a strong professional/technical bias, local businesses will view the arrival of heavy industry with some trepidation. In addition, as occupational composition shifts so too does the ability of the city to attract different sorts of workers. Accordingly local economic development policy can be a matter of some concern to major employers. More generally this concern for socialization into particular workplace aptitudes is also reflected in the interest business takes in the schools.

Housing costs can be another issue. The wages that workers demand reflect the cost of living that they have to meet and housing is a major element in that cost. In expensive housing markets workers may make adjustments for increasing rents and housing prices by moving further afield but this is likely to increase their commuting costs, and these too will be reflected in wage demands. Housing costs have been a major issue for employers and for local governments in the Silicon Valley area. The zoning policies of local governments have tended to be generous towards industrial and commercial land uses and frugal with respect to residential. The result has been perverse: more employment as a result of increasing industrial and commercial uses but insufficient housing for the workers. Major employers have had trouble hiring workers from outside the area owing to the high housing costs there and have been faced with increasing wage demands from existing employees. This led to the formation of a major lobbying group in the area, The Santa Clara Manufacturing Group, with a view to intervening in the land use planning process and increasing the supply of land for residential purposes.

But what goes on in the living place also feeds not just into the labor process and the cost of living and therefore wage levels, but also into the politics of the workplace. Factories led to large-scale urbanization and urbanization has always been problematic for business. This is because of the way in which it brought together large numbers of workers in the same place; for this in turn had a number of important implications for the growth of an organized working class. In particular it allowed the achievement of thresholds for organization and for the publication of newsletters and newspapers that would, through their effects on the working class's self-understanding, underpin that organization. Organizing a political party or a labor union requires resources: not just people but also membership fees. Numbers allowed the achievement of that critical mass that would generate the resources necessary for (e.g.) a full time secretariat, publications, campaigns for union recognition, legal expenses, the payment of bail in case of arrest, and so on.

To some degree industrial capitalists tried to forge their own solutions to this problem. In the nineteenth century many evinced interest in the formation of *planned communities*. These would be built around or next to their factories, and through their internal arrangements and their location, they would help to sequester their own workers against the growing tide of industrial unrest. There are numerous examples of these planned communities. In the United States they included George Pullman's creation, Pullman on the south side of Chicago. Gary, at a substantially greater distance from the center of Chicago, was another instance, established in the opening years of the twentieth century by US Steel. In England examples include Sir Titus Salt's Saltaire just outside the northern city of Bradford; and Bournville, established by the Cadbury chocolate company on the southern outskirts of Birmingham (figure 4.4).

Isolation from existing urban centers was a major feature of these projects. And while it might be argued that this was necessary if an entirely new settlement was to be constructed it also had other attractions: in particular, isolation from the hotbeds of labor unionism in the city itself. The same end of facilitating labor control was apparent in the internal design of these communities. Pullman built churches and libraries but excluded taverns. Churches and libraries, he believed, would counter the development of working class consciousness: churches through spiritual uplift and showing an alternative way to salvation; and libraries through education. Taverns, on the other hand, merely led the worker to waste his or her money on drink, leaving less for essential needs, and so provoking demands for increased wages.

Dense concentrations of working class people living together were therefore seen as a threat to industrial peace and stability: a threat to the balance of political forces in the workplace, in other words. To some extent businesses tried to cope with this through their own locations. David Gordon (1977) has noted how, in the early years of the twentieth century, industrialists evinced growing interest in sites away from, but accessible to, the major industrial centers: in short, industrial satellite towns or suburbs. Around Chicago these included Chicago Heights and Hammond; Lackawanna next to Buffalo was another one; as was Norristown next to Philadelphia. This has continued. New centers of capitalist development have continued to come about in areas that are virgin from the standpoint of labor organization – areas like Silicon Valley and Utah's Software Valley or the sites of Japanese automobile production in the US – and this may be an important source of their competitive success. To the extent that it is possible firms may decentralize their production functions into the less militant environments of small towns and hitherto agricultural areas: an important source of the rural turnaround and the rise of the Sunbelt in the US.

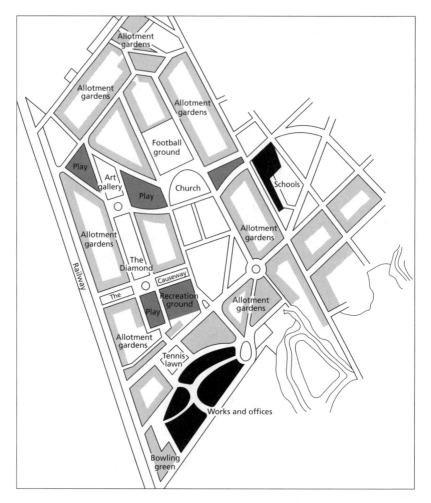

Figure 4.4 The planned community of Port Sunlight. Port Sunlight was created by Lever Brothers (the antecedent to the present-day Unilever). Note the provision for the morally uplifting in the form of churches, an art gallery, allotment gardens instead of taverns, and liberal – extremely liberal! – allowance for recreational space.

Source: P. Jackson (1989) *Maps of Meaning*. London: Routledge, p. 83.

Part of what is involved here is what Dick Walker (1978, 1981) has called "the suburban solution": that through suburbanization into immediately adjoining areas as well as into extended urban regions labor militancy could be held in check. But in addition to the arguments set out above, and as Walker has made clear, suburbanization has broken up the old, dense concentrations of working class people and helped to produce new forms of social life. These center more on the family and family-based consumption and less on the classical loci of working class life like the saloon, the working men's club, and the union local.[14]

14   See, for example, Alt (1976) and Rosser and Harris (1965).

But if geographic arrangements are important to the tractability of workers, to labor peace, so too are they important in the actual physical process of production. As I argued earlier, geographic arrangement is a productive force. Different land uses exist in relations of complementarity with one another. If the productivity of workers is to grow, then businesses need to expand in order to take advantage of economies of scale; they need to locate close to one another to take advantage of various external economies such as the pooling of labor reserves and the deepening of the division of labor between them, and this in turn means expanded public facilities – new freeways, expanded, possibly new, airports, new reservoirs to assure the water supply; in some instances houses may have to be cleared away in order to allow factories to expand.[15] In short, in the interests of the heightened productivity that underpins increased profitability, cities have to be physically restructured: room has to be found for these various functions.

This clearly has the potential for an antagonistic relation with people where they live. Any newspaper in the advanced industrial societies is full of stories about land use conflicts: about the opposition of various neighborhood groups to new freeways, new industrial parks, expanded airports, and the like. And to the extent that the geography of the city is reorganized in these ways, there are indirect effects which can further impact people in their living places. As central cities shift from a retail to a corporate headquarter/office function, for example, waves of gentrification have been induced in the immediately surrounding residential areas. As we saw earlier this in turn has generated controversy over residential displacement.

The forces pushing for these land use changes, moreover, are not just particular firms, developers, or local governments. To the extent that the land use changes are brought about, so the city can expand and this is of intense interest to the developers, the utilities, the local banks, and all those belonging to the local growth coalitions whose activities were discussed in chapter 3. They will almost certainly bring their full weight to bear on what are perceived to be major land use projects that facilitate the growth of the local economy, therefore, and will use all the resources at their disposal to push them through. These include the money they can draw on to fight referenda, and to fund the elections of those who will be friendly to their plans and projects. There will inevitably be attempts to influence wider publics through discourse; arguments about jobs, property taxes, the future of "our city."

They will not win all their battles. But to the extent that they fail to transform the city's geographic organization, its efficacy as a productive force, then investment will shift elsewhere, perhaps to greenfield sites where there will be no opposition, perhaps to cities where by virtue of their limited prospects residents can be persuaded to make the sacrifices. And as this happens so the balance of opinion in cities, by now branded as development-hostile, may begin to shift. There are, in short, important forces tending to the subordination of the geography of living places in the city to the city's function as a workshop.

---

15   See, for example, the Poletown case and the clearance of a large area for a new Chrysler plant in Detroit in Jones and Bachelor (1984).

The living place as source of profit

The living place is connected to the workplace in another way: it itself has to be produced and in a capitalist society its production is a source of profit. Many businesses and therefore many workers depend on the living place as commodity: for example, all those who construct the houses, install the streets and the utility lines that allow the houses to be built, build the shopping centers to serve the new residents, finance the construction, etc. Real estate development is big business. Much of the dynamics of living place conditions comes precisely from this fact. There are two things to consider here: (a) the cycle of construction and destruction in the living place; and (b) the low-density form of residential development. We consider each of these in turn.

The cycle of construction and destruction in the living place

To start with reflect on the dynamics of the residential geography of American cities. On the one hand there is very active construction of new residential subdivisions on the edge of the city – what we understand as the immediate precondition for suburbanization. Less noticed is a process of abandonment at the center of the city. This is more apparent in some cities than in others and depends in part on conditions in the local real estate market. In booming markets like San Diego or Boston it will be less apparent than in cities that have recently grown much less rapidly in terms of population: Detroit, St Louis, Buffalo, Cleveland would be instances. But residential development does not obey a simple demand-side logic. It is not as if the construction spigot can be turned off easily when fewer people decide that they want to live in a particular city. There are large numbers who depend on the vitality of a city's real estate market: not just the developers, the land speculators, and the savings and loans but also the firms that do much of the subcontracting – the plumbers, the electricians, the carpenters, the bricklayers, the innumerable backhoe firms. So instead of waiting for the demand to materialize developers make it happen, not necessarily by attracting new demand into the city, though they will certainly support efforts along those lines, but through developing new types of housing designed to titillate the consumer's palate.

Think of the way the American city is structured. As you go towards the edge of the city housing becomes more "up-to-date," incorporating new ideas: from houses with no garages, to ones with one-car garages to those with two-car garages at the very edge of the city; from houses with small lots to ones with larger lots. Subdivision design also changes: the grid pattern of streets is displaced by winding streets and cul-de-sacs; developments start incorporating golf courses and lakes as additional attractions to would-be buyers. Innovation in real estate products is a way of life for developers, and necessarily so.

For it is driven by competition. Developers and builders have to innovate or die. This is not just the competition of one developer with another as they seek the consumer's money. It is also with existing homeowners and landlords. Buyers can buy a new house or an existing one. New houses will usually be more expensive since the costs of labor – and housing construction is labor-intensive – keep on increasing. But developers and builders compensate for this by giving their products something new that older houses don't have. If abandonment of property at the center of the city has a cause, this is it. Supply can leap ahead of demand because developers and builders have to keep on "supplying." As a result there will be too much housing for the demand and some will be taken off the market, possibly cleared for some other use – a freeway, a downtown arena, perhaps – or simply abandoned. It is like a game of musical chairs; when the music stops it is the homeowner or the slum landlord who is left without a chair while developers and builders remain in the game.

Think and Learn

Based on our earlier discussion of some of the factors affecting property development in Britain, would you expect there to be more abandonment of housing there than in the US, or less? Why do you think that?

The low-density form of residential development

The residential development that occurs on the edge of the city – suburbanization – always seems to be lower density than what preceded it. Lots tend to be bigger; houses are complemented by lakes, golf courses, and other "community facilities"; houses get bigger too, and so more space gets consumed per resident. Yet people need to move around. Given the separation of living place and workplace, above all they need to get to work. And since we are not talking about self-sufficient peasants who live and work at the same place they also need to get to the shops. The vast majority, of course, do so through the automobile. And it could not be otherwise.

Residential densities are far too low to make mass transit – buses or light rail – profitable. Only with more geographically concentrated populations can you fill buses or subway cars. If people are spread out then it would take a vast number of transit lines with very few people patronizing any one route in order to serve them; and while they might be served, the result would not be profitable for the providers. So everybody uses the car and the form of real estate development assumes that they will: not just the house on a large lot but the large shopping center or hospitality–entertainment complexes with their acres of parking space and which are virtually inaccessible without a car. In short, reducing dependence on the automobile would require residential densification: the tearing down of existing structures and their replacement with high rises for a start; weaning people away from having lots and being happy instead with a balcony; replacing the single garage with a communal one in the basement of the high rise. This would involve a cultural transformation if not, perhaps, a revolution. But the biggest obstacle probably lies elsewhere: major businesses have important economic stakes in keeping things as they are.

An automobilized society is an immense source of profit, not just for the auto producers but for the gasoline companies. In consequence they are far from friendly to proposals for shifting resources away from, say, road construction, to the provision of mass transit or to an increase in the gasoline price that might signal a real commitment to reducing air pollution in cities and easing some of the pressure on the world's fossil fuels. In addition, the auto producers have important supporters.[16] Auto production is a huge employer in the US. It is a propulsive sector for the economy as a whole because of the demands it makes on other industries – the glass, rubber, plastic and aluminum producers for a start – and the secondary effects it has on (e.g.) those who repair automobiles, the concrete mixers who depend on the market provided by highway repair and construction, and the insurance industry which insures against injury, damage, and theft. It is part of the very weft and woof of the economy as a whole so that it can always call on a vast amount of support at the slightest possibility that there might be a challenge to its supremacy as *the* way to move around.

And there are other allies. Low-density development at the edge of the city represents a very large amount of debt. Banks and savings and loans hold that debt. If, for example, there was a thoroughly radical increase in gasoline taxes to, say, the levels prevailing in Western Europe, then property values at the edge of the city would suffer greatly. Far fewer would want to buy there. Existing owners would find that the value of the house would be exceeded by the remaining debt on the mortgage; at which point some would certainly be convinced that there were better investments for their housing money elsewhere – closer in to the center of the city – and simply walk away. The financial agencies would then foreclose and try to get what they could out of it, which would in most cases be considerably less than the value of the outstanding debt.[17]

In short, low-density development corresponds to a particular form of living place. But it is far more than that. It is quite simply a cash machine. And that connects the living place back to the workplace.

Summary

In chapter 3 we talked about issues that define a politics of the workplace under capitalism. But there is also a politics of the living place, and necessarily so. This is separate from the politics of the workplace because with capitalism there is a separation of living in the sense of eating, sleeping, relaxing, meeting neighbors, from the place where one works: typically the factory or the office. Accordingly the issues are different ones.

---

16 See, for example, the discussion of this in Whitt (1975).
17 On the general topic of the relation between urban form and energy consumption see the excellent paper by Walker and Large (1975).

Historically a major living place issue was housing. To some degree this continues. Disputes about gentrification are a case in point. But the availability of housing space has been largely displaced as an issue by concerns about neighborhood. Neighborhood or resident organizations come into being, particularly in areas of owner-occupied housing, to contest rezonings, planning decisions that they see as threats to their amenities, their open space, their schools, and their property values. Towards the center of the city other tensions emerge as a result of the mortgage lending practices of banks. This is the problem of redlining. Banks are wary of lending money for the purchase of homes in areas where home values are declining since from their standpoint such loans are poor investments. This, however, can leave existing homeowners stranded since if they want to move out to a different neighborhood they have difficulty finding buyers for their houses.

In the United States another layer of concern is added as a result of the widespread fragmentation of local government in urban areas. As a result of this, local taxes can be a major issue and many local governments, often with resident support, structure their land use policies so as to enhance the local tax base and also minimize the demand on it in the form of expenditures. So shopping centers are attractive investments because they increase the flow of government revenue without increasing the need for some expenditures, like those on education.

What we confront in discussing the politics of the living place, therefore, is another set of territorial strategies implemented/supported by those who happen to be embedded in various ways in particular neighborhoods or local government jurisdictions. This accounts for the prominence of homeowners in the politics of the living place. A major concern for them is the flux of land use that characterizes the property markets of urban areas. Urban areas are characterized by continual movements of people and employment from one part of the city to another, breaching neighborhood and local government boundaries alike and often posing threats of various sorts to existing residents.

Major instruments of this change, of the movement of money which underlies it, are the developers, the financial agencies, and the landlords. Accordingly these are often anathema to the resident and neighborhood associations as they invest in new areas and disinvest, as in the case of redlining, from others. This is not to say that these forms of capital are always mobile so that if development plans don't work out the investments made can be liquidated. What makes developers and their financial backers dig in their heels in land use disputes is often the fact that they have already put significant amounts of money into the ground, devoted resources to planning the project, resources that cannot easily be transferred to other projects elsewhere. And failure to secure the necessary land use permits can result in loss, particularly if the price originally given for the land reflected confidence that the permits would be forthcoming.

It is not always the case that the developers and their allies are opposed. Some of their plans may be welcome ones, or at least some may be more welcome than others: shopping centers instead of residential and single family as opposed to rental housing, for instance. Part of what is at stake is fiscal:

how the new developments will affect the balance of local government revenues and expenditures. But there are also questions of property values. The
result is that neighborhoods and local governments, sometimes in coalition
with developers, landlords and financial agencies, often find themselves in
competition one with another to attract in the wanted and to push the
unwanted on to others.

These processes, however, unfold against a background of national difference: of difference in fiscal systems, in land use regulation, in short in the
incentive frameworks confronting landowners, residents, the financial agencies, landlords, developers, and local governments. Education is a major
concern of parents in both the US and Britain, but it plays itself out in different ways in the two countries. This is partly to do with a more geographically
egalitarian way of funding schools in the British case but also with the greater
prevalence of private schooling there. For the effect of the latter is to divorce
for many of the wealthier the question of schools from that of neighborhood
or indeed local government. Likewise urban housing markets in Britain are
different in their outcomes. For various reasons – greenbelt legislation, the fact
that agricultural land is untaxed – the addition of new housing at the edge of
the city is much less vigorous and this has tended to preserve the vitality of
inner city housing markets. As a result redlining has been relatively muffled
as an issue.

But to talk about a politics of the living place in abstraction from a politics
of the workplace – something we are in a sense invited to do by the separation of home from work – is limiting. There are important connections between
the two. In a number of different respects the distinctions between living and
working and between their associated politics are false dichotomies. What
happens in the workplace affects what happens in the living place and vice
versa.

We look for things in the living place that we are denied in the workplace:
a less alienated relation to nature, a less instrumental approach to others,
among other things. In other instances the relationship between living and
working is more complementary. Parents are anxious about neighborhood
schools because of what it implies for the work prospects of their children.
And on their side, employers have always worried about the living arrangements of their workers because of their belief that they can carry over into the
workplace in the form of worker resistance. This helps explain the nineteenth-
century interest in the creation of model communities and the ongoing
attempts of some firms to escape the big cities for what they regard as the less
class polarizing conditions of smaller towns.

Finally, of course, the living place is itself a commodity or, through its geographic form, mediates the sale of other commodities: it is, in short, a tremendous money-spinner. Developers develop residential neighborhoods, banks
lend money for that purpose and then to the buyers of the housing, because
it is profitable. The form of the subsequent development has likewise been
important for other businesses. For while the emergence of the low-density
suburb helped spell the demise of mass transit it created a demand for the
automobile, for the gasoline to fuel it, and for the highways on which it is

driven that has been an immense fillip to national economies, particularly that of the United States where the "low" in low density is about as low as it gets.

REFERENCES

Alt, J. (1976) "Beyond Class: The Decline of Industrial Labor and Leisure." *Telos*, 28, 55–80.

Gordon, D. M. (1977) "Class Struggle and the Stages of American Urban Development." In D. C. Perry and A. J. Watkins (eds) *The Rise of the Sunbelt Cities*. Beverly Hills, CA: Sage Publications.

Jones, B. D. and Bachelor, L. W. (1984) "Local Policy Discretion and the Corporate Surplus." In R. D. Bingham and J. P. Blair (eds), *Urban Economic Development*. Beverly Hills, CA: Sage Publications, chapter 13.

Rosser, C. and Harris, C. C. (1965) *The Family and Social Change*. London: Routledge and Kegan Paul.

Walker, R. A. (1978) "The Transformation of Urban Structure in the Nineteenth Century and the Beginnings of Suburbanization." In K. R. Cox (ed.), *Urbanization and Conflict in Market Societies*. Chicago: Maaroufa Press, chapter 8.

Walker, R. A. (1981) "A Theory of Suburbanization: Capitalism and the Construction of Urban Space in the United States." In M. Dear and A. J. Scott (eds), *Urbanization and Urban Planning in Capitalist Society*. London: Methuen, chapter 15.

Walker, R. A. and Large, D. (1975) "The Economics of Energy Extravagance." *Ecology Law Quarterly*, 4, 963–85.

Whitt, A. (1975) "Californians, Cars and Technological Death." In H. I. Safa and G. Levitas (eds), *Social Problems in Corporate America*. New York: Harper & Row, pp. 17–23.

FURTHER READING

A classic on the mechanisms of neighborhood change, and still worth reading, is C. L. Leven, J. T. Little, H. O. Nourse, and R. B. Read (1976) *Neighborhood Change*. New York: Praeger. It should, however, be supplemented with studies of the politics of neighborhood change. The more useful of these, though again old, include: H. Molotch (1972) *Managed Integration: Dilemmas of Doing Good in the City* (Berkeley and Los Angeles: University of California Press); and Y. Ginsberg (1975) *Jews in a Changing Neighborhood* (New York: Free Press).

The literature on gentrification is huge. A start can be made with N. Smith and P. Williams (eds) (1986) *Gentrification and the City* (London: Allen and Unwin), but websites would also be helpful.

On the neighborhood activism of home owners, an excellent case study of home owner organizations in the Los Angeles area is provided by Mike Davis's "Homegrown Revolution," chapter 3 in M. Davis (1992) *City of Quartz* (New York: Vintage Books). For a more analytic approach see: K. R. Cox and J. J. McCarthy (1982) "Neighborhood Activism as a Politics of Turf," in K. R. Cox and R. J. Johnston (eds), *Conflict, Politics and the Urban Scene* (London: Longman). A more advanced treatment that also extends to the organization of tenants can be found in J. E. Davis (1991) *Contested*

Ground (Ithaca, NY, and London: Cornell University Press). Davis also sheds light on fights around gentrification.

Useful materials on US–British contrasts are not as plentiful as they should be. Two of the more notable offerings, even though one is now quite old, are: M. Clawson and P. Hall (1973) *Planning and Urban Growth* (Baltimore: Johns Hopkins University Press); and H. Wolman and M. Goldsmith (1992) *Urban Politics and Policy* (Oxford: Blackwell).

On the politics of suburbanization see papers by Andrew Jonas: "Busing, 'White Flight,' and the Continuous Suburbanization of Franklin County, Ohio." *Urban Affairs Review*, 34(2), 340–58 (1998); and "Urban Growth Coalitions and Urban Development Policy: Postwar Growth and the Politics of Annexation in Metropolitan Columbus." *Urban Geography*, 12(3), 197–225 (1991). Also useful are: K. R. Cox and A. E. G. Jonas (1993) "Urban Development, Collective Consumption and the Politics of Metropolitan Fragmentation." *Political Geography*, 12(1), 8–37; William V. Ackerman (1999) "Growth Control versus the Growth Machine in Redlands, California: Conflict in Urban Land Use." *Urban Geography*, 20(2), 146–67; and Stephanie Pincetl (1992) "The Politics of Growth Control: Struggles in Pasadena, California." *Urban Geography*, 13(5), 450–67. But for a provocative, if occasionally bilious, critique of suburban no-growth policies and the residents' groups that support them see Bernard Frieden (1979) "The New Regulation Comes to Suburbia." *The Public Interest*, 55, 15–28. More measured on the topic of suburban no-growth is M. J. White (1978) "Self-interest in the Suburbs: The Trend toward No-growth Zoning." *Policy Analysis*, 4(2).

On urban housing markets and the way they function, including neighborhood change, suburban development, and exclusion and redlining, the reader by Roger Montgomery and Daniel Mandelker (1979) *Housing in America: Problems and Perspectives* (Indianapolis: Bobbs-Merrill), although over twenty years old, is still useful. See, in particular, chapters 5 ("Housing Market Dynamics"), 9 ("Neighborhood Revitalization"), and 11 ("Opening up the Suburbs").

Finally, the relation between the geographic form of urban development and the stakes that particular branches of the economy have in it, particularly the oil and auto companies and the real estate industry, see the masterful discussion by Dick Walker and David Large (1975), which, although now quite old, remains highly relevant: "The Economics of Energy Extravagance." *Ecology Law Quarterly*, 4, 963–85.

Part II

Territory and the Politics of Difference

Chapter 5

Difference, Identity, and Political Geography

Context

Thus far our emphasis in this book has been on people's material objectives: earning a wage, making a profit, levying taxes, bringing about local economic development, redistributing income via the welfare state. We now need to broaden our horizon. For life cannot be reduced to material relations. The material relations of human beings, their interactions in a division of labor, say, presuppose an ability to *communicate*. That ability to communicate in turn assumes a common set of *meanings*. Likewise, in interacting with nature, we try to understand our interactions, how nature reacts, so that we can enhance our material control, and this too demands *understandings* and so meanings.

This would suggest that imposing meanings on our experiences is quite as important to us as earning our bread; indeed, if we couldn't determine the meanings of our various relations, both with nature and with other people, it would be very difficult to survive. As human beings, understanding our world, understanding our relations with others, knowing what to expect, how to act in particular circumstances is of central significance. Unlike other animals we assign meanings to our experiences, to the objects we encounter, to the people we interact with. It is in terms of these meanings that we structure our activities. When meanings are undergoing rapid change we may be struck by the meaning-dependent nature of our activities. What it means to be male or female has changed greatly over the past few years and has had important effects on our social relations.

So we all need *interpretive frameworks:* sets of conceptual pigeon holes into which we can insert our experiences and, on that basis, decide how we should act. These interpretive frameworks have both *cognitive* and *normative* implications. They tell us how to differentiate one object or activity from another (the cognitive); and also how we should act towards them (the normative) and why we should act in those particular ways. These normative aspects may be underpinned by more or less understood social philosophies of the liberal or conservative variety or by the codes of behavior promulgated by the organized religions. Meanings are justified in this way, including the meaning that we ourselves have as individual people to others.

These interpretive frameworks are shared with those of others. This is necessarily so since if we did not share meanings with other people communication and therefore interaction would be impossible: nobody would know what anyone else was referring to. This sharing provides us with a degree of *security* and a sense of familiarity. We act in accordance with what our interpretive frameworks tell us and, by and large, our expectations are confirmed. Imagine how disorienting it would be if we could *never* predict how someone would react to a gesture, an instruction, an expression of disapproval, etc.

Meaning systems are social in their character: they are meanings that we share with others and can anticipate as being shared by others. They are also social in their origins. Meaning systems are socially *constructed*. People are inducted into certain meanings by others and pressure is applied to make them accept those meanings. We can see this in the simple socialization processes that we all experience as children. Our parents teach us certain things and then make sure we remember them. Chastisement or withdrawal of favor is the punishment for failure to learn those essential meanings. This does not mean that accepting new meanings is purely a matter of others exercising power over us. The meanings that we acquire have to work for us in some, at least minimal, sort of way. We learn as children that sneezing at the table is not a good idea not just because we will be told off but because we learn that it is to our disadvantage if others sneeze at the table when *they* have a cold. The same applies to other widely accepted norms, like those of punctuality and control of bodily functions: we recognize the desirability of these things because we want *others* to be equally punctual and self controlled.

The Politics of Difference

Interpretive frameworks provide a social mapping. They generate expectations which, by and large, tend to be realized. They therefore provide us with a sense of security and familiarity that would otherwise be lacking and which is important to us: something we quickly become aware of when we travel and encounter people who do *not* share the interpretive framework we developed in the US, in Britain, in Canada, or wherever. And the social definitions of others, which we constantly verify through our actions as students, professors, short order cooks, etc., provide us with a sense of who we are: a sense of identity. Under capitalism, however, senses of identity get reworked and not everyone is left with something that they can feel good about. Invidious distinctions start creeping in and concepts of relative social worth develop. Feelings of inadequacy take root, of inferiority. In turn these can lay the foundations for social movements whose goal is to rework identities so that people no longer have these feelings, as in the feminist movement.

These are classifications in which there are top dogs and bottom dogs, good and bad, clean and dirty, civilized and barbaric, capable and inefficient, rational and emotional, strong and weak, hardworking and idle; a world of "us" and "them." As one constructs others in this way, defining not just as different but also as inferior, so one simultaneously constructs oneself as superior, deserving, altogether meritorious: a veritable paragon of human virtue. If others are different and not deserving, then the person making the claim must be deserving, civilized, hardworking, rational, or whatever the opposite happens to be. This is not just a matter of self congratulation. It is also a way of explaining to oneself why one is indeed privileged: why indeed it is *not* a matter of (e.g.) anonymous social processes with which, by the good fortune of being born into the family that we were, we were able to connect on favorable terms. Rather it turns out to have been a matter of personal merit: taking control of one's own life, exercising foresight and self discipline, working hard, obeying the law.

This particular example also helps us understand what drives the construction of these moral hierarchies. It is not just that people want to feel good about themselves in a world where status insecurity is pervasive, though that is important. Defining others as different, as lacking, and disparaging them is part of the project of reproducing or achieving material advantage. In other words, part of the logic of difference is the pursuit of material interests.

It cannot be emphasized sufficiently that we live in a world of inequality. The rules that structure society, the rules of private property in the case of capitalism, are ones that tend to perpetuate inequality. It would seem that to those who hath shall be given! But things are not quite that simple since it is always possible that the rules that produce such unequal outcomes will be challenged. The problem is that while capitalism generates material inequalities of an extraordinary magnitude, it also holds out the possibility of upward mobility. Unlike in feudalism, you too can be one of the advantaged. But not all can be advantaged at the same time. Not all can be capitalists in a capitalist society since who would produce the goods on which profits are to be made? Which means, in turn, that the rules that produce that inequality are likely to get challenged. To forestall that challenge those who benefit from them need to justify the inequality from which they so clearly benefit. In part this has been accomplished through assertions, rhetorics, of Difference. The other person (the generalized "Other") is defined as somehow lacking, as deficient in some way, and therefore as responsible for his or her own subordination, his or her status as employee rather than employer, as unemployed rather than employed, and so on.[1]

---

1   On the other hand, this discourse of Difference is often supplemented by arguments that the Other, while the worse off in the relationship, is actually better off than if the relation did not exist! Sometimes these stories are told in such a way that the Other even appears to be the one advantaged by the relationship, as benefiting from the self-denying actions of those who, we had thought all along, had actually been the beneficiaries. This is the world of the abstaining capitalist who resists the temptation of consuming in order to invest and so provide

But this sort of disparagement, this discourse of moral superiority and inferiority and of rights, can also be used where material inequality is not evident; where in fact the goal is to *create* it. Competition in job markets, in housing markets, is often of this form. There are large numbers of people who qualify for a particular job or who can afford particular sorts of housing. But by excluding some from the competition, the number of competitors can be reduced. Those who "win" in such cases can benefit from an enhanced scarcity of workers in the case of job markets or an enhanced scarcity of buyers in the case of housing markets. In the one case the result is higher wages or even a job as opposed to none at all; and in the other, lower housing prices, or if housing is in particularly short supply, just a roof over one's head.

This is the world of racializing, gendering, ethnicizing, and nationalizing. The competition of equals or near equals for jobs or housing can, in short, produce a discursive jostling to define the other as somehow inferior, undeserving, inappropriate, incapable and so justify an exclusion from which they can benefit. In South Africa, and for many, many years, blacks were not allowed to take the examination for the relatively well paid position of blaster in the gold mines on the spurious grounds that they could not be trusted with explosives. It is also the world of gender exclusion in job markets, something which has a long history.

In short these distinctions often tend to get naturalized. People are defined as *naturally*, *biologically* superior or inferior, however mistaken or prejudicial that might be. This has been and continues to be the basis of a good deal of racist and sexist argument. Likewise, the untouchables of the Indian caste system are born; their untouchableness is beyond their control; nothing they can do, other than escaping to the city where they will go unrecognized, can affect the way they are defined. Even when racial and gender differences are not part of the field of contrast, as in the case of underclasses, it is still not uncommon to hear it explained as a matter of "breeding."

Think and Learn

We have been talking about inequality and cultural hegemony. What, if anything, do you think these have to do with concepts of social justice? Is social justice something that is contested? Are there different concepts of social justice according to the race or gender of the person for whom it is a concept? What do your conclusions suggest about the defensibility of the idea of *absolute* standards of social justice?

---

employment for others; or that of the white man's burden, the selfless bringing of light and the benefits of civilization to the benighted masses of Africa. The result is on the one hand to underpin the privileged material position of those articulating this discourse of (justified) inequality. And on the other, to define themselves as thoroughly creditable specimens: as people who not only deserve to occupy the positions they do, but at the same time behave in such a way as to uplift their inferiors.

Even so, we should be careful not to see the emergence of a sense of social hierarchy as purely a matter of a means to the end of justifying material inequality or obtaining a material advantage over others. Recognition, respect, a sense of social worth become desirable goals in themselves. Holding on to a particular position in the status hierarchy then becomes a matter of distancing oneself from those lower down; it is as if proximity might contaminate. This becomes particularly clear in the case of reactions to poor, unskilled immigrants who may present no threat in job or housing markets to those relatively high up but who, by virtue of their colonial origin perhaps, have been pre-interpreted as inferior. Maintaining one's image with others, one's respect, then depends on a refusal to accept them as equals, which in turn means a careful regulation of contact, the avoidance of informal relations, intermarriage, rejection of the other as a neighbor.[2]

This is not to say that there is any one single dimension of moral worth in these constructions. Identities are multiple. People aren't just working class; they are also white, male, and American as well, even proud New Yorkers or, in Britain, Scousers or Geordies. People have multiple identities which they draw on according to circumstance. Working class people are often oppressed by a sense of cultural inferiority, of exclusion,[3] as being defined by the middle class as boorish, lacking manners, sophistication, and good taste. But at least given existing cultural codes some believe they can take pride in being white or American. And indeed the fact that they can so compensate for a sense of cultural inferiority, of not having the right address, the correct vocabulary, is what helps account for the virulence of their racism and nationalism. They may be towards the bottom of the heap but at least they can rationalize to themselves that they still have a long way to go to reach the bottom-most point. And as long as despised racial minorities and nationalities are kept there they won't.

Geographies of Identity

We have now seen that concepts of difference are either rooted in social relations of inequality or are designed to produce them. They tend to produce/reproduce inequality and that is the intention of those who propagate them. But social relations, regardless of whether or not they are ones of inequality, are also, and invariably, *spatial* relations. In order to relate people have to make contact, they have to connect. At the same time, they may come

2   On the other hand, it is hard to keep market relations out of the picture entirely. Individual British people may indeed accept West Indians as social equals. But to the extent that the view of them as inferior, as dirty, as not good housekeepers, as criminal, and so not deserving of informal association and contact is a prevalent one, then concern over property values can intrude into the functioning of housing markets.

3   Merely check out the soap operas on American TV and their respective casts of besuited doctors, lawyers, executive class men, and well coiffed suburban matrons. And they're not exactly flattering to women either.

to know their *place*, in quite literally geographic as well as social terms. In this section these ideas are explored through two different themes: one of these discusses the relation between place and identity where it intersects with relations of social inequality. The second is about what can happen when people from outside encounter people who have lived in a place a long time: in other words, a case of "insiders" versus "outsiders."

Place, identity and inequality

Stigmatized places

Capitalism brings people together in cities. Urbanization is part of that social-ization of production through which the productivity of labor develops under capitalism, and which we discussed in chapter 2. But people come together in relations of inequality. In particular there are, as we have seen, very consid-erable variations in income. This affects where people live in the city. For in the city, and as we discussed in chapter 4, people are subject to a residential sorting process. Money is social power and no less power over where you live. It allows you to live in "more desirable" neighborhoods or, if you can't afford it, it excludes you from them. Social stratification is at the same time, there-fore, a spatial stratification.

This residential allocation of people is the context for the creation of a moral geography: a geography of "good" and "bad" neighborhoods, "upmarket" and "downmarket" suburbs. Places get stigmatized and this is justified in terms of the personal characteristics of people. Those who live in the exclu-sive suburbs "deserve to" because they have worked hard and saved and so reaped the rewards of that hard work and foresight. Those living in "problem" council estates, the "projects" with their stereotypical drug trafficking, family disintegration, and vandalized stairwells and elevators, or even on "the other side of the tracks," likewise deserve to because they lack the desirable char-acteristics of those who, not uncoincidentally, are making these judgments.

This is not to say that it is their view against the views of those they are defining as society's outcasts. Rather there is a more general process of social definition which tends to work its effects on those who are its immediate targets. This is a process that works behind people's backs in a very subtle way. It is one that works through media images and discourses, through the role models of the "soaps," through socialization in family and school, and through the public statements of "our leaders." And given the nature of the beast it could not be otherwise. How could capitalism possibly be reproduced if those who earnt the most money and lived in the best neighborhoods were castigated for the unfair advantage that (e.g.) inherited wealth had given them, for the fact that their stock earnings were produced by the sweat of someone else's brow (typically, and ironically enough, someone they despise as a social category)?

Within this moral socio-spatial hierarchy residents jostle further to redefine their spaces, their neighborhoods, in some way which will enhance their sense of social worth. In stigmatized areas people will always claim that the problems for which their neighborhood has become known happen somewhere else. They appeal, therefore, to a sense of geographic scale, to the idea that the moral geographies that are represented in the minds of some are too coarse to pick up the local variations that, of course, allow those making the appeal to dredge some sense of self respect out of their situation.[4] Not that this is exclusively a tactic for those living in stigmatized areas. It is all a matter of degree. Residents who are not that much better off may try to redefine their neighborhood so as to give them a degree of status and sense of social worth they might not otherwise have. And even those who thought themselves safe can be in danger of stigmatization (see box 5.1).

Where we live is something we carry with us and may want to conceal: " 'We don't tell people we come from Bethnal Green,' said one woman, 'You get the scum of the earth there' " (Willmott and Young, 1975, p. 289). In Paris dealing with this sort of residential stigma is something that is habitual for the residents of the suburban public housing projects: what the media and they themselves refer to as a "dumpster," "the garbage can of Paris" (Wacquant, 1993, p. 369):

> When asked where they reside, many of those who work in Paris say vaguely that they live in the Northern suburbs rather than reveal their address in La Courneuve. Some will walk to the nearby police station when they call taxicabs to avoid the humiliation of being picked up at the doorstep of their building. . . . Residential discrimination hampers the search for jobs and contributes to entrench local unemployment since inhabitants of the Quatre Mille encounter additional distrust and reticence among employers as soon as they mention their place of residence. . . . All youths recount the change of attitude of policemen when they notice their address during identity checks, for to be from a *cité* carries with it a reflex suspicion of deviance if not of outright guilt. (Wacquant, 1993, p. 371)

The fact that the residents themselves refer to the *cités* as the "dumpster" or "the garbage can of Paris" indicates how deeply the prevailing social definitions have bitten; how, that is, people accept a dominant meaning system whose ultimate effect is to justify and so protect inequality by blaming the victim.

But not only do the stigmatized accept these designations; their acceptance intensifies their subordination. This is because the lack of respect accorded by others results in a lack of *self* respect. Being defined as a loser turns out to be a self fulfilling prophecy. People do not believe in their own abilities to make a difference. Any attempt to mobilize residents around self help programs designed to improve living conditions is therefore likely to fall on deaf ears.

---

4   In the public housing projects on the edge of Paris: "What appears from the outside to be a monolithic ensemble is seen by its members as a finely differentiated congery of 'micro-locales': those from the northern cluster of the project, in particular, want nothing to do with their counterparts of the southern section whom they consider to be 'hoodlums' . . . and vice-versa" (Wacquant, 1993, p. 369).

Box 5.1 *Redefining Neighborhoods*

Upwards

In his book on Los Angeles, *City of Quartz*, Mike Davis (1992) reports on a case in which residents tried to redefine their neighborhood, or at least public consumption of what their neighborhood was, in an attempt to redefine their position in the city's moral geography. Anyone who has driven around Los Angeles will recall the way in which different neighborhoods announce themselves with signs at their self-designated boundaries: Holmby Hills, Baldwin Hills, West Hollywood, etc. The case that Mike Davis discusses concerned an attempt to break away, separate from, the area known as Canoga Park: "more than three thousand homeowners in the foothills of western Canoga Park petitioned . . . in early 1987 to redesignate their area as 'West Hills.' The members of the West Hills Property Owners Association complained that they were forced to look down from the patios of their hilltop $400,000 homes on mere $200,000 hovels in the flatlands (below) . . . the secessionists whined that Canoga Park was 'bad . . . very slummish'" (p. 154).

And downwards

"The proposal was not for a toxic waste dump or a new train line through the town center. It was for a charter school, a new public school paid for with local tax dollars but run independently. But in the middle-class community of Glen Cove on the North Shore of Long Island, where pride in the public schools goes a long way toward buttressing property values and *self-esteem* [emphasis added], it touched off a bitter dispute" (Kate Zernike, "Suburbs Face Tests as Charter Schools Continue to Spread." *New York Times*, December 18, 2000, p. A1).

Charter schools are schools which are privately run but with public money. As the children enrol in a charter school, so the tax monies that would have gone to the public school system in a particular locality are diverted to the charter school. They were originally conceived as a way of improving education in relatively disadvantaged urban areas, particularly in the inner cities, and this has colored perceptions of them. It is this perception that seems to have been at work in this particular instance. But in addition, and more recently, charter schools have been justified on the grounds of the enhanced competition they bring for the public schools: they offer the possibility of breaking the monopoly of publicly run schools on (publicly funded) education. Even so, the old definition endures and is aided and abetted by the way in which some State agencies define them, in consequence intensifying the apprehensions of people like those of Glen Cove. Thus, according to the same article, "the [New York] state chartering agency refused a charter to neighboring Great Neck last year, saying the schools there were good enough that parents did not need options." In other words: in the State's view, charter schools are for disadvantaged areas, and if they locate in your area, then that is a step down the road to residential stigmatization.

Think and Learn

Consider the standard form of land use zoning in the US. The land uses are arranged hierarchically. Single-family housing can be built in areas zoned for apartments but not vice versa. Low-density single-family can be built in areas zoned for high-density single-family but not vice versa. What does this suggest about the state's views on the desirability of different forms of residential land use? What is its implicit ranking and how do you think that affects people's images of "good" and "bad" neighborhoods?

Think and Learn Again

A common way in which people differentiate between "good" and "bad" neighborhoods is in terms of the extent of crime: property crimes, vandalism, in particular. But what if we knew more about the residential geography of those committing so-called "white collar" crimes: tax evasion, bankruptcy? Would our views change? Given that we do not know as much about this geography as we do about "blue collar" crime, what does this suggest about the relative degrees of public opprobrium accorded "blue collar" and "white collar" crime? And do you think those relative weights are fair? And why or why not?

Ossis and Wessis

Similar sorts of moral geographies can occur at the regional scale: the contempt Northern Italians feel for Southerners, for example. But in cases like these the traffic is likely to be two-way: the derided, those who are held in contempt, have their own stories to tell, ones that reverse the relation of superiority/inferiority. Highly illustrative of the processes involved is what unfolded subsequent to the reunification of West and East Germany over ten years ago after about 45 years of cultural divergence between the two areas. It pits in hostile embrace *Wessis* or the West Germans and *Ossis*, former residents of East Germany:

> To simplify outrageously the *Wessis* think that the *Ossis* have spent too long living in their own little world and are naive and unsophisticated; that they lack self-confidence, apologizing constantly; that they are unused to hard work because in the old GDR [German Democratic Republic] days they were always running short of materials; that they are no good as managers or entrepreneurs because they have no experience of a market economy; and that, far from being grateful for the benefits that unity has brought with them, they are always complaining. . . . The *Ossis*, for their part, reckon that the *Wessis* are insufferably arrogant, often without justification; that relentless competition has made them hard as nails; that quite a few of them are crooks who have grown rich on selling dud

insurance to trusting easterners, or snapping up eastern assets at far below their real value; and that to cap it all, they have pinched the best jobs in the east. At first all this just made the *Ossis* feel depressed and helpless, but now some are becoming resentful. . . . *Wessihass* (hatred of westerners) has become fashionable. (*The Economist*, Survey of Germany, November 9, 1996, p. 6)

For just short of half a century the people of Germany were divided in two. There was West Germany and there was East Germany: the so-called German Democratic Republic or GDR. Contact between the peoples of the two new states was very limited. Both developed in very different directions. In West Germany a vigorous process of capitalist development unfolded from the early fifties on, making the country one of the richest in the world. In East Germany state planning ruled the roost and market relations assumed a very subordinate role in economic life. Norms of competition and the pursuit of material gain were accordingly only very weakly developed. On the other hand, the social safety net was much more extensive in the East. Rents and utilities were highly subsidized. Women in particular benefited as a result of the extensive availability of childcare, a liberal abortion regime, and worker training programs that were as generous for women as for men.

Reunification, however, has occurred on the terms of West Germany. The rule of the market has been extended to East Germany. The East German political class was displaced as a result of the overthrow of the communist regime there and the country was reabsorbed into a state that was through and through West German. This is not to say that East Germans objected to this. In the beginning all this meant "freedom." But they have since discovered that "freedom" on West German terms is Janus-faced.

Reunification has been far from painless. New senses of difference have emerged as the two Germanies have come together. Both East Germans and West Germans define themselves in terms of the Other and find the Other wanting, as inferior, as indeed is apparent from the quote above. Significantly enough, and according to the same article from which that quotation was drawn, in 1996 two-thirds of those in the former East Germany considered themselves first and foremost "east German" and just less than a third of those from the former West Germany considered themselves "west German."

For those of the former GDR or those who call themselves, in contrast to those from West Germany, *Ossis*, the West Germans or *Wessis* are a threat to a way of life for which they now find themselves nostalgic. They miss the more developed welfare state. They miss the feeling of warmth, of being together and needing each other to survive under difficult circumstances. They also miss the low key work life. The *Wessis* are the threat to this because they, in the form of employers and civil servants, are seen as the ones imposing a new way of life of competition, consumerism, and self reliance and also one that is less gender-neutral in its effects. There is an emergent sense of a colonial relation with the *Wessis*, a feeling of being second class citizens. They are unable to compete in job markets on the same terms as West Germans. They are treated as inferior by West German businesspeople, as less able.

Out of this experience both *Ossis* and *Wessis* draw senses of themselves, in contrast to the Other, that are flattering. In contrast to the *Wessis*, *Ossis* see themselves as more humane, more socially sensitive, including gender-sensitive. But what *Wessis* see is lack of initiative and self reliance, an adherence to the past and a reluctance to embrace the new. The implicit contrast, of course, is one that is flattering to themselves. This is because they are the ones who, by their own definition, take the initiative, are self reliant, and forward looking.[5]

Insiders and outsiders

Ethnicizing in apartheid South Africa

> I didn't have many friends. One can't really get used to the East London people because they feel: What are the Cape people coming to live here for? They always had something bad to say about us. They called us the Amalawu which means Bushmen. My children couldn't speak Afrikaans as well as we could when we were children at Upington, because in Nyanga they didn't hear it so much. But they could understand it. And they spoke it to the people of De Aar and Beaufort West and Mossel Bay and Richmond who were also resettled in Mdantsane. So we were called the Bushmen. Some didn't say it to your face but behind your back. They said their children wouldn't have a place to live, because the Cape people were being given all the houses. And it wasn't easy to ask someone to mend something for you, or to put on a door, because you had no one of your own. Ag ja, they were not bad people, but you didn't feel at home with them, so you kept to yourself and your children. The children got on well, they all went to school together and the East London children wanted to hear about Cape Town, but the adults were different. (Joubert, 1980, p. 206)

This is an extract from a biography. It is the biography of an African woman and narrates her, always difficult, sometimes traumatic, experience under apartheid. Even so, the relations and processes depicted here have a degree of universality. The fact that lines of Difference are being drawn by and among blacks rather than, as one might expect under apartheid, by whites between themselves and blacks, is indicative of this.

At this particular point in the biography the woman in question, Poppie Nongena, has been relocated (the term used in the extract is "resettled") by the South African government from Cape Town (the quote mentions Nyanga, which is a black township in Cape Town) to a new township, Mdantsane, in the homeland of the Ciskei. She has not gone of her own accord and her

---

5   But bolstering the discursive resistance of the *Ossis*, an unwillingness to accept the definitions of the *Wessis*, is a high level of consciousness of being working class, an interesting product of the GDR years: clearly a *pride* in being working class. In a recent survey, for example, 61 percent of East Germans identified themselves as working class in contrast to only 29 percent of West Germans. Among West Germans almost as many as those among East Germans who identified themselves as working class, identified themselves as middle class (*Informationen zur politischen Bildung*, 269 (2000), p. 60).

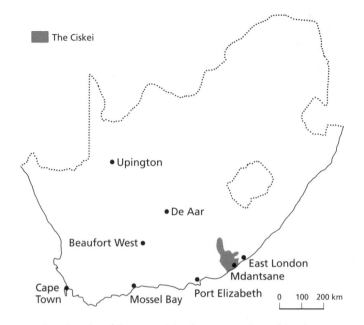

Figure 5.1 Poppie's South African world: places mentioned in the extract from Elsa Joubert's *Poppie Nongena*.

relocation is part of the South African apartheid government's grand scheme at that time of "whitening" South Africa. This, they intended to do, by giving blacks their own, ultimately independent, homelands and eventually relocating all blacks into them (see chapter 8). In Mdantsane she encounters blacks from other parts of what was then the Cape, including some from the nearby city of East London. Figure 5.1 indicates the different place names mentioned in the quotation.

The quote starts off by alerting the reader to the presence of a particular differentiation that is being made in Mdantsane, not just by Poppie but by those she is encountering from East London: there are the Cape people and the East London people. The "Cape people" are from a diversity of locations throughout the Cape (see figure 5.1) but what they all have in common is that they are *not* from East London, even though East London too is in the Cape.

The second point to note is that the Cape people are being defined by the East London people in derogatory terms: as Amalawu, meaning Bushmen. "Bushman" in this case implies primitiveness, backwardness, as anyone who has seen the movie *The Gods Must Be Crazy* will understand. There is, however, a logic to this apart from the fact that the East London people have, as we will see, an agenda: this is the fact that the children speak Afrikaans, which is also spoken by those Bushmen who have become urbanized. "So we were called the Bushmen," concludes Poppie. The agenda of the "East London people" is clearly one of housing, which is allocated by the state. They are concerned that there will not be houses for their children. For this reason they do not want the "Cape people" around. They would like to see them excluded. And if the government won't do it for them, then they will shame the "Cape people" into leaving, by being made to feel inferior, unwanted, not "at home."

From island chauvinism and pigmentocracy to Caribbeans[6]

[T]he harsh lessons of British racism have helped to create an identity among Afro-Caribbeans living in Britain commensurate with their concrete situation and historical experience. The crude binary classification of ethnic groups within Britain – black/white – has broken down the absurd and deeply offensive hierarchy of shades which has long vitiated the Caribbean psyche. (James, 1992, p. 17)

In the Caribbean English-speaking blacks identify with their islands of origin – Barbados, Jamaica, the Bahamas, etc. – in displays of island chauvinism. But in Britain, and in a context of quite severe racial discrimination, these differences assume much less importance than the racial difference between all of them, regardless of island of origin, and the white majority. So it is in Britain that blacks have forged a Caribbean identity to replace that of the different islands from which they come.

More specifically, in the Caribbean Britain had numerous island colonies. These included Jamaica, Barbados, and Bahamas, along with smaller islands like Nevis, St Kitts and St Lucia. On those islands black people identified with their island of origin in so-called *island chauvinism*. They also identified, however, with their position in a ranking of skin pigments. This was a highly fragmented ordering into numerous layers, with position in the social and political hierarchy defined according to lightness of skin: lighter skins on top and darker ones beneath, forming what was known as a *pigmentocracy*. After the Second World War, and particularly during the 1950s, large numbers of black West Indians left their islands and migrated to Britain in search of work. This has transformed the identities of those migrating.

Settlement in Britain has had several important effects. One has been the confrontation with British racism. Since all the migrants from the West Indies were treated the same in this regard, whether they were of a lighter or a darker shade, their non-whiteness became more important to them as something they shared than those variations in skin pigmentation which had been so important in the West Indies. Likewise they were all treated in Britain as West Indians rather than as Jamaicans or people from Barbados. So island of origin became less important than seeing themselves as all West Indian.

The rejection of variations in skin pigmentation as a basis for social hierarchy and sense of self worth among themselves was not just an effect of British racism, however. Before coming to England they had never seen a working class white or a poor one for that matter. This was an important reason for the strong value they put on degrees of whiteness. But in England the myth of whiteness was exploded for they saw there, for the first time, poor, working

---

6   This section draws heavily on the work of Winston James (1992).

class whites suggesting that whites were not somehow naturally ordained to occupy superordinate positions. This in turn led to a re-evaluation of blackness. Being black could be positive and no less contributory to a sense of self worth than being white.

Their racist treatment in England has also led to a re-evaluation of their sense of themselves as West Indian. "West Indian" is a colonial term as in the "British West Indies." The link with Britain has lost its luster as a result of the job, police, and housing discrimination that West Indians have faced. They now tend to see themselves as "from the Caribbean" since this term does not have the colonial connotations that "West Indian" does (James, 1992). In other words, while they were defined by British whites as the same and indeed, as a result of their common oppression, they came to see themselves as similar, it was not exactly in the terms the British whites were defining them. Rather they wanted to dredge some elements of self-worth out of their situation. So defining themselves as Caribbean rather than West Indian was an act of defiance, albeit a symbolic one.

Reflection on these two case studies, some comparison and contrast, allows us to make some generalizations and observations of a wider applicability. The first point to make here is the extraordinary importance assumed by length of residence, of provenance, in popular perceptions of rights; the widespread view that outsiders are not so deserving as insiders or the long established. This is clearest in the case of Poppie and the reaction of the "East London people," relocated only a short distance to Mdantsane, and the "Cape people" coming from outside the area.

But it was almost certainly at work in the British case[7] and helps to explain the way in which West Indians became residentially segregated, often in very overcrowded housing. This is because of the way in which British public housing was allocated: priority was given to the long-term resident.[8] And so too is it elsewhere. In Western Europe a common plaint from long-term residents faced with rural gentrification is that the newcomers, through their purchases, are displacing locals, including their adult children: a highly territorialized conception of social justice, in other words.

---

7   The sense of territorial rights with respect to West Indians is strikingly apparent in the views of a white dairy worker from the Midlands recorded in an interview by Thomas Cottle: "It is a crime, all of it. First they come here where they don't belong. . . . And would you believe, the government lets them have anything they wish, at any time they wish. But help those of us already here? No, they haven't an ear for that. I suppose the next step will be the government telling us *we* don't belong here anymore because we're sixth generation, or tenth generation" (*Black Testimony* (1978), quoted in Arthur Marwick (1982) *British Society Since 1945*. Harmondsworth: Penguin).

8   Compare Elizabeth Burney writing in 1968: "Few, if any, housing authorities of any importance refuse altogether to accept applicants who are not native sons within the sacred boundary born and bred. But they often look at them askance, and either apply general residential rules which are by their nature weighted against newcomers, or apply a different set of rules which deliberately handicap strangers or, sometimes specifically, immigrants from abroad" (p. 61). So while these rules were not originally devised as a way of limiting the access of immigrants to this desirable form of rental housing, they came to serve that function.

Think and Learn

Sometime in the 1970s a group of students from the University of Leeds went to the town of Leamington Spa, over one hundred miles to the south, to raise money for a Leeds charity. The following week there was a series of letters to the weekly newspaper in Leamington. Can you imagine how Leamingtonians reacted to this charity drive? Positively or negatively? Why, do you think? And do you think that their viewpoint was justified? Why or why not?

A second point is that with movement from one place to another, with convergence of different people from different places on a particular city or country, identities can change; new identities emerge to replace the old. This is most obvious in the case of those migrating from the West Indies to Britain where old attachments to island and shades of darkness are cast off and replaced by a sense of being, simply, Caribbean. Something similar is going on in the case of Poppie. It is unlikely that Poppie thought in terms of "Cape people" and "East London people" before her move to Mdantsane. But as in the case of the West Indians in Britain, she finds that she is defined by others as belonging to a larger category of person: as being differentiated out not as a Capetonian, which she is, but as a Cape person, which is a much broader designation.

At the same time, and this is the final point, she is willing to accept this definition, to find certain redeeming values in it, just as do the West Indians in Britain. This exemplifies the fact that identities are formed in opposition – something that should also have been apparent in the *Ossi/Wessi* case. The designation by the British of those from Jamaica, Barbados, and so forth as West Indian is not a friendly one. It is grouping together all those who share certain similarities that touch (un)sympathetic chords in the British mind: that they are all black, but also that they are all equally a threat – to jobs, to housing opportunities, supposedly to taxes since (the perception is) they will have to be supported when they are unemployed, and so on. So too in the case of Poppie. She is identified by the East London people with all those others who, while sharing a *general* location, much as do the West Indians, are seen as a threat to the chances of their adult children also gaining housing in Mdantsane. So accepting the definition, albeit with qualifications as in the case where the West Indians assume the identity of Caribbeans, is part of an emerging politics of identity. Accepting it is tantamount to strength in numbers, to achieving a solidarity in the face of a common threat to life chances.

Think and Learn

Think of the complexity of social hierarchies, the construction of senses of relative social worth. Outsiders may be defined as less deserving, but what if they are also a racial minority? And what if that racial minority has been subject to pre-interpretation, has been defined in schools and other media of socialization as inferior?

New Social Movements in Spatial Context

These relations of domination and subordination are unstable ones. Difference becomes a political issue as various groups push for recognition, construct stories about themselves, their history and experience, which explain why they should not be demeaned, why they should be accorded more respect, included in the life of the nation on equal terms. This helps us understand the emergence of what have been called "new social movements," the most prominent of which have been the black civil rights movement and the women's movement in their respective and various forms. But other movements also belong, at least in part, to this politics of difference. In Western Europe so-called "guest workers" have joined together to put pressure on national governments there to accord them citizen rights and to be naturalized as Germans[9] or Danes. Anti-colonial struggles also fall into this category.

Consider the resultant politics in terms of its characteristic geographies. In the first place, if effective organization of the stigmatized, "inferior," culturally oppressed and excluded is to occur then home bases are required: sites of interaction, places, where they can *re*define the contrast between themselves and those who oppress them, and build alternative meaning systems. These may amount to little more than women's support groups, gay bars, or – important in the black US South – one's own churches. More obviously it may mean segregated neighborhoods with their own newspapers and cultural organizations: places, in other words, where people can have the respect that they are denied outside of their barrios and ghettoes and organize to change broader meaning systems.

Alternatively, meaning systems and the inequalities they either entail or seek to justify can incite escape. Often this means a move to the city or to more developed areas where contact can be made with other "refugees" and the nucleus of a women's movement or a civil rights movement, a national movement in the case of colonial societies, formed. Emancipatory movements, therefore, often acquire a distinct metropolitan, urban flavor and this means that as the message is brought back into the hinterlands in an attempt to further the emancipatory process, so the subsequent politics can get confused with one of outsiders versus insiders, cosmopolitans versus locals, urban versus rural (see box 5.2).

Integral to achieving the respect that the members of emancipatory movements seek, their acceptance as full members of society, is a project of reclaiming space or claiming a space of their own. In some instances the goal is integration, inclusion, into that society of which they are defined as a lesser element or part and with respect to which they feel marginalized. They incorporate demands for freedom of mobility on the same terms as everyone else: the elimination of the barriers that confine racial minorities to particular neighborhoods and schools or which confine women to the home or more recently

---

9   This issue has been an especially aggravated one in Germany owing to the longtime German insistence that German nationality only be accorded to those "of German blood."

Box 5.2   *Local Identities, Outsiders and Non-racialism*

A useful example of outsider/insider, cosmopolitan/local tensions associated
with the intrusion of ideas of social emancipation comes from Peebles, a small
town in Scotland about 24 miles to the south of Edinburgh in what is known
as the Borders region. It involves conflicts between the long-term residents of
the town on the one hand and, on the other, newcomers and the more
national cultural ideals of non-racialism espoused and advocated by them (see
Smith, 1993). The immediate cause of controversy was an annual event, the
Beltane Festival. A major element of this is a street parade featuring people
in fancy dress: cross dressing, exchanges of status, and, most significantly, the
use of golliwog costumes. For many years golliwog dolls were a pervasive
feature of British culture; few who were children more than fifty years ago
will not have encountered them as cherished toys. But as caricatures of blacks,
with dark skins, frizzy hair, and prominent red lips, they later acquired the
opprobrium of civil rights groups and, it is fair to say, became seen as symbols
of a racist culture.

It was only recently, however (early nineties), that their use in the Beltane
attracted critical attention. What triggered it were the complaints of an
Edinburgh schoolteacher, formerly resident in Peebles, to schoolteachers in
that town, who then made a public stand against the practice. This injection
of some opposition into the preparations for the Beltane was taken up by the
local media and subsequently by the national media, both newspapers and
television. In the latter instance at least the interpretations were critical and
did indeed identify the practice as racist.

The dominant local response and that of the organizers was one of
resistance. On the one hand they defended the use of the costumes in the
Beltane as entirely without racial significance. On the other hand there was an
anti-outsider rhetoric, a defence of a local way of life and local traditions. This
was concretized not so much by the continued use of the golliwog costumes,
indeed there were more of those than usual, but by a systematic drawing of
attention to the criticism that had been made of it. One character in the
parade, for example, dressed up as the Edinburgh schoolteacher in question
with a fishing net over his (her?)[10] shoulder carrying a captured golliwog.

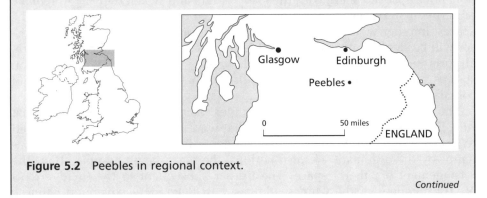

Figure 5.2   Peebles in regional context.

Continued

10   The teacher in question was indeed female but given the practice of cross-dressing in the
parade it is unclear whether it was a man or a woman who performed the caricature.

What seems to have been at stake here is an identification with Peebles. The festival was always an insider affair: a celebration of the town of Peebles and its community from which people drew a sense of self worth. Peebleans celebrated their community feeling, their one-ness, and their difference from non-Peebleans. The attribution of racism by outsiders intensified this identification. Peebles people were being criticized, made to feel bad, and unjustly, by an outside world based partly in the cities of England with their racial politics and partly in the cosmopolitan world of Edinburgh with its middle class values. Their response was to reassert their difference from the English and the lowland Scottish from around Edinburgh. Not only were they different, and proud of it, they were not racist either.

It is likely that the golliwog incident was a trigger for resentments that had been building for some time. Like many small towns of less than ten thousand people Peebles has been undergoing serious social change for some time. Unable to hang on to its young people through lack of jobs it has on the other hand attracted a considerable number of Edinburgh commuters and retired people. Outsiders, in other words, have come to live in Peebles and, in all likelihood, brought very different values and a different way of life with them.

to the so-called "pink collar ghetto." At the same time they may demand that the national space express their equality with others, that their great men and women be celebrated through national holidays, the naming of highways and airports, and the erection of statues – and in symbolically central places too, like town centers. Where the movement for recognition is on behalf of a linguistic minority, then recognition may come in the form of dual language street signs or a stipulation that the government documents that circulate among the citizenry and tie them to the government – applications for drivers' licenses, or passports and tax forms – be dual language.

In yet other instances, however, the demand is for separation, decolonization: not a space to share with their oppressors and which they try to remake on terms that are less demeaning, but a space of their own. These are to be spaces, in other words, in which signs of the former presence of the oppressor will be significantly undone: statues of colonial heroes will be torn down, airports renamed, new currencies will circulate, new images will appear on stamps, and the police will acquire uniforms that set them apart from the old colonial past.

The obvious examples are the struggles for independence from imperial rule. But even among those countries that were never formally subordinated to imperial rule, but whose peoples and governments were treated with contempt by the dominant Western nations, there can be a paranoia about the infringement of "their" space. The uproar subsequent to the (inadvertent) NATO bombing of the Chinese Embassy in Belgrade – sovereign Chinese territory – in 1999 bears witness to this. This is not to say that rhetorics of colonial domination are confined to nations that experienced imperial subordination in its classic form. It has been a common rallying call for separatist movements of diverse stripes, including that pressing for an independent Quebec and the movement for an independent Corsica.

Interestingly, in the subsequent struggle, a common reaction on the part of local power elites, those for whom the meanings work, will be to define the movement as the work of outside agitators: to assimilate it, in other words, to a set of meanings which opposes the innate goodness of the locals, the natives, to the duplicity and heinousness of foreign troublemakers out to deceive the (local) innocents. So (e.g.) the problem in Iran was defined not as one of women fighting the oppression of Islam but as originating in the United States in the form of movies filling them with evil ideas. And when blacks in the South became restive in the early 1960s the reason, so local white elites argued, was that wrong headed liberal ideas, often brought by white, sometimes Jewish, students from the hated and despised "North," were filling black minds with falsehoods: ideas which they had difficulty dealing with due to their own (supposed) simplicity.

Summary

In order to engage practically with the world, with other people, with nature as we engage in labor processes, we need understandings, and owing to the fact that practical engagement with the world is always social, these understandings have to be shared. An interest in material interventions in the world in order to satisfy material needs, the sorts of interest with which we were concerned in chapters 3 and 4, has to be complemented therefore by the development of modes of communication with others – speech, writing, body language – which allow us to negotiate with others a set of shared meanings: a meaning system or interpretive framework.

No matter where we are in the world or when, some meaning system will be operative. But they do change over time and space. Given that one is socialized from birth into a space–time specific meaning system, confrontation with a different set of meanings can be highly disorienting. There is, therefore, a very close relation between meaning systems and a sense of personal security, of familiarity.

Meaning systems or interpretive frameworks are closely bound up with the question of identity. Who we believe ourselves to be, how we believe we are supposed to act in certain circumstances, is always a matter of social definition. We occupy different social positions – husband, student, daughter, grandmother, policeman – and certain social expectations attach to those particular roles. In order to be a policeman, in order to feel competent as a policeman, one must be aware of how one is expected to act and this is a matter of social definition.

But under capitalism identity becomes highly political, producing a politics of identity; one in which, that is, the demeaned organize to achieve the recognition that they believe is being unfairly withheld from them. For now, in contrast to pre-capitalist societies, social positions – lawyer, laborer, employed, unemployed, homeowner, renter – are compared in invidious ways. Some are made to feel bad about themselves, while others, not uncoincidentally, are able to define themselves as good, as altogether exemplars of human virtue and rectitude. This is not to say that there were no hierarchies in precapitalist society, an ordering of social positions in terms of desirability. But these orderings were justified in terms beyond human mitigation. It was a question of God's will or a deistic retribution for the transgressions of ancestors. But always the possibility was held out of salvation in an afterlife or of a happier reincarnation. Under capitalism there are no such consolations for a life of oppression. Rather, to be relegated to less desirable social positions becomes a matter of individual failure, a failure to measure up to the trials of competition.

The reason for this is the emergence of material inequality as something to be struggled over: a struggle for material privilege which is possible because, unlike in precapitalist societies, that privilege is no longer justified in terms of the primordial terms of gender, kinship, age, or (noble) birth. Privilege now has to be justified in other terms. And since under capitalism everyone is supposed to be free to be what they are, since whether one is wealthy or poor is a matter of choice, then no one has the right to complain: it is the individual's responsibility and he or she has no one else to blame. If they don't (e.g.) work hard, save, invest in their future and that of their children, then too bad.

So at the same time as those lower in the social scale are demeaned those higher up can congratulate themselves on their success: they did it all themselves. But a moment's reflection will of course make clear that this is not at all the case and that, for example, variations in inherited wealth are immensely important in allowing some to take advantage of opportunity and denying it to others. But then if one's inferiors were to press their claims along those lines, to argue for a radical redistribution of wealth, one that might truly instigate the equality of opportunity so evidently compromised – i.e. to alter the rules under which competition occurs – where would the privileged be? From their standpoint, therefore, it *has* to be a matter of individual choice and drive.

This is not to say that there are not other discourses that have been and continue to be drawn on in order to justify material advantage or to make claims to it. Racist and sexist arguments have been extraordinarily common and have helped secure for white males a lion's share, if not a monopoly, of the more materially advantageous positions in society. Arguments about "a woman's place," what their – biologically determined – role is, have proven extraordinarily powerful in limiting their access to paid work, as well as in keeping them out of the political arenas in which they might have been able to challenge the laws that have circumscribed their choices. And so too has it been the case with racial minorities in the advanced capitalist societies, particularly blacks. By virtue of their supposed biology they have been defined as good only for the most menial of tasks.

On the other hand, while the drive to justify material advantage is at the root of these differentiations, the differential distribution of social worth can also become something struggled over for its own sake. Status insecurity is rampant. No one wants to be demeaned by "associating with the wrong types" or "living in the wrong neighborhoods." And indeed the issue of "living in the wrong neighborhoods" underlines the fact that the distribution of social worth has a geography and has in turn geographic preconditions. There is, for example, a moral geography which we explored through the idea of stigmatized neighborhoods. There are also geographic preconditions in the sense of people from different places making contact with one another. This produces the insider/outsider problem and we drew attention to the quite extraordinary strength of the belief that insiders have rights that should be denied to those from elsewhere.

The denigrated, the despised, the patronized, the excluded have not necessarily taken it lying down, however. Rather they have joined together in what have become known as new social movements in order to achieve some moral equality, some equality in terms of a sense of social worth. Through revisiting their histories, telling stories about themselves, they have appealed both to themselves and to an outside audience in an attempt to achieve both a greater sense of confidence in their own abilities and a greater degree of acceptance of what they believe themselves to be.

Social movements have their geographies. Among other things they need bases from which to organize. These typically are urban where the thresholds for achieving strength in numbers are more likely to be reached. The movement then diffuses its ideas back into smaller towns and rural hinterlands, though the media can also be complicit in achieving this. Their goals can also be depicted as geographic in character. They represent claims to space by either achieving a greater sense of inclusion or expelling the moral oppressors.

In the following two chapters we explore these themes of identity and difference. The next chapter focuses on the question of national identity, which in its politicized form of nationalism proved such a deadly force in the twentieth century. Chapter 7 brings into greater relief the politics of difference that has been generated around race and gender and in the former colonies.

REFERENCES

Burney, E. (1968) *Housing on Trial: A Study of Immigrants and Local Government*. London: Oxford University Press.

Davis, M. (1992) *City of Quartz*. New York: Vintage Books.

James, W. (1992) "Migration, Racism and Identity: The Caribbean Experience in Britain." *New Left Review*, 193, 15–55.

Joubert, E. (1980) *Poppie Nongena*. New York: Henry Holt.

Smith, S. J. (1993) "Bounding the Borders: Claiming Space and Making Place in Rural Scotland." *Transactions, Institute of British Geographers* NS, 18(3), 291–308.

Wacquant, L. J. D. (1993) "Urban Outcasts: Stigma and Division in the Black American Ghetto and the French Urban Periphery." *International Journal of Urban and Regional Research*, 17(3), 366–83.

Willmott, P. and Young, M. (1975) "Profile of a Suburb." In C. Lambert and D. Weir (eds), *Cities in Modern Britain*. London: Fontana.

FURTHER READING

There are few better supplements to the ideas presented in this chapter than three chapters in the book edited by Doreen Massey and Pat Jess (1995) and entitled *A Place in the World?* (Oxford: Oxford University Press and the Open University): chapter 3 by Gillian Rose ("Place and Identity: A Sense of Place"); chapter 4 by Pat Jess and Doreen Massey ("The Contestation of Place"); and chapter 5 by Stuart Hall ("New Cultures for Old"). Beyond these much of the reading on this topic is relatively advanced. For those willing to engage in depth with it, however, rewarding references include: R. Fincher and J. M. Jacobs (eds) (1998) *Cities of Difference* (New York: Guilford Press); and P. Jackson (1989) *Maps of Meaning* (London: Unwin Hyman).

Chapter 6

Political Geographies of Imagined Communities: The Nation

Context

This chapter is about nations and the associated ideas of national identity and national movements. In order to dramatize some of the processes involved here I want to start the chapter by relating some of the changes occurring over the past decade in what has come to be known as Taiwan. This is because Taiwan, for a long time known as the Republic of China in contrast to the mainland *People's* Republic of China, has recently been the site of rather striking shifts in national identity. But in order to understand this a little history is useful. The Republic of China came into being subsequent to the civil war in China which culminated in the victory of the communists in 1949. The defeated nationalist forces under Chiang Kai-Shek – the so-called Kuomintang or KMT – then retreated to an island off the mainland (figure 6.1). This island, the island of Formosa or what later came to be known as Taiwan, had formerly been a Japanese colony that had been returned to China subsequent to the ending of the Second World War. This was to be the site of the headquarters of *the* Chinese government as the nationalists defined it, in opposition to the "usurpers and oppressors on the mainland," and the base from which it would eventually be liberated from the communists.

But in order that it so function, its Chinese-ness or sinic quality, what the mainland and Taiwan had in common, had to be emphasized. This was problematic since at the time the island was occupied by the native Taiwanese complete with their own language, folklore, and traditions, quite separate from those of the nationalist Chinese. The nationalist Chinese therefore set about "sinicizing" the island. The goal was to replace the local version with specifically Chinese culture, language, and history. So, for example, the government subsidized Beijing opera – *the* art form of China – and declared Mandarin Chinese to be the only legal language in schools, thus prohibiting the use of Taiwanese. Place names from mainland China were used to rename the streets of the capital city of Taipei (Leitner and Kang, 1999).

Figure 6.1 Taiwan in regional context.

More recently, though, there has been a retreat from this vision of Taiwan's future as a province of a reunited China under nationalist tutelage. For despite the recent economic development of mainland China living standards have diverged considerably. Taiwan, after all, was one of the original newly industrializing countries. On top of that it has recently undergone considerable democratization, and reunification would almost certainly be under communist rather than the nationalist rule originally planned. Very possibly reunification would also mean a deterioration in living standards for in all likelihood the Taiwanese population would be taxed so as to raise standards of public provision on the – less well provided for – mainland. Accordingly the Taiwanese state is now tilting in the direction of an independent future. And this re-visioning of the island's future has allowed old senses of difference to re-emerge, often with government connivance and even support. In other words, it is the non-Chinese nature of the country that is now being emphasized.

As a result, and for example, government subsidy for Beijing opera is now out, and subsidies to Taiwanese puppetry are in. There is a rediscovery of Taiwan's unique history. The fifty years of Japanese occupation now feature as a source of difference from China, for example. Japanese rule is looked on fondly and many older people like to speak Japanese. This is because Japanese rule was altogether more benign than on the mainland where the Nanking massacre symbolizes a very different sort of occupation. Similarly there is now a rediscovery not just of the island's Taiwanese past, its language and traditions, but also of the aboriginal cultures that predate even the Taiwanese.

I relate this at the beginning of a chapter on the nation and the sense of nationality since it captures so much of the processes behind them. For a start the formation of national identity is certainly about difference and fields of contrast. The nationalist Chinese initially tried to stamp out difference in order to further their project of Taiwan as a province of China with Chinese traditions and language. More recently the differences in question – differences in the colonial legacy of the Japanese, differences in language and history – have reasserted themselves. But, and this is the second point, the idea of "stamping out difference" suggests that national identity is somehow malleable. It is not, in other words, something immutable that we are born into. Rather it is constructed. Third, in this act of construction the state is often an important agent. In this instance it tried to erase difference in order to construct an identity as Chinese. As the dangers of reunification have become evident, however, the state has moved to resurrect old senses of difference in order to justify the existence of an independent state of Taiwan. Fourth, and finally, behind the desire of the Taiwanese to create an independent state are questions of economics. Taiwan is a much wealthier country than mainland China and reunification would threaten the high standard of living that its people presently enjoy.

I am going to start out the rest of this chapter by talking about the idea of social construction that is so evident in the Taiwan case and the raw materials from which nations are constructed: the fields of contrast that are called on by national movements, states, and lobbying groups, as they seek to build followings. In the second part I turn to what I regard as the two major contexts within which, and through which, a sense of national identity is constructed: the state and capitalism. Finally, and in order to illustrate these arguments, we will look at some case studies of nation formation.

Nations as Social Constructions

We have made Italy. Now we must make Italians[1]

The idea that nations are constructed and by people acting on and through each other – i.e. "socially" – is a very odd, even bizarre, idea. For we are used to thinking of our nationality as something we are born with, something that is given to us like our gender, skin color, or our genetic inheritance *in toto*. Nation is conceived by most people in almost biological terms, as something we inherit and about which we can do little, though the process of

---

1   Massimo D'Azeglio subsequent to the Risorgimento and the creation of a united Italy.

"naturalization" stands in clear opposition to that assumption.[2] Indeed in the nineteenth century "nation" was often used interchangeably with the term "race." The British referred to themselves as "an island race," different from "the Irish race," etc.

But strange as it might seem, the common view today, at least among social scientists, is that nation and the sense of belonging to a particular nation are socially constructed. And after all, this is the gist of such commonly held notions as the US as a "melting pot": the idea, that is, that Americans are made and not born, that they are socialized through school, fellow workers, friends, the media, etc. into becoming Americans. Likewise, if nationhood, the sense of belonging to a nation, was inherited, how could we explain the fact that what it *means* to be (e.g.) a Canadian or a French person has changed over time; and that for some peoples, like the Quebecois, the pretension to nationhood is actually very recent and can be documented as something that emerged over time rather than originating at the dawn of civilization?[3]

But clearly it is not just politicians like D'Azeglio who are involved in "making nations." We are *all* involved, though some more so than others: those who write "national histories," those who identify and cultivate "national literatures," the intellectuals who codify and act as the guardians of "the national language," those who give us a sense of our shared experience and destiny by invading and oppressing us, the journalists who interpret events and give them some meaning. But not any meanings will do. Rather these are things that must resonate with the people to whom they are addressed. If they are being appealed to as sharing certain *national* predicaments, certain *national* experiences, then that particular audience must feel that they are indeed *its* predicaments and *its* experiences and that they are national in character rather than, say, religious, racial, or merely ethnic.

So precisely what are the raw materials out of which nations are constructed? We talked in chapter 5 of fields of contrast and the definitions of others as crucial to the assumption of some identity, whether it be that of "Caribbean" or "Cape people." So how do those ideas apply to specifically *national* identity? Clearly as members of a nation we differentiate ourselves from others in many different ways. We see ourselves as more similar to each other than to others along numerous dimensions, and these axes of variation typically tend to reinforce one another.

As Americans, Canadians, British, or French, we recognize that we share certain practices and activities with some – our fellow nationals – but not with others: language, customs, voluntary organizations, sports, religion, styles of

2   Even so, the term "naturalization," as in "making natural," is revealing of how nationality is perceived.
3   There is an identity that overlaps with that of Quebecois and which antedates it. This is the idea of being French Canadian. French-speaking people are found throughout Canada. Indeed in a province that neighbors Quebec – New Brunswick – they comprise about 30 percent of the total population. But the drive by the Quebec national movement for an independent state has involved de-emphasizing what they have in common with the French Canadians elsewhere and emphasizing what is unique to themselves: in particular the fact that they are residents of Quebec.

dress, or domestic architecture, perhaps. We come to recognize that we share a way of life with some but not with others: an American or British way of life. We obtain this awareness in diverse ways but in general through patterns of interaction. The field of contrast with respect to which we are situating ourselves is, in other words, geographic and it is through contact over space – vicarious as through the media or school textbooks with their stereotyped images[4] or immediate as when we come into contact with "foreigners" – that we define that field. We read about other peoples and their "strange" and "different" ways in the newspapers, we encounter something of their exoticness in the movies or perhaps while reading a novel, and the extent of international tourism and migration today means that few of us have not come into contact with a "foreigner": somebody whose native language is not ours, who wears a turban or a sari, and attends a mosque.

A good sense of this differentiation process is often provided by travel writers. Typically their audience consists of fellow nationals – American travel writers are writing for Americans, British travel writers for the British, etc. – and they write from the standpoint of their audience. But they are also writing from their own standpoint as Americans, British writers, etc.: they are looking at the "other" through American or British eyes. Consider the following passage from the American writer Paul Theroux about the English and observe the way he proceeds through the play of contrast:

> "Mustn't grumble" was the most English of expressions. English patience was mingled inertia and despair. What was the use? But Americans did nothing but grumble! Americans also boasted. "I do some pretty incredible things," was not an English expression. "I'm fairly keen," was not American. Americans were show-offs – it was part of our innocence – we often fell on our faces; the English seldom showed off, so they seldom looked like fools. The English liked especially to mock the qualities in other people they admitted they didn't have themselves. And sometimes they found us truly maddening. In America you were admired for getting ahead, elbowing forward, rising, pushing in. In England this behavior was hated – it was the way the wops acted, it was "Chinese fire-drill," it was disorder. But making a quick buck was also a form of queue-jumping, and getting ahead was a form of rudeness – a "bounder" was a person who had moved out of his class. It was not a question of forgiving such things; it was, simply, that they were never forgotten. The English had long, merciless memories. (Theroux, 1983, p. 15)

This is not to say that we sense ourselves as unlike others in *all* respects. Americans share the same language with many others around the world. As far as religious belief is concerned, while Judaism is a distinguishing trait of the people of Israel,[5] Roman Catholicism is not a unique characteristic of Italians

---

4   The Dutch, their clogs, their windmills, and their canals, the French with their chic women, their men with their berets, puffing on Gauloise cigarettes, and drinking wine, while watching a bicycle race go by.
5   Not exactly, of course, since there are the Israeli Arabs of Northern Israel. But Judaism *is* the state religion.

and nor can the Russians define themselves relative to others through the Orthodox Church. On the other hand French cooking is *French* and Chinese cooking *Chinese*. And just as Judaism differentiates Israelis from most people elsewhere so too is it the case with Hinduism for most Indians.[6] There are also more nuanced cases. Italians aren't the only lovers of opera in the world but it is more a part of their national tradition than would be the case anywhere else; the same goes for Spaniards and the national sport of bullfighting, since bull fights can also be found in parts of Latin America as well as in parts of southeastern France. But just as opera is more central to the lives of Italians than to the lives of others so too is bullfighting only an essential aspect of Spanish-ness as opposed to French-ness or Mexican-ness. Likewise fields of contrast can change. The French are the only ones among whom the use of the French language is dominant but one wonders what the creation of an independent Quebec would do for this particular sense of difference.

We also differentiate ourselves from others in terms of our history. Through school textbooks, through the public pronouncements of politicians, we obtain a sense that our history has been different from that of others. In fact we will learn that we have often been the object, perhaps unwilling object, of the history of others: exploited by them, invaded and oppressed by them; while at the same time they have been the object of our (historical) activity as we resisted their attempts to oppress and enslave us. And we will certainly learn that our history is something of which we can be proud. In this way we arrive at the conclusion that we have shared certain experiences: colonialism, an oppressive state, the conquest of the wilderness, the building of a worldwide empire, the liberation of others from tyranny.

Think and Learn

Consider the problem of writing English or American history as *national* history. What would you include and what might you exclude? Why might that be? What might your answer have to do with the fact that, reputedly, the teaching of French history stops just before the Second World War, even though the textbooks provide a survey up to the present day?

And so too is it with geography. We share a geography with our fellow nationals. This is the homeland. Everybody has a homeland, or so it is believed, but "they" have different ones from "ours." This is the sphere of pre-dictability. It is where everything is familiar. Beyond the homeland lies uncertainty, something to be learnt if we are ever to be comfortable in an everyday life there. Familiarity in this case generates not contempt but affect. We feel for our country as a place in a way we can't feel for other countries. The land-

---

6 Which helps account for the emergence of so-called "Hindu nationalism" in India.

scape is not just familiar and predictable – the stark geometry of fields that characterizes so much of the American landscape, the repetitive features of the built environment like the commercial strip or the skyscraper, the yellow taxis of New York City – it is loved.

Think and Learn

Thinking of your own country, what landscape features do you regard as typically English, Canadian, Scottish, American, etc.? In trying to arrive at an answer it might help to think in terms of the politics of landscape: what landscape (or townscape) features might be the object of preservation for the various heritage societies that have sprung up? Alternatively you might get clues from thinking of the various photo collections with titles like *Beautiful Britain* that are produced for coffee tables.

Our country is something we can long for when abroad and separated from it. And part of the place is the way of life discussed above. The English person on holiday can long for a good cup of tea or a friendly English policeman just as an American abroad is likely to find the local hamburger no substitute for the "real" thing and the local newspaper cannot replace an American one, if only because it features so little about the United States!

It is not just as a place that we differentiate "our" country from "theirs." Our familiarity extends to the internal geography of "our" country and its diversity of landscapes. Americans will obviously have a stronger sense of "the South" or "the Midwest" than would an English person; while for an English person "the North" has resonances that would be beyond most Americans. Similarly an English person will recognize the landscape of drystone walls dividing up the smooth and swelling contours of a Northern moorland as something only found in England and therefore distinctively English. This means that in our "own" country we can bring to bear on our – geographically differentiated – experiences an interpretive framework that stands some chance of success while elsewhere our ignorance makes us tentative and unsure (figure 6.2).

On the other hand, the various "geometries" of difference are a little more complicated than might have been suggested elsewhere. People can share some things, some relations with each other, while at the same time also preserving some areas of life, some elements of history perhaps, that are uniquely theirs. In other words, senses of difference often have a geographically nested form. The Welsh, the Scottish, the English, each have their own cultural traditions and histories, but as Britons they also have much in common. There are not simply separate Welsh, Scottish, and English national traditions but also a more geographically encompassing *British* tradition to which the Welsh, Scottish, and English have all contributed, and appropriated from it, positive

Figure 6.2   An essentially English landscape? For many, and perhaps more so in the past, this would have been regarded as a very English scene. But it is not just a landscape, it is a man-scape, suggesting the ways in which the national is more often than not highly gendered.

Source: G. Manley (1952) *Climate and the British Scene*. London: Collins, plate XIb, p. 158.

meaning.[7] The empire was, after all, the *British* Empire. British victories in two world wars were emphatically *British* accomplishments, even if individual regiments often betrayed a national affiliation: the Highland Light Infantry or the Welsh Guards, for example. And Scottish, Welsh, and English all suffered in the same way from Hitler's bombs. Scottish heroes like the explorer Livingstone, the inventor James Watt, or the writers Robert Louis Stevenson and Sir Walter Scott, the Irishman George Bernard Shaw[8] or the Welshman Lloyd-George are also notable Britons, and so on.[9] But in other cases the

7   This effect may now be widening to include an embryonic European identity. So as the following quotation reveals, the identity drawn on depends on the particular state entity involved: "English football fans might wave the flag of Saint George, but they also love to belt out choruses of 'Rule Britannia.' And those who roar their support for the English soccer team one week may be equally content to cheer on the European team in the Ryder Cup golf tournament with the United States, or support the British Lions rugby team, which still includes players from Ireland, as well as Scotland, Wales and England" (*The Economist*, October 3, 1998, p. 65).

8   Albeit "Irish" prior to the granting of independence to Ireland in 1922.

9   However, not all Scottish, Welsh, and English heroes or notables are appropriated by British nationalism. Robert Burns is much more Scottish than British and Dylan Thomas is obviously

attempt to construct a more all-embracing sense of nationality has been much more difficult. Canada has become a state with strong centrifugal pulls: from Quebec, from Newfoundland, from the Western provinces, etc. For whatever reason Canada has never been able to construct a sense of itself sufficiently positive and sufficiently unifying to make a Quebecker or a Newfie Canadian first.

Think and Learn

With the development of the EU attention is turning to the degree to which it can be a focus of popular identification: to what degree do the French, the Germans, the Dutch, etc., identify themselves as Europeans? This has in turn generated a search for symbols of Europeanness that can express this embryonic sense of nationhood; that could, for example, appear on the euro when it became a widely circulated currency. Think of the dilemmas involved in developing such symbols and suggest ones that might just work.

There is something similar to the British case in the US. There are Italian-Americans, Irish-Americans, Polish-Americans, Cuban-Americans, Hispanic-Americans, as well as Americans who claim a – state-sanctioned – dual citizenship with Israel. Each of these subscribes to American traditions, understandings and senses of difference from other peoples. They all celebrate the national holidays, for instance. They all vote in American elections. The vast majority will use the English language as their first language. They will all learn the same history in high school, be party to the same national political controversies, etc. On the other hand, this is not exactly like the British case since the territorial element is missing, or rather it is geographically external to the US. Instead of the US divided into compact areas of Italian, Jewish, or Irish settlement as per the case of the English, Scottish, and Welsh, the territories which are still important in their frames of self-reference are elsewhere.[10]

So we are continually involved in bounding: in defining lines of difference between "us" and "them." But in this process not all differences get selected in. Several "rules" seem to be at work here. First, differences that are relevant to what is to become specifically *national* cross-cut numerous other social divisions. Nationals divide themselves from those who define themselves as belonging to other nations. But, and for example, there is no such thing as a

---

very Welsh. This suggests that there are distinct Scottish, Welsh, and English cultural traditions that have yet to be merged, and which may never be merged, in a distinctly British one.

10   The idea of "Scandinavia" is another interesting concept of identity worth drawing attention to in this regard. Danes have Denmark, Norwegians have Norway, and Swedes have Sweden. But they all have Scandinavia and identify with it to some degree, though not, one suspects, to the extent that the English, Scottish, and Welsh identify with being British, if only because Scandinavia has no political expression.

capitalist nation, a working class nation, a female or a male nation, nor yet one just for landowners or homeowners. In other words, and perhaps this is relevant to its appeal, the national abstracts from lots of things that divide people. It is rather what unites men and women, old and young, employer and employee in their imagined or indirect interactions with each other that provides the raw materials out of which senses of national distinctiveness, national identity, are forged.

Second, we tend to define ourselves in opposition to those with whom our contacts have been less than agreeable, with whom we feel some rivalry perhaps, or who threaten us in some way. The antipathy at work here can be of a very mild nature. The Canadians are an interesting case in point since it can hardly be said that their relation to Americans is one of hostility. Even so, many feel threatened by the presence of such a powerful neighbor: powerful economically as well as culturally. Accordingly, and to some degree, they tend to define themselves in contrast to Americans, or more accurately in contrast to their *perceptions* of Americans. It is common for Canadians to see Americans as less concerned for the underdog, for instance, and therefore themselves as more so, and this may be justified by reference to the fact that access to health care in Canada is not a matter of money but one of simply being a citizen. Canadians often see Americans as violence-prone in contrast to themselves and they will document this by reference to comparative murder rates and the relative freedom with which Americans can own guns. The question of Cuba has likewise served as a way of distancing themselves from the Americans, since while the US has pursued a policy of trying to bring Castro down by confrontation, the Canadian attitude has been that the only way of achieving change is by gradually establishing business and tourist relations with the country: in other words, and in the common Canadian view, Americans try to achieve their ends on the world stage by strong arm tactics while Canadians tend to be less confrontationalist. There are many other instances: Canadians tend to congratulate themselves on the question of race, for instance – perhaps unjustifiably – and so differentiate themselves from Americans with their clearly more tortured history in that regard. In short, much of our sense of difference is grasped through contrasts with others; and obviously contrasts which make "us" look good relative to "them."

And this is the third point with respect to this selecting-in and selecting-out process. We do indeed tend to differentiate ourselves from others in ways that we believe are flattering. The Canadians can congratulate themselves on their national health insurance scheme as evidence of the way they care for their fellow-citizens, but the American retort would be that such schemes are socialistic and that their own adherence to the private provision of health care demonstrates their commitment to norms of efficiency and to the development of health care and health care standards that results when market forces are unrestricted. By the same token what is unflattering is not a matter for differentiation. In their self understanding, the contrasts that they draw with others, Americans don't dwell on the extermination of the Indian (or the buffalo for that matter), just as the Germans are silent on their militaristic and genocidal past. And the Israelis are much more inclined to identify themselves as the vigilant defender of one of the world's most persecuted populations than as displacing a people from their homeland and oppressing those who remain under their rule.

In other words we construct ourselves with respect to others not just cognitively: how we differ in terms that others would accept as "factual." Rather we construct ourselves normatively, morally: how we are not just different but also better, more worthy, more deserving of the respect of others. We look to our differences from others as a source of personal significance and sense of self worth. We define ourselves as superior to other peoples. The pecking order of peoples varies from one people to another, but those doing the ranking usually come out as "number one." Even those peoples which, arguably, and by common knowledge, have been guilty of the most heinous crimes of genocide, racism, and imperialism can find some trait which establishes them as "number one," e.g. the "hardworking" Germans and Japanese, the "hospitable, easy-going, not-standing-on-ceremony" Americans, the "stolid" British bringing "civilization" to the natives, the "artistic and stylish" French spreading to "their" natives their particular brand of civilization, etc. And in the same way we can find fault with even the most commendable nation if it happens not to be our own.

What is interesting about this process is the way it is bringing different peoples together into a sort of global moral economy. Peoples can define themselves as "number one" but how much better if others accept the scale of values that you are drawing on: that everybody agrees it is better to have a free market economy just like the Americans rather than one in which the state plays a more prominent role; that everybody agrees that the military prowess shown by the British in reoccupying the Falklands is something that others should emulate, etc. Obviously many of these things are contested. But that doesn't mean to say that the attempt to make them hegemonic ceases. As a frequent reading of *The Wall Street Journal* would attest, imposing a particular way of economic life on the rest of the world is still a worthy project for many Americans and one that is ongoing through the offices of the International Monetary Fund, the World Bank, and the like. And this is still contested, as in the attempts of the Japanese to impose *their* particular concept of how economic life should be organized, their particular version of "freedom," in the Far East.

Contexts and Agents

Set against the backdrop of the full span of world history, the idea of the national is relatively recent. There is no evidence that the ancient Britons or the ancient Germanic tribes actually thought of themselves as British or German. As Hobsbawm (1990, p. 3) has written: "Nations, we now know . . . are not, as Bagehot thought, 'as old as history.' The modern sense of the word is no older than the eighteenth century, give or take the odd predecessor." In this particular timing it coincides with two other recent developments: the rise of the centralized state and of capitalism. The purpose of this section of the chapter is to demonstrate just how state and capitalism have been implicated with the rise of the idea of the national.

Both state and capitalism have been crucial contexts for the construction of nations. They have also, or more accurately those acting out state or capitalist[11] roles, featured as agents in that construction. States have formed the crucible for the formation of nations. Nations have been formed as peoples reacted to, acted against, states, as in the revolt against colonialism. But states have also intervened in the formation of nations for their own purposes. They have been active in the construction of difference, as indeed we saw in the discussion of Taiwan at the beginning of this chapter.

By the same token capitalism has also featured as a major context for the formation of nations. It has been through the creation of markets on ever larger geographic scales that people have become aware of their similarities to, and differences from, others. Through the material insecurities it has resulted in, it has also given people a motive to differentiate between the insiders, or the embryonic nationals in this case, and the outsiders: the immigrants, the foreign firms that threaten to take away "American jobs," and so forth. At the same time the unevenness of market outcomes, the creation of stratification systems coinciding with those emergent senses of difference, has often been a catalyst for national movements aimed at either overthrowing the system (if you are an underdog) or, if you are a top-dog, preserving it.

We will start out by considering the state and then discuss the importance of capitalism. But note how this is something of a false compartmentalization. For it is often the strains and tensions emanating from the unevenness of capitalist development that are the context for the formation of peoples demanding their own state. This is because it is through their own state that they hope to achieve a reversal of their (economic) fortunes.

The state

In the first place consider how the state provides for people a common set of experiences in the form of taxes, conscription, and public policy in general, and the diverse ways in which this can generate senses of difference and shared experience. Perhaps most obviously the common nature of these policies can generate a sense of unity within a country and of difference from those in other countries. People are united around discussion of the same policy alternatives and around nation-specific policies – like national health insurance in Canada – that separate them off from those in other countries like the United States. They face similar existential dilemmas like how to minimize taxable income, the answer to which will depend on which country they live in. And what is happening in Washington, in Ottawa, in Paris, or wherever provides a common point of departure for conversation. Most critically, at time of war the state of which one happens to be a citizen defines who your friends and enemies are.

11 Including the working class as a class found under capitalism.

So far so good. However, the effects of living with the same policies may have other seemingly contradictory effects. In the first place, what if the policies being implemented are experienced not just as something to talk about, to adapt to and live with, but as oppressive, a failure, as exploitative, even enslaving? This can create a sense of unity perhaps where it did not exist before and cross-cutting other differences like those of class, religion, ethnicity, tribe. This was the case with anti-colonial movements in the 1930s, 1940s, and 1950s in countries like India, and later Indonesia and Algeria, as the majority came together around a program of throwing the imperialists out. The view was that the colonies were being run not for them but for the colonial powers and their white settler allies. Rather the experience of colonial rule for the indigenous majority was one of racial discrimination, forced labor or at least the risk of it, and generally arbitrary rule.

Alternatively it may not be a matter of getting rid of the specifically colonial state and replacing it with an independent one. Rather the state in question may be "independent" but not representative of the wishes of the vast majority. It may implement policies – like the colonial state – for a small minority; or make such a hash of its responsibilities that it spreads nothing but misery. The exploitation of the *ancien régime*, the way it privileged the aristocracy, united the vast majority of the French and led to its overthrow in the French Revolution and its replacement with a republic. The Russian revolution of 1917 was able to succeed because Russians of all ethnicities, classes, and religions were frustrated by the failures of the Russian state in prosecuting the First World War.

State policy, by virtue of its inadequacies, therefore, can unite those subject to it and counter a sense of diversity and differences of interest that would otherwise make united action difficult. But it can also, and paradoxically, generate a sense of difference among its citizens – rather than between its citizens and the citizens of other countries – in circumstances where a sense of difference either did not exist before or was in a very rudimentary, unarticulated form. This is typically the prelude to the emergence of separatist movements demanding their own separate state and hence the partition of the existing one.

States tend to be homogenizing. The law is the same for everyone. But given some difference among the population, and depending on the law in question, this can be felt as highly discriminatory. In the interests of national unity states have been prone, for example, to define a particular language as *the* national language: the only language, in other words, which is acceptable in any business involving the state, including the highly sensitive areas of education and state broadcasting. This can make language an issue, a difference that people are sensitive to, where it was not an issue before. Because of this some may be excluded from highly lucrative positions in state employment, their culture can be threatened as their children lose the ability to speak in their native tongue. The symbolic effect of public notices written in the state language rather than the local one – highway signs, public conveniences, place names on maps – can add insult to injury. Such was the history of Germanization in what had been the western part of Poland in the nineteenth and early twentieth centuries. And so too is it with economic policy. To the extent that government policy is the same for all regions, to the degree that the state takes no measures to reverse geographically uneven development, for example, it can be defined as complicit in the backwardness of some regions and this also can generate a sense of difference, particularly if the state has set itself up as guardian of the economic fortunes of its citizens.

But if nations seek states through overthrowing existing ones, states also seek nations. States are quite crucial players in the development of a sense of nationhood. With the development of democracy this has become more urgent. Governments can't just tax and conscript at will. Their acts have to be justified to those on whose behalf they are supposed to be acting. This is typically in terms of some "national interest" which we "all" share. Moreover, the societies which states regulate are extraordinarily divided: there are class divisions, racial divisions, religious divisions in some countries, and cleavages along gender lines. Accordingly anything the state does will be interpreted in terms of the advantage it provides for someone else. How, therefore, to reconcile people to this except in terms of some notion of national solidarity: that, for example, it is right that the government should take the taxes of all in order to provide health care for the elderly and the poor because they are people we should care about as a result of their being, like us, British or Canadian.

Creating this sense of a shared identity, of belonging to the same nation, of being united with one another and against other nations, is something that states do because they have to, therefore. States build nations; they are active in their construction. Some have clearly been quite successful at it – France (see box 6.1), Germany, Britain – though others, like Canada, Spain, or

---

Box 6.1 *States and Nations: The Case of France*

In the contemporary world France seems to represent a paragon of national integration. It has a level of national unity, an immunity to centrifugal forces, that is the envy of many other nations. Underlying this is a strong homogeneity of language, religion, and – seemingly – cultural tradition. Separatist movements such as those of Brittany or Corsica are relatively weak. To the extent that there are tears in the national fabric they are vertical in form rather than horizontal: divisions of class and around the role of the Catholic Church in French life, though the latter is of much less significance than it used to be. But this high level of national integration is actually quite recent. And in this process the state played an important role. In the mid-nineteenth century, for example, France exhibited considerable diversity in its linguistic geography. The thoroughly French-speaking areas were confined to the north. In the far east and south and in Brittany local dialects which would be hard to classify as French prevailed (figures 6.3, 6.4): an interesting fact when set beside the high level of importance that attaches to the French language as a national symbol. At the same time this linguistic geography was matched by the maps of what we might reasonably call unpatriotic behavior. Figure 6.5 indicates the geography of resistance to military conscription; while figure 6.6 gives us some idea of those areas of France where people were

Figure 6.3 Largely or wholly French-speaking areas, 1885.

Source: After E. Weber (1976) *Peasants into Frenchmen: The Modernization of Rural France 1870–1914.* Stanford, CA: Stanford University Press, p. 68. Copyright 1976 by the Board of Trustees of the Leland Stanford Junior University.

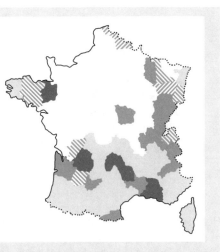

Figure 6.4 Largely or wholly non-French-speaking areas, 1863.

Source: After E. Weber (1976) *Peasants into Frenchmen: The Modernization of Rural France 1870–1914.* Stanford, CA: Stanford University Press, p. 68. Copyright 1976 by the Board of Trustees of the Leland Stanford Junior University.

Figure 6.5 Lack of patriotism as reflected in attempts to avoid military service, 1819–1826: draft evaders and self-mutilation to avoid service.

Source: After E. Weber (1976) *Peasants into Frenchmen: The Modernization of Rural France 1870–1914.* Stanford, CA: Stanford University Press, p. 106. Copyright 1976 by the Board of Trustees of the Leland Stanford Junior University.

Figure 6.6 Lack of patriotism as reflected in the cost of collecting taxes, 1834: more than 4 francs per 1,000 francs collected.

Source: After E. Weber (1976) *Peasants into Frenchmen: The Modernization of Rural France 1870–1914.* Stanford, CA: Stanford University Press, p. 106. Copyright 1976 by the Board of Trustees of the Leland Stanford Junior University.

Continued

most resistant to paying taxes. Southern France was obviously the area where national feeling, on these indices at least, was weakest.

A crucial area of state intervention which had as one of its effects an enhanced degree of national integration was that of education. The school reforms of Jules Ferry towards the end of the nineteenth century are regarded as particularly important. Thus in 1882 enrollment in public or private schools was made compulsory. In 1886 an elementary teaching program was instituted along with elaborate provision for inspection and control.

Belgium, somewhat less so. In the less developed countries the process is at even earlier stages. In many African states a sense of being, say, Nigerian, Zimbabwean, or Angolan has to compete with other identities such as that with a tribal grouping. So while nation-building is important, it is not necessarily easy and straightforward.

The object is to facilitate the creation of a sense of unity among those living within national boundaries and against those various unities found beyond those boundaries. Some of this is clearly an unintended by-product of the general insertion of the state into the fabric of society. This is because the mediation of social relations by the state provides a sense of similarity of circumstance among citizens, and of difference from the citizens of other countries: contact with the state through social services, clinics, schools, pensions, and common experiences like paying taxes, voting, serving in the armed forces. But states are also active in more intentional ways, trying to inculcate in the population at large what the nation is about, what it stands for, its values, its history, as indeed we saw in the case with which this chapter opened, that of Taiwan.

In this regard the state assumes broadly educational roles. The state school system is immensely important here for it gives the state a captive audience at an early and impressionable age. The state prescribes the structure of the school day and makes sure that it exposes the children to potent national practices and symbols: the oath of allegiance, worship according to the doctrines of a state church, salutes to the flag, possibly pictures of great national statesmen or war heroes on the schoolroom walls. The state also prescribes the curriculum. *National* history will, of course, be emphasized along with the *national* geography and the *national* literature. In countries where there is linguistic diversity the state may try to bring that to an end by insisting on instruction in the *national* language.

Especially important is national history. The histories told in school rooms are built around heroes who peculiarly exemplify national attributes, or at least those attributes which can be given a positive interpretation: the invention of Thomas Edison, the heroism of Nelson at Trafalgar, Abraham Lincoln's pursuit of equal rights for those of all races, Teddy Roosevelt and the devotion of Americans to the wilderness and its preservation, the music of the great

German composers, Bismarck as founder of the German welfare state. Other nations and peoples are drawn into those histories to likewise highlight national virtue: the peace loving British confronted with German aggression; the innocent, well meaning Americans, ever thinking the best of people, and the treachery of the Japanese at Pearl Harbor. The British Empire was not exploitative but was a happy welding together by the British of different peoples among whom they spread the advantages of civil order, education, and Christianity. And all is progress: Franklin Roosevelt undid past injustices to America's poor; post-war German governments have redeemed the nation's past, and so on (see box 6.2).[12]

Box 6.2 *National Interpretations of American History*

An illuminating example of the way particular themes in the history of a nation are abstracted and identified as the essence of the nation and of its people comes from an article that appeared in *The Wall Street Journal* (November 21, 1990, p. A12). In this article the writer, Peggy Noonan, addressed the problem of converting contemporary immigrants to the United States into Americans. In particular she tried to identify the critical stories and celebrations that needed to be communicated to migrants, mainly through the schools, as "what America is, what we believe in, what club they've joined and, by extension, what dues they owe."
    She identified what she called seven unifying myths. They were:

1   The coming of the Pilgrims: how they came to America in search of freedom to practice their religion; and the trials and tribulations that they suffered as colonists in order to make a better world for themselves.
2   The American Revolution: this symbolizes the commitment of Americans to democracy and to the individual. It was the first time in history that people had come together to produce a constitution; and it endowed the individual with certain inalienable rights.
3   The Civil War: a moral struggle in which a nation went to war with itself and in which right prevailed. No one has the right to own another. The freedom of individuals was affirmed.

continued

---

12   Much of this, of course, is of a strongly mythic character based on simplifications or in some cases, actual falsehoods. The idea of the French resistance in the Second World War is one example of a powerful national myth: "The romantic image of selfless French men and women in berets and leather jackets blowing up bridges and ambushing columns of German soldiers has become one of the most persistent of legends." But according to the same article "even those few French who risked their lives to gather military information for the Allies or downed airmen to safety often did it more for the money than out of patriotism. The standard payment for getting an escaped Allied prisoner out of France, across the Pyrenees down into Spain was $5,000, or about $50,000 today. A number of people grew rich out of the resistance" (Bernard D. Kaplan, "Author: French Resistance a Big Lie." *Columbus Dispatch*, January 21, 1996, p. 6A). The myth, apparently, was promoted by De Gaulle who needed it to bolster his position in London with the allies.

4   The Winning of the West: illustrates American willingness to suffer great hardship and loneliness in order to make progress and invent a new world. It exemplifes a continuing tradition to leave a place of safety and move on, a tradition reaffirmed in the voyage to the moon.

5   The Great Immigration of 1840–1920: how the immigrants learned the system, beat the system, and now run America.

6   The exploration of space: an ordering of talent and commitment and courage, not for conquest but to satisfy human curiosity.

7   The civil rights struggle: a struggle on behalf of the emancipation of the individual from arbitrary constraint; it succeeded because America had a conscience to which an appeal could be made.

Perusing this list a number of points of a general character seem clear. The first is that all of these themes flatter Americans: Americans are moral people (7), not interested in dominating other peoples (6), willing to sacrifice for a higher purpose (1, 3, 4), and willing to give individual ability its due (2, 5). Nationalist histories are invariably selective and by embellishing in this way they reinforce identification: which is precisely their purpose, of course, as the author of this piece recognizes (i.e. she is concerned with how to convert immigrants into Americans: regaling them with stories of atrocities against Indians or more recent atrocities against the Vietnamese, of the elimination of native species like the buffalo, the assault on nature in orgies of soil erosion, or the hideous exploitation of people on slave plantations, in the factories of the late nineteenth century and in the sweatshops of contemporary Los Angeles and New York, or even the present day gun mania, is hardly likely to do the trick, particularly if not set against the context of "progress" and an impulse to "reform" which is also seen as peculiarly American).

The second point is the way various national traits are disclosed by a comparison of the seven "myths" one with another. There are, for instance, repeated commitments to: (a) the individual and individual rights (2, 3, 5, 7); (b) progress and to the hard work and suffering that have to be endured if "progress" is to be brought about (4, 6); (c) higher ideals like "curiosity" (6), "freedom" (1) and what is "right" (3, 7).

This is not to pillory these peculiarly American beliefs. The point is that every nation has them. They're just different, though as in the American case they invariably have strong positive valence: things people can identify with. One can imagine what the British would emphasize, for example: the so-called "bulldog" spirit and the refusal to give in, even when things look bleak (e.g. the early days of the Second World War), which can be assimilated to the "stiff upper lip" or imperturbability in the face of any crisis; British inventiveness (e.g. the "mother" of the industrial revolution, the disproportionately large number of British scientists who have won Nobel Prizes); and a commitment to "quality" which makes the British resistant to the claims of popular (read "American") culture; and which of course incites counter-charges of "elitism" from Americans. So nationalist histories are inevitably selective and so neglect the negative: events, themes perhaps, that people would have great difficulty in identifying with, that would give them no cause for pride.

The state's educational mission extends beyond these rather obvious expressions, however. For many, education is extended by a period of military service. This provides scope for further exposure to emotionally charged symbols like the flag and the national emblem as well as for indoctrination into who the *national* enemies are. But not all "education" is compulsory. States also establish and finance *national* museums, military and otherwise, *national* galleries, *national* theaters, *national* parks, *national* festivals and exhibitions, in the context of which citizens are provided with other opportunities for learning about, appreciating, their *national* heritage.

Complementary to this educational activity is the creation or preservation and nurturing of national symbols: symbols of national unity, of what the nation means, of the glorious national history, etc. These assume a wide variety of forms, including:

- War memorials and tombs of the unknown soldier ("unknown" and therefore representing *all* those who have fallen in war).
- Public holidays in honor of "national" events or people who contributed to the life of the nation and to the fulfillment of national ideals. These can be days of remembrance, as in Memorial Day and Martin Luther King Day in the US, Armistice Day in Britain, or of celebration: Bastille Day in France, the countless Independence Days in so many countries. In some instances these occasions are obviously intended to extend a sense of significance to those who might reasonably be in some doubt as to whether or not their contributions to the nation were valued: Labor Day in the US is a case in point, but Martin Luther King Day is also important.
- Ritualized practices: the Presidential Inauguration, the lighting of the national Christmas tree in Washington, coronations, the opening of Parliament, and (certainly at one time) royal weddings.
- Practices underlining the role of the state as itself symbolic of the unity of the nation: these would include the formal openings of anything that the state has been involved in and which contributes to the life of the nation – new freeways, new major bridges, new public hospitals (in countries where the hospitals are state-owned) or public universities, new "national" parks, even new (nationally, as opposed to locally, owned) airports.
- The images selected for postage stamps and for tokens of the national currency: monarchs, "great" presidents, those regarded as contributors to the national life.

All these serve the national project by simply drawing attention to, endowing with feelings of reverence, objects of thought and feeling that are held to be of national significance. Each of them symbolizes belonging; and by the same token the exclusion of those who don't "belong." They bring together people in communities of an imagined nature; while at the same time excluding those (non-nationals) for whom it is not meaningful and therefore who don't belong. "We" understand; "they" don't. The Presidential Inauguration brings together all Americans watching it on TV, just as on Independence Day as people watch the fireworks they think it perfectly normal that people everywhere else in the

US – regardless of race, creed, gender, or class – are watching the fireworks at the same time. In Britain the monarchy was revitalized to fulfill this purpose: the development of practices like the coronation, trooping the color, or royalty opening new hospitals, which reinforce in the minds of the individual Briton the unity of the nation as symbolized by the Queen (though note in each of these instances the role of the media – TV, radio, newspapers – in (literally) mediating the community nature of these events, in facilitating mass participation in them: see box 6.3). Unity is being forged not only among a people but against other peoples. The nation belongs to "us" and not to "others."

Of course a common life, a sense of unity, can be threatened from within as minorities or otherwise marginalized groups flex their muscles. Who has to be recognized as part of the national life if the state is to retain its legitimacy as representing "the nation" changes over time: new groups to be integrated into the nation and made to feel part of it. The selection of the faces to appear on stamps, coins, and banknotes is especially interesting. A number of black American jazz musicians – Louis Armstrong, Miles Davis, Duke Ellington,

---

Box 6.3   *The Role of the Media*

It has been argued that it was print-capitalism, the mass availability of printed material readily sold and bought, that made the national *possible* (Anderson, 1983). This suggests in turn that at one point in time at least, prior to other forms of media like radio and television, literacy was a sine qua non for the development of a sense of common nationality; which puts the educational reforms of Jules Ferry which we reviewed in box 6.1 in a slightly different light. The fundamental reason for this is that it would have been through printed materials, especially newspapers and magazines, that people first learnt of what they shared with others elsewhere, beyond the immediate locality, that is, and what they did not share with the putatively non-nationals. Through the newspaper, for example, they would have learnt of the activities of a shared monarch, of the unveiling of national monuments, of dilemmas shared with other (putative) nationals elswhere, and through maps to illustrate points, something of the geography of the country. The newspaper itself would have had a broadly educative purpose in this regard, distinguishing, for example, between "National News" and "International News" and so helping people interpret "the news" in these terms.

Earlier in this chapter I referred to the role of travel writers in alerting us to a sense of difference from people elsewhere. But the non-print media clearly also play an important role in this regard. It is through photographic images, for example, through cinema newsreels (now, in an age of TV, no longer shown) and foreign movies, that we learn or learnt something of what distinguishes in physical appearance – hair style, modes of dress, even eyeglass style – a French person from an American. And in this regard, ponder on the huge importance of magazines like *National Geographic* (note the title) in conveying how different the world is beyond American shores.

Figure 6.7 Black recognition: the years subsequent to the civil rights movement in the US have seen increasing recognition of black achievements by the government. One of the ways this has been done is through the images on postage stamps. These particular stamps are also revealing for the way in which a neglected artform, jazz, is recognized here as not only black – for out of the total series of ten stamps, none bore a white image – but also as essentially American. For example, while jazz is popular in Europe the idea of France or Germany commemorating jazz musicians is faintly ludicrous since it is obviously an adopted art form there.

Thelonious Monk, and Charlie Parker, for instance – have all appeared on American stamps within the past ten years (figure 6.7); but for a black face to have appeared on a stamp just fifty years ago would have been unthinkable.[13] Likewise there is the female face on the dollar coin: again, a highly unlikely event prior to the women's movement which took off in the sixties. And a few years ago the image of Billie Holiday brought together both the excluded gender and an excluded race.

But if all this is to work, if the nation is to succeed in its educational and symbolic practices in forging the sense of a common life, history, and destiny which differentiates them from the citizens of other countries, then the educational practices, the symbols that the state offers, have to be not only acceptable but acceptable with enthusiasm: they *have to resonate*. For example, it is no good introducing something like the Queen's Christmas Broadcast as,

13   The new found respectability of jazz, its recognition as a major American art form and one that has contributed to the distinctiveness and vitality of the culture of the United States – and arguably its greatest contribution to world culture – is also significant.

potentially at least, a unifying symbol if the monarchy is hated. Indeed one might argue that with the development of a widespread scepticism about the future of the British monarchy that particular tradition no longer works the magic it once did. Likewise, and with reference to the symbolic first pitch from the President to open the baseball season, it would do little good for American identity if baseball was not a consuming passion for large fractions of the population. Again, recent changes in attitude towards baseball as a professional sport suggest that this too may be a tradition which is losing its effectiveness. Likewise where there are deep divisions in the population then the symbols chosen must not be divisive. This is why, subsequent to the realization of black majority rule in that country, the South African flag was redesigned: the emphatic use of the color "orange" generated negative feelings among the majority of that country's population since it was so much associated with the Afrikaners and the apartheid state, though significantly the color orange has not been entirely excluded.

So too is it with respect to the state's broadly educational role. The state has to tread carefully if it is to seduce, persuade people that they share more than what separates them. It is no good forcing a common language on everyone where there are strong cultural traditions surrounding the particular languages found in a country. Better yet would be an emphasis on the contributions representatives of those different cultural communities have made to the national life, the writing – perhaps rewriting – of national history so that everyone feels a part of it.

Recent events in South Africa are especially interesting here. Until very recently blacks were excluded both formally and morally from the life of the state. Only whites enjoyed the vote and it wasn't clear that blacks were citizens at all. And this was despite the fact that whites were a very small minority (15 percent) of the total population. This has now changed. South Africa has a black majority government. Symbols of black exclusion can no longer continue and new symbols expressive of the multiracial character of the country have to be found. The first part has been easy. The faces of prominent white statesmen of centuries past have been removed from the country's bank notes (figure 6.8). The second part has been more difficult. For given the still fraught character of race relations in South Africa the South African government has drawn back from replacing white faces with black. Instead, in the interests of national unity it has retreated into something more anodyne: not people as representative of South Africa's unity, but wildlife!

On the other hand, it is important not to exaggerate the state's power in nation building. The experience of Eastern Europe during the late 1980s and early 1990s provides an important object lesson. Although the area had a history of nationalism of the most chauvinistic kind, after 1945 communist regimes seemed to have succeeded in eliminating those feelings. Fraternal peace between the individual nations and indeed within multinational states like Czechoslovakia, Yugoslavia, and Bulgaria seemed to have been consolidated. But in retrospect the attempt of those regimes to eradicate nationalist feelings has to be judged a failure. For with their collapse there has been a clear and very strong resurgence of nationalist sentiment. Multinational states

Figure 6.8   Changing national symbols in South Africa. The top image is of a banknote from the apartheid era of white domination. It depicts an early white settler, Jan Van Riebeck, first governor of the Cape. The bottom two images are from postapartheid banknotes. The black majority government has studiously tried to tiptoe along a line on one side of which it would offend whites by including black faces, and on the other side of which it would offend blacks by retaining white faces.

like the USSR, Yugoslavia, and Czechoslovakia have disintegrated, often with much bloodshed. Quite how this has been possible needs careful investigation. It may be, for example, that the return of capitalism makes a difference and that there is a logic which leads from the pressures of capitalist development to nationalist interpretation. Alternatively it is conceivable that nationalist feeling was preserved through vehicles, like the family and church, which were driven underground. And elsewhere the emigrés kept the national idea alive.

In understanding the development of nationhood, therefore, it is important to balance a consideration of the top-down forces represented by the state with the bottom-up forces emanating from civil society. Striking instances of the power of the latter come from states in Africa and South Asia and the Caribbean that became independent in the post-1945 period. In each instance the drive for independence was led by a nationalist movement that the colonial states attempted to suppress. Far from wanting national integration the colonial states would have preferred, and indeed often tried to foster, a national *dis*integration that would have prolonged their rule. Similarly there are the various separatist movements that have led to the creation of new states. Senses of nationhood have been nurtured among some at variance with the one that the government has tried to create and this can result in national movements demanding the partition of existing states. Violence has often been necessary to bring this about as in the case of Yugoslavia, and it may yet occur in the case of Taiwan since, as we have seen, it now seems bent on a state separate from that of China. But violence is not always necessary: the division of Czechoslovakia into the Czech Republic and Slovakia was almost entirely peaceful in character.

Nation states strive for "purity": a unity between state boundaries and nation. Within respective boundaries states try to construct a common sense of nationhood. If, for whatever reason, they fail, then the emergence of separatist movements and subsequent partition may be the result. But partition is only possible if the senses of nationality apply to geographically discrete populations – as they did in the Czechoslovakian case. A major source of the violence in Yugoslavia was that they didn't. So a Croatian state was a threat to the Serb minority as was a Bosnian state. The result was the attempt by the Serbs to create their own spaces, the territorial base for their own states, by expelling Croats and Bosnians from areas which they dominated numerically. This is so-called "ethnic cleansing" and it is by no means new.

"Ethnic cleansing" is a solution that arises in particular geographic situations: where the dissident fractions are spatially interspersed among the national and where some wholesale geographic excising of the dissidents through secession is not a feasible option. Moreover, the "ethnic cleansing" which occurred in Yugoslavia in the summer of 1992 as (e.g.) Serbs expelled Croats and Bosnians from certain areas in order to create homogeneously Serb areas was by no means new. The Germans expelled Poles from the half-German, half-Polish areas of Western Poland during the Second World War and replaced them with Germans. Likewise under apartheid the South African state tried to create a homogeneously white nation by forcibly relocating blacks out of South Africa into areas that were set aside as future independent black states; what were to be the independent homelands.

Not that ethnic cleansing is always of this clearly forcible character. Many whites are now leaving South Africa of their own accord since they do not feel part of the sort of nation that is now being constructed there. Likewise states can stop short of force but still aim at "ethnic cleansing." There are still important elements in the Israeli state, for example, which would like to see Palestinians so discouraged about living under Israeli rule that they will simply leave for places like Jordan. Their places can then be taken by Israelis. In such ways can states remove those who have no sense of identification with the nation and replace them with those who do.

But to return to the "success" of the state in generating a sense of a common life, there is much that the state does in this regard that is quite inadvert. For what it often provides is simply the territorial frame for other activities that provide a sense of community, or shared social life, among those participating in them. The history of sport from this standpoint has yet to be written. But how can one doubt the sense of community, albeit gender-biased, of a common national enterprise, provided by the National Football League or major league baseball in the US or by soccer in England and Scotland? The use of nationally charged symbols, the singing of the national anthem, the single game or series of games which brings the season to a close and unites, if only through the media of newspaper and television, insistently informs people that they share something, that they belong together, and addresses them as fellow nationals.

Think and Learn

Reference was made in the paragraph immediately above to the conception of "a common national enterprise" as "gender-biased." What other evidence can you think of that supports the claim that nation construction has been a highly genderized process? Are there, for example, as many national heroines as heroes? How might you explain this?

Soccer leagues, just like the National Football League of the United States, are decidedly *national* institutions. There are, for example, England's Premier League, Scotland's Scottish Premier League, Germany's Bundesliga, and Italy's Serie A. The nations provide the essential arenas for these competitions. The focus of the press, of the fans, is defined nationally. A sense of a common social life is created on a national basis. As the British historian Eric Hobsbawm (1983) remarked in discussing the significance of professional soccer in Britain around the turn of the twentieth century: "the topic of the day's matches would provide common ground for conversation between virtually any two male [*sic*] workers in England or Scotland, and a few score celebrated players provided *a point of common reference for all*" (pp. 288–9, emphasis added).

One qualification that should be entered here, however, is that while indeed the nation provides a territorial frame for these activities that bring many people together and divide along national lines, they may also divide in other ways. Sport *is* national but, as far as popular and media attention are concerned it would also seem to be highly masculine. Women don't get much of a look in, and where they do, as in tennis or soccer, they are still defined as second best.[14] Alternatively, there may be divisions that correspond to the social stratification or racial composition of the population. In England rugby union has historically been the game of the better off – of those who send their children to private schools, for example – while soccer has clearly been more proletarian in its appeal. In South Africa the divisions are, as might be expected, racial: soccer for black Africans and rugby for whites.

Capitalism

Capitalism has been an extraordinarily important context for the formation of national identity. This is because of the way it installs a regime of extreme material insecurity, a struggle for material advantage, and an enmeshing of people in geographically extended webs of market exchange: a world of labor migrations, imports, exports, financial flows, geographically expanding markets everywhere and for everything. As we should recall from chapter 2, these tendencies are all related, all mutually entailing. The creation of markets entails competition which in turn results in a sense of material insecurity. That anxiety, that fear of the future, of bankruptcy, unemployment, foreclosure, of liens on property, in turn produces a drive for accumulation on the part of businesses, for upward mobility on the part of workers; and drives to keep others in their place for fear of the challenge that their competition might mean to material advantages, to nest eggs and to corporate surpluses as hedges against material insecurity, already achieved. At the same time the urge to accumulate, to obtain a higher wage, unleashes a geographic expansion of markets: of labor markets, product markets, money markets, which in turn threaten the advantages businesses, workers, and residents enjoy in particular places.

This has been the essential context for the formation of new identities linking large numbers of people together on a territorial basis and in opposition to people elsewhere. We have already noted this sort of process at work in the cases of gender and race in chapter 5, and they will be taken up further in chapter 7. But so too is it the case with the idea of national identity. People are pulled together into relations of competition with distant others, as in product markets, or those from other places as in the case of the migration of workers, or, for some local businesspeople, perhaps, inward investment. In the context of the insecurity this provokes it is not difficult to imagine them starting to act as peoples, to construct each other as peoples in opposition to one another: to differentiate themselves in terms of what they share in their history, culture, perhaps, from others who are singled out as threatening to material positions – but furthermore, to differentiate themselves on the basis

---

14   This is something that in some instances, like the Wimbledon tennis tournament, they have tried to combat by pressing for equality of prizes.

not just of the cultural, the historical, but also of some sense of proprietary right with respect to a particular space. This is the idea of living in a particular place as mandating exclusive rights not available to newcomers, to outsiders; a tendency that we first encountered in chapter 5, and mediating the territorial character of the identities being formed. This emergent sense of peoplehood is then used as the basis for calls on states to regulate markets to their advantage: to usurp, to defend, to achieve positions of advantage as we noted in the case of Padania in chapter 1.

In this process of social construction Hobsbawm (1990) has argued for the importance of what he has called "proto-national bonds." By these he means the existence of certain feelings of collective belonging which could operate on a scale above that of the purely local. He argues further that the most decisive criterion of proto-national bonds is the consciousness of belonging/ having belonged to a lasting political entity. England and France were political entities, they were both monarchies, for example, before there was a sense of Englishness or of Frenchness. But, and on the other hand, this is clearly not a necessary precondition for the formation of national identities. There was no "lasting political entity" prior to the formation of a sense of American identity, unless one is willing to accept the original colonies as providing it; but they were *separate* political entities.

All this, however, is quite abstract as an understanding of the development of national identities. The discussion, therefore, needs to be complemented by one that identifies the more concrete contexts within which senses of national identity have been formed. Two in particular are cities and the colonies. For on the one hand, capitalist development unleashes processes of urbanization on a hitherto unprecedented scale. These in turn bring people of very different cultural formation together. At the same time it has been associated with empire and the formation of colonies. These again brought into contact peoples of very different cultural background, and in a relationship of domination and subordination, which in turn meant that the identities that would be constructed would inevitably be ones of opposition.

The formation of capitalist economies has invariably taken place within the context of centralizing states but also against a background of geographical variation: variation in language or dialect, custom, historical tradition, old political authorities like dukedoms, kingdoms,[15] or, in the African case, tribal chieftaincies. The development of the economy, the establishment of mobility of capital and labor on a national scale, brings the representatives of different cultures into contact with one another, as in cities. This creates new fields of contrast, new bases for attributing similarity and difference and hence identification. At the same time it brings them into competition with one another. And to the degree that competitive advantage/disadvantage seems to congeal around these newly forming identities, so that formation process is given a further thrust.

An important consequence has been a degree of aggregation into larger proto-national or even national groupings. The Italians who came to North

15   As in pre-1871 Germany.

America in the late nineteenth century came with a very limited sense of being Italian if they had that sense at all. They were (e.g.) Sicilians, Neapolitans, or Venetians first and foremost. But in the very different cultural context of the North American city, linguistically, sometimes religiously, and in terms of eating habits, they were forced in on themselves. They had to interact with one another. They confronted a spectrum of difference with, for a Neapolitan, for example, Sicilians and Genoese much more similar than the Irish, the native born Americans or whomever. They had to rely on each other in the matter of securing work, housing, establishing churches, supporting cafes, and other vehicles of collective life, therefore. In consequence they began to see themselves as having a good deal in common with each other, regardless of the fact that they came from different parts of Italy. They began to see themselves as Italians, and probably for the first time. And the fact that they were identified as Italian by the native born Americans – since they must have all looked and sounded pretty much the same to their untutored senses – only reinforced that effect. One can imagine similar processes in the multi-ethnic Habsburg Empire with Czechs, Croats, Serbs, Romanians, Slovakians, etc. first realizing the cultural affinities – if not exact, certainly close – with other Czechs, Croats, Serbs, Romanians, Slovakians in such cities as Vienna, Trieste, Prague, Bratislava, Budapest, and Belgrade.

Think and Learn

How does this process of making Italians in the context of the American city compare with the process of making Caribbeans that we reviewed in chapter 5? In what ways are they similar and in what ways are they different?

Moreover, capitalist development is extremely uneven. Some sections of the population are more developed than others in the sense of their capacities to participate in the more demanding[16] aspects of the labor process, or even organize it, etc. These differences often mirror other differences of a more cultural nature or simply in who was there first, among the different populations coming into contact within urban or national labor markets. The larger cultural groupings that emerge in the cities often vary in the degree to which they can compete for jobs. This may be a result of variations in education between the different parts of a country. The attempt to standardize language, to create a language of business or of the state, can also leave particular groups feeling that opportunities across the different proto-national groupings are unequal.

It is this sense of inequality that can transform the proto-national into the national. The formation of a Quebecois nationality is a case in point, as we

16   "Demanding" in the sense of requiring lengthy training and/or on the job experience.

will see below, for a major stimulus was the language question: the fact that the language of business in the major city of Montreal was English and this served to block the upward mobility of the French-speaking. In South Africa the formation of an Afrikaner nation is owing in part to similar circumstances. The Afrikaners who spoke a language derived in part from Dutch and German were for many years relegated to a subordinate position in the stratification system of white South Africa: they were over-represented among the poor and under-represented in higher paying jobs in business and state. Their relative absence from state employment was especially galling since it was in part due to the fact that for many years the South African civil service did not recognize Afrikaans as a language for purposes of state business; Afrikaners were therefore automatically excluded on language grounds.

The other context, apart from the urban, has been the colonial. We will review this at greater length in the next chapter. Suffice it to say here that the cultural divide between colonial administrator or white settler on the one hand and the indigenous population on the other was accompanied by a yawning material gulf. This was one, moreover, that could readily be constructed by the indigenes as highly exploitative in character. The resultant identities – the Europeans and the nationalist forces – were constructed in opposition one to another, therefore.

Class and nation

We have seen that the development of capitalism is part of the backdrop, the essential context, for the formation of nations. But capitalism also and simultaneously generates a division of the population into classes. The principal classes are those who own the means of production, i.e. the capitalist class, and those who lack the means of production and therefore have to sell their labor power to the capitalists in order to obtain access to means of subsistence through a wage, i.e. the working class. There are also intermediate classes consisting of the owners of very small businesses in which the owner also contributes to the labor through which its characteristic product or service is produced.

In chapter 2 we saw how this polarization between business and labor forms the principle cleavage between the major political parties, though support for the right-wing parties that tend to be pro-business in their positions is also considerably bolstered by the support of the better-off fractions of the working class: the supervisors, the more highly skilled, those with professional qualifications of some sort. But this type of polarization, or rather resistance, is something business could do without. The policies it prefers – ones of liberal labor law, low levels of corporate taxation and of progressivity in the income tax, a weak welfare state – are ones that historically at least have been vigorously opposed by political parties of the left. The profitability of business has therefore been challenged by national labor movements.

But at least in their rhetoric labor movements, or at least the more radical of them,[17] have also had an important international dimension. The view has been that since business organizes itself on an international scale through its investments so too must labor: in short, "Workers of the World Unite." In turn, however, this internationalism has created an opening for business. This is because it has allowed them to depict labor leaders as traitors to the nation;[18] and relatedly to drive a wedge between the more class conscious leadership of the labor movement and its less class conscious, more nationally conscious membership.[19] In short, in nationalism business has found a useful tool for countering the challenge of labor.

But in point of fact the international character of their brotherhood has always been problematic for labor. Attempts to form labor unions on an international basis have never been outstanding successes. The labor unions of the more developed countries are often more concerned about the jobs of their own members than about raising the standard of living of their fellow workers in less developed countries, as in fact we saw in chapter 3: or at least that is how it would seem judging from their support for (e.g.) tariff protection against the exports of those countries. In fact, in their formation class and nation have interacted with each other in important ways. On the side of class the formation of classes through the development of labor unions, employers' organizations, representative political parties, institutions, and traditions has clearly been on a nationwide basis. People have come to identify themselves as members not simply of the working class but (e.g.) of the *French* working class. And not simply as business but as *British* business. At the same time the nation and national identity provided a context for class formation. It was with the plight of fellow nationals that the working class in Britain identified, and not unreasonably since it was experienced much more immediately than say the plight of working class people in Germany or France. Likewise it was with the specifically British working class that British capitalism had to deal in legislative battles.

This contextualization of class and class formation by the sense of nationhood and national difference and hence identity posed a real threat to the ability of the working class to press its claims and to retain its sense of opposition. This is because it has constantly had to fight the claims of the national

---

17   This would include therefore the French, Communist Party-backed *Confédération Générale du Travail* but not the American AFL-CIO.

18   In writing about the attempt to draw on nationalist themes to counter the growing support for left-wing parties in Europe in the 1920s and 1930s, Michael Mann has written that the direction of aggression was mainly "inward against those 'disloyal' to the nation, weakening its collective capacity to deal with the profound crises of the period. Socialism and anarcho-syndicalism were obvious targets: both proclaimed internationalism, socialism was pro-Russian and anarcho-syndicalists denounced the nation-state. The terms 'Bolshevik' and 'anarchist' were perennial terms of abuse, along with the simpler 'Red', all conveying a sense of foreignness and disorder. . . . British Conservatives contrasted their own patriotism with the divisiveness and 'foreign origin' of socialism. Labour failed to be 'Britons first, and Socialists only second'" (Mann, 1995, p. 32).

19   This is not to argue that the leaderships of labor movements are always more class conscious than their members.

interest, a national interest which was usually defined in terms of the subordination of working class interests to such policy goals as national efficiency and competitiveness or national military strength.

Yet one hesitates to be too definitive about this. It is true that the nation state has tended to fragment the class struggle, isolating it within national boundaries, and then making the working class vulnerable to appeals couched in terms of the national interest. But the effectiveness with which unification around national symbols could extinguish working class feeling and polarization *vis-à-vis* the capitalist class has clearly varied a great deal. Ruling classes which have been able to point to a record of great national triumphs, of imperial expansion, for example, were obviously in a much better position to subvert working class feeling than the defeated or those simply unable to project themselves on the world stage. Sweden, in international terms a relatively weak country, was for many years regarded as very socialist in its domestic policies. This also extended to its foreign policy. It was one of the few non-communist countries, for example, to provide financial aid to the African National Congress in its struggle against the apartheid regime of South Africa. New Zealand is another instance of a relatively weak player on the world stage but historically labor-dominated and which has sought to carve out foreign policy positions at variance with the militaristic stances of the so-called Great Powers.

Likewise the inequalities from which nation abstracts cannot be too extreme if nation is to work its magic, if its symbols are to be experienced as truly positive for all and consequently equalizing. In countries like Mexico or Peru, where there is intense social polarization and where differences of race and standard of living reinforce one another, it would be surprising if the existing national symbols were not regarded by the poor with a good degree of cynicism: whose nation is it, after all? Likewise it may be that nation works best as a set of uniting symbols if there is a sense of social fluidity so that (e.g.) anyone can indeed rise to become president. The British–American contrast is interesting here for the British commonly experience Americans as much more intensely nationalistic, but British social structure has the image, at least, of being much less fluid, as one where families can be locked into subordinate positions for generations.

Case Studies

National movements are often coalitions of different sets of interests, some more cultural in character, some more economic. Useful case studies with some provocative similarities and contrasts are provided by Quebecois and Afrikaner nationalism respectively. Both Quebecois and Afrikaners regard themselves as stepchildren of the British Empire. The Dutch settlers of South Africa – ultimately to become Afrikaners – were under Dutch rule until 1806 when the British took over. In the case of Quebec, British rule came almost a half-century earlier in 1763 when they displaced the French as the hegemonic imperial power in what was to become Canada. In both instances the sense of national difference has been with respect to the English and, again in both cases, the language question has been to the fore. Likewise an economic stratification that worked to their disadvantage has been an important issue. On the other hand, given the geography of the respective cases, there was no way in which there could have been an Afrikaner state similar to that of a Quebec state.

Quebec separatism

The Canadian province of Quebec is dominantly French-speaking: over four-fifths of the population speaks French in the home. This is of significance to those professions which, by virtue of language, see the francophones as "their market." The obvious cases in point are the journalists and media people, the teachers – both in schools and universities – the lawyers, and the clergy. Their major point is that the French language is the linchpin of a particular way of life that is worth nurturing and should therefore be defended, though clearly this is not disinterested on their part. French is their medium and any serious anglicization of the Quebec population would be a threat to the exclusivity of access they currently enjoy. In other words: culture is economic!

The appeal of the Quebec nationalists has been much wider than this, however. While many have been moved by the appeal to defending a distinct way of life and culture against the threat of anglicization others have been driven by a concern to widen[20] their career prospects. For a long time there has been an economic stratification in Quebec: business has been dominated by the English and English has been the language of business. In the late 1970s, among the 105 largest corporations in Quebec only 14 had a majority of French-speaking directors. Of the other 91 only 9 percent of the directors were French-speaking: and this in a province that was 80 percent francophone. The result has been a stratification of power and wealth in which the English-speaking were at the top and the French-speaking were at the bottom: a clear ethnic or national stratification. This has been a source of resentment for the francophones, for only by adopting the English language did it seem that they could be upwardly mobile.

What has united the different strands of Quebec nationalism, therefore, has been the language question. The Parti Quebecois, the major vehicle of Quebec nationalism, came to power in the province in the mid-seventies and perhaps its most crucial act was the introduction in 1977 of Bill 101. This made French the only official language and banned the use of English in the workplace. It also limited the right of English-speaking immigrants to send their children to English-speaking schools. In this way they hoped both to put a stop to creeping anglicization, particularly in Montreal where there was a large English-speaking minority and where immigrants from non-English-speaking

20 As opposed to "conserving," which appears to have been the goal of the lawyers, the teachers, and the French media people.

countries were opting for English; and also to open up more higher-level employment for the French-speaking.[21]

On the other hand, the drive for an independent state of Quebec has had an additional rationale. For it would require an expansion of state employment in the province and so open up additional high-paying jobs for the French-speaking. In this way it could further mitigate their position of economic inferiority with respect to the anglophones. Historically the francophone university-educated have shown a bias to state employment in order to avoid the problems they would face in the private sector. It is this, for example, which helps to explain the taking into state-ownership of several English-Canadian owned electric utilities in the 1960s: opening up jobs for more French-speaking civil servants.

In retrospect a crucial precondition for Quebecois nationalism has been the incorporation of Quebec into broader economic structures: national and global markets in particular. The agents for this incorporation have been the primarily anglophone-owned and managed firms located in Montreal. The expansion of anglophone-dominated businesses – manufacturing firms, banks, insurance companies – led to the urbanization of large numbers of French-speaking Quebecois and their subsequent contact with those speaking English. This had two effects. On the one hand it created the sharp economic stratification between anglophones and francophones to which I drew attention above. And on the other, due to the anglophone cultural dominance in Montreal, it also brought with it an uncertainty about whether a unique way of life could endure. Accordingly Montreal became the epicenter of Quebec nationalism.

Afrikaner nationalism and its changing political goals

In South Africa today whites are outnumbered by black people (Africans, Coloreds, and Indians) by a factor of about seven to one, though it is only recently (1994) that a non-racial franchise was introduced. Historically the state in South Africa has been white, both in the composition of legislators and senior civil service and in those holding the franchise. Whites, however, are divided into two clear cultural or, according to some views, ethnic groups. Well over half the white population identify themselves as Afrikaners. They are the descendants of Dutch, German, and French Huguenot settlers in South Africa. They speak a language, Afrikaans, which is related to Dutch but which today is far from identical to it. They have their own church, the Dutch Reformed Church, their own universities and schools, their own intellectuals, and a

---

21   An initial effect of this was the relocation of some English-dominated businesses out of Quebec, suggesting that nationalism had a calculable dollar cost. But for yet other businesses (the spatially entrapped?) its effect was to increase the demand for French-speaking managers. There has also been some "ethnic cleansing" since many anglophones have left the province: the 10.8 percent who spoke English at home in 1999 compares with 15 percent in the early 1970s.

distinct cultural life. The remaining whites are English-speaking and are the descendants of settlers from Britain. They are culturally much less self-conscious than the Afrikaners, though Afrikaner identity has been formed significantly in opposition to the British and what they have represented for the Afrikaners. From 1948 to 1994 the Afrikaners dominated the South African state and were the implementers of apartheid policies, though from the formation of South Africa until that point the ascendancy was for the most part that of the English-speaking.

The origins of Afrikaner nationalism are complex, though as indicated above, it has been formed to a substantial degree in opposition to the English-speaking and to Britain as an external force on the subcontinent. Of course, it is hard to be sure whether the depictions by Afrikaner nationalists of their history and its intertwining with that of the English-speaking and the British Empire are accurate or reformulated for nationalist consumption. But at any rate the history of contact with the English looms large in the stories that Afrikaner nationalists tell.

Three particular historical experiences exemplify what the Afrikaners believe they have suffered at the hands of the English:

1   The Great Trek. Until 1806 the Dutch settlers were governed by representatives of the Dutch East Indies Company. After that, however, the Cape, which was where most of the whites were concentrated at that time, became a colonial possession of the British. It is commonly believed among the Afrikaners that the British were less than sympathetic to their interests. In particular, they abolished slavery, which had been a common practice among the Afrikaner farmers. This helped trigger off the movement of many settlers east and northeast out of the Cape Colony into the African interior and away from the representatives of the British colonial government. Symbolically the Great Trek has contributed a great deal to Afrikaner nationalism. According to the conventional Afrikaner narratives great privations were experienced, particularly in the form of hostile African tribes but, and in testimony to their admirable qualities, the Afrikaner trek Boers, as they became known, prevailed and ultimately established their own republics in the (aptly named) Orange Free State and in the Transvaal. They endured much but, and according to nationalist doctrine, that testifies to the repressiveness of the British and to their own desire for freedom (figure 6.9).

2   The second great event is the Boer War (1899–1902). In Afrikaner nationalism this has become a potent symbol of the hostility of the British to the Afrikaners, of their resistance, and of the cruelties they experienced at British hands. Much of what they believe about the Boer War, however, is true. It is true that the British wanted to overthrow the independent Boer republics, particularly the Transvaal. This was because British mining interests that had started exploiting gold after 1886 in the vicinity of what was to become Johannesburg were thoroughly dissatisfied with the way the Afrikaner or Boer state of the Transvaal catered to their interests (figure 6.10). The Boer War was a very imperialist war, therefore. The Boer resis-

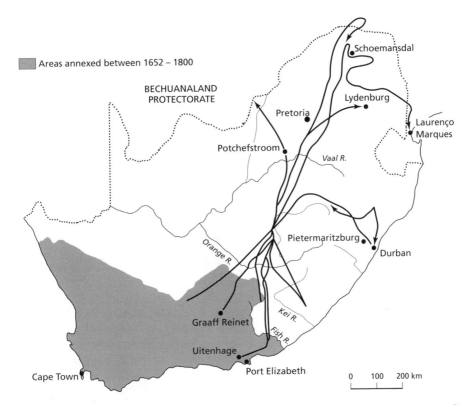

Figure 6.9 The Great Trek. The major routes followed by the "voortrekkers" (i.e. "those who trekked before") traced trajectories that started in the Eastern Cape, which was already under British rule in 1806, north-northeast into what was later to become the Transvaal, the Orange Free State, and Natal.

Source: After figure 1.5, p. 15, in A. J. Christopher (1994) *The Atlas of Apartheid*. London and New York: Routledge; and figure 5, p. 51, in T. R. H. Davenport (1987) *South Africa: A Modern History*. Toronto and Buffalo: University of Toronto Press.

tance was indeed of an heroic magnitude. Vastly outnumbered, men and young boys adopted guerilla tactics and kept the British army at bay for much longer than they had anticipated. The army responded by laying waste the crops and livestock of the Afrikaners and herding the women and young children into concentration camps where large numbers died in piteous conditions.

3   Language policy. Subsequent to the conversion of the Orange Free State and the Transvaal to the status of colonies in the British Empire the British government, in an attempt to heal the wounds, gave the four colonies of which South Africa consisted at that time (the Cape and Natal were the other two) independence as the Union of South Africa. Unfortunately in the newly independent state English was declared as the state language. This excluded many Afrikaners from state employment and meant that

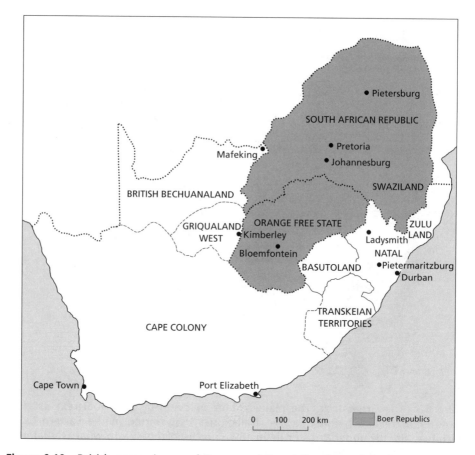

Figure 6.10 British possessions and Boer republics at the time of the Boer War.

Source: After A. J. Christopher (1994) *The Atlas of Apartheid*. London and New York: Routledge, figure 1.6, p. 16.

Afrikaner children could not learn their native tongue in the schools of South Africa. This produced more resentment.

Many Afrikaners believe, therefore, that they have suffered at the hands of the English. This sense of oppression was heightened by the different positions they occupied in the South African stratification system. The English-speaking were always better off. They were especially strong in the owner-ship and management of the mining and manufacturing industries and also in the higher levels of the country's civil service. Mining, moreover, was polit-ically strong and this served to give it privileges lacking to agriculture, which was where most Afrikaner business enterprise was concentrated. There was a white working class but most of it was Afrikaner in origin. This recalls the sort of cultural stratification of the economy that has characterized Quebec and which became such an issue there.

It was in this context that Afrikaner nationalism emerged in the 1920s, giving birth to a succession of so-called National Parties. In its development a marginalized professional middle class of teachers, clerics, lawyers, and journalists played a key role. These were people for whom an Afrikaner-speaking constituency was important: again, note the similarities to Quebecois nationalism. Both lawyers and teachers suffered from the proscription of Afrikaans as a state language. Journalists and clerics, on the other hand, were concerned about the urbanization of the Afrikaner population and the way it was resulting in intermarriage with English-speaking people so that old cultural habits were being discarded. It was this group that codified the Afrikaans language and facilitated its dissemination through the sponsoring of books and stories glorifying Afrikaner history and embellishing on the crucial events in their historical formation – events like the Great Trek, the Boer War, and the Battle of Blood River at which the trek Boers wreaked horrible vengeance on the Zulus.[22] They also formed a political party and a brains trust – the Broderbond – to plan for the policies they would institute when they came to power.

Their mass appeal, however, was primarily economic. They played on the sense of Afrikaner inferiority with respect to the English-speaking in this regard. Among other things they pushed forward plans for the economic uplift of Afrikaners as a whole. They also called for, and to a considerable degree achieved, a diversion of Afrikaner money into Afrikaner businesses: Afrikaner savings into Afrikaner banks, Afrikaner insurance premiums into Afrikaner insurance companies, for example. This proved particularly effective during the 1930s when poorer whites – dominantly Afrikaners – bore a disproportionate part of (white) misery.

This is not to say that the political goals of the Afrikaner nationalists have been single minded. Rather they have varied according to what would allow them to achieve enhanced positions in the South African stratification system and to protect their unique cultural life. Initially their goal was domination of the white South African state and displacement of the English-speaking as the ruling class. This they accomplished with the election of a National Party-dominated government in 1948. The state was used to facilitate their upward mobility – much, again, as it was in Quebec: the civil service became primarily Afrikaner; state contracts went to Afrikaner firms; and other firms were set up by the state and given over for their management to yet other Afrikaners.

However, there are also important differences from the Quebec case. The relation of the English-speaking whites and the Afrikaners was different both numerically and geographically. The English-speaking comprise about 40 percent of the white population compared with the 15 percent they comprised in Quebec when the secessionist movement there got seriously under way in the 1970s. The geography is also different. In Quebec the French presence has been dominant almost everywhere, including in the most anglophone part of

---

22   This is an event that until recently was celebrated as a national holiday in South Africa, testifying to the Afrikaner dominance of the South African state.

the province, the city of Montreal. In South Africa, however, the Afrikaners have lived interspersed with English-speaking whites. There was no Afrikaner heartland which they dominated numerically. Any Afrikaner state would have to, as indeed it did, include large numbers of English-speaking whites.

Besides which, the Afrikaners *needed* the anglophones. This was because whites were in a minority and, therefore, Afrikaner-speaking whites were in a very small minority indeed: and this in a context of black demands for the franchise and therefore the loss of white privilege, economic as well as political. For while the Afrikaners may have closed the economic gap with the anglophones after 1948 as a result of the various policies which the National Party introduced, black majority rule would clearly place these gains in jeopardy. Always in Afrikaner calculations, therefore, has been the development of a "white" South African nationalism which would appeal to Afrikaner and anglophone white alike as a bulwark against black claims. With the stiffening of black resistance after the Soweto riots in 1976 this tended to displace the more raw, Afrikaner nationalism of earlier years. Accordingly the National Party became a much more catholic "church" and began to attract a majority of the votes of anglophones.[23]

Nevertheless, in the end black majority rule came to South Africa. From 1990 onwards the National Party entered into negotiations with black representative groups, particularly the African National Congress (ANC), around a new constitution for South Africa; a new constitution that would enfranchise the country's black majority. For some Afrikaners this was clearly a threat, though the mix of cultural and economic concerns has varied a great deal from one to another.[24] As a result there were, and still are, calls by some for the creation of a distinct Afrikaner state.

One proposal has called for the establishment of an Afrikaner state in the western part of South Africa. This seems to be primarily aimed at maintaining the cultural integrity of the Afrikaners. Few people live there at present, and that is one of its virtues since it would be politically more acceptable to black Africans. It is also economically one of the less developed areas of South Africa. The call for a Boerestaat, on the other hand, seems more directed at the preservation of economic privilege. It would include, for example, Johannesburg, Pretoria, and most of the country's mineral wealth. Blacks could continue to live there as migrant workers, and would, presumably, be a majority of the labor force, but they would not have political rights (see figure 6.11).

Finally, and in summary, note once again the considerable similarities that exist between Afrikaner nationalism and that of the Quebecois. In both

23   At the expense, however, of losing many of the most ardent Afrikaner nationalists to a new political party, the Conservative Party, after 1983, though this breakaway was in part because it was believed the National Party was making too many concessions to blacks.

24   Black majority rule, for example, would be a threat to lower-level civil servants who could easily be replaced by blacks. So while many Afrikaners would be able to hold their own economically in the "new South Africa" others saw the writing on the wall. There were also cultural concerns. Under National Party rule *all* school children had had to learn Afrikaans, for example, and this seemed unlikely under ANC rule.

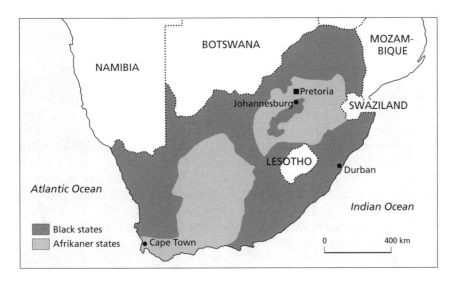

Figure 6.11   Proposals for Afrikaner homelands.
Source: After *The Economist*, August 22, 1993, p. 203.

instances the threat to a cultural identity from anglicization has been important. In both instances anglicization has been linked to an English-speaking
political hegemony, so it is against the English-speaking that Afrikaner and
Quebecois identities have been formed. Resentments have been intensified as
a result of subordinate positions in respective social stratification systems: like
the Afrikaners the Quebecois have been penalized as a result of their language.
And in both instances what brought issues of cultural, economic, and hence
political subordination to the fore has been urbanization since it was as a result
of urbanization that the dominant and subordinate were thrown into everyday contact and the subordinate became aware of barriers to their upward
mobility.

Summary

Nations are constructed; more accurately they are constructed by people
working with and through each other. They are in other words *socially* constructed. People aren't born English, French, or German; they are *made*. In a
sense making nations and senses of national-ness can be viewed as a labor
process. Some are involved as agents more than others though no one is totally
uninvolved. And like other labor processes those doing the constructing work
with "raw materials." These comprise senses of difference from others elsewhere: senses of different ways of life, different histories, different geographies, different homelands. Not all senses of difference, however: rather those
that flatter relative to others, that distinguish "us" in positive ways from particular "thems" with whom "we" have an uneasy, even hostile, relationship

and differences that abstract from those that actually divide "us" among ourselves – differences of class, gender, generation, for example.

The idea of the nation and the national is in historical terms very recent. It has coincided with two other transformations of momentous significance. These have been the rise of the centralized state and the emergence of capitalism to a position of dominance in economic life. Not surprisingly, both state and capitalism have figured prominently in the emergence of particular nations and senses of national difference. In some cases it is in their capacity as state agents or as market actors that people have taken the lead in constructing senses of nationhood and separateness. In the case of Taiwan reviewed at the beginning of this chapter the state was clearly implicated in shifting the senses of national difference prevailing on that island. The case of the Northern League and Padania that we discussed in chapter 1, however, is quite different. This is a bottom-up movement directed *against* rather than *from* the state and motivated by a desire for a more advantageous positioning in wider markets; and so much so that, in contrast to Afrikaner nationalism which also had a strong element of furthering the economic cause of a particular fragment of the population, the sense of specifically "national" difference – a difference in history and social life from the rest of Italy – is very weak indeed.

Nations abstract from other senses of difference: differences of race, gender, and, in particular, of class. To the extent that these differences threaten the sense of national unity then the idea of the nation and its distinctiveness has to be reconstructed. Women, blacks, and labor union leaders have to be recognized on stamps and bank notes, through public holidays, and with national honors,[25] for example. There are, in other words, other forms of identity around which distinctive forms of politics develop. It is to some of these that we turn our attention in the next chapter.

REFERENCES

Anderson, B. (1983) *Imagined Communities*. London: Verso Press.

Hobsbawm, E. J. (1983) "Mass-producing Traditions: Europe 1870–1914." In E. J. Hobsbawm and T. Ranger (eds), *The Invention of Tradition*. Cambridge: Cambridge University Press.

Hobsbawm, E. J. (1990). *Nations and Nationalism Since 1780*. Cambridge: Cambridge University Press.

Leitner, H. and Kang, P. (1999) "Contested Urban Landscapes of Nationalism: The Case of Taipei." *Ecumene*, 6(2), 214–33.

25   In Britain the national Honours List, which is published annually and which prescribes those to be honored by the Queen through various forms of recognition – knighthoods, MBEs, CBEs, etc. – would form an interesting object of study. To what extent, for example, has the gender composition of the list changed over recent years? How has the rise of Scottish and Welsh nationalism affected the representation of Scots and Welsh people respectively? And what has been the history of labor union recognition?

Mann, M. (1995) "Sources of Variation in Working-class Movements in Twentieth-century Europe." *New Left Review*, 212, 14–54.

Theroux, P. (1983) *The Kingdom by the Sea*. Harmondsworth: Penguin Books.

FURTHER READING

An excellent survey that came to my attention after I had written this chapter is by the sociologist David McCrone (1998): *The Sociology of Nationalism* (London: Routledge). But this should be read alongside Hobsbawm's *Nations and Nationalism Since 1780* (see references) and Ernest Gellner's (1983) *Nations and Nationalism* (Oxford: Blackwell). A good general statement from a geographer is James Anderson's (1988) "Nationalist Ideology and Territory," chapter 2 in R. J. Johnston, D. Knight and E. Kofman (eds), *Nationalism, Self-determination and Political Geography* (London: Croom Helm).

There are some good edited collections by political geographers, including David Hooson (ed.) (1994) *Geography and National Identity* (Oxford: Blackwell) and Guntram Herb and David Kaplan (eds) (1999) *Nested Identities: Nationalism, Territory and Scale* (Lanham, MD: Roman and Littlefield).

Excellent case studies of nationalist movements by political geographers include: J. Penrose (1990) "Frisian Nationalism: A Response to Cultural and Political Hegemony." *Environment and Planning D: Society and Space*, 8(4), 427–48; J. Anderson (1990) "Separation and Devolution: The Basques in Spain," pp. 135–56 in M. Chisholm and D. Smith (eds), *Shared Space, Divided Space: Essays on Conflict and Territorial Organization* (London: Unwin Hyman); and J. A. Agnew (1984) "Place and Political Behavior: The Geography of Scottish Nationalism." *Political Geography Quarterly*, 3(3), 191–206. Also useful in terms of its relevance to the process of nation-building is one on Singapore by Lily Kong and Brenda Yeoh (1997) "The Construction of National Identity through the Production of Ritual and Spectacle." *Political Geography*, 16(3), 213–39.

On more specialized issues: Dan O'Meara's (1983) *Volkskapitalisme: Class, Capital and Ideology in the Development of Afrikaner Nationalism, 1934–1948* (Cambridge: Cambridge University Press) is *the* book on Afrikaner nationalism.

Chapter 7

A World of Difference

Introduction

As objects of identification nations can provide a false sense of unity; it is as if the world were divided geographically into a patchwork of mutually exclusive identities. But within nations there are typically quite severe stratifications which often take as their justification, though not exclusively, concepts of biologically based difference. The most commonly encountered of these relate to race and gender. Indeed, concepts of the nation are deeply gendered and racialized. According to the title of a nineties book on the subject of British identity, "There Ain't No Black in the Union Jack." And we remarked in chapter 6 on the gendering of the nation. These stratifications can become the object of struggles, struggles in which space is an important factor. In a very real sense claims for recognition are also claims to space.

This chapter explores the politics of these struggles and how geography has been implicated both in the creation of polarized and polarizing identities and in subsequent struggles for recognition. It proceeds through a series of case studies of issues important to contemporary debates about identity and identity politics.

Colonists and Colonized

> Take up the White Man's burden–
>   Send forth the best ye breed–
> Go bind your sons to exile
>   To serve your captives' need
> To wait in heavy harness
>   On fluttered folk and wild–
> Your new-caught, sullen peoples,
>   Half devil and half child.

Take up the White Man's burden–
   In patience to abide,
To veil the threat of terror
   And check the show of pride;
By open speech and simple,
   An hundred times made plain,
To seek another's profit,
   And work another's gain.

(First two verses of Rudyard Kipling's
"The White Man's Burden," 1899)

An important product of the various European imperialisms of the past four centuries was the settler society: societies in which the primary social cleavage was that between the settlers, usually from the imperial country, and the indigenous population. Included here are the settler colonies of the Spanish, British, French, and Portuguese empires in particular. But in addition there are some less obvious instances that need to be taken into account: Northern Ireland, Israel, Taiwan, and odd cases like Liberia and Sierra Leone.

A major feature of these societies, one which was used to legitimate the economic and political dominance of the settlers, was the development of a set of meanings which, while originating in other imperial contacts like those of the explorers, colonial administrators, and missionaries, was taken up enthusiastically by those who came to settle. These meanings counterpose, above all, a culturally superior Westerner or European to a culturally inferior native.

Depending on the context the following distinctions were central to the way the Other, the native, was constructed in the colonies:

settler/native	European/non-European
white/dark skinned	progressive/traditional
nation/tribal	science/magic
rational/irrational	controlled/emotional
civilised/barbaric	moral/immoral
orderly/violent	knowledge/ignorance
mature/immature	sophisticated/primitive
clean/dirty	Christian/non-Christian
innocent/cunning	

These are dualisms or conceptual oppositions that tended to "leak" into one another like those implicit in Kipling's poem: "half devil and half child." Cleanliness would be associated with civilization, and science, maturity, and modernity with whiteness. Perhaps without exception settler colonies were appallingly racist. But not all the dualisms applied in any particular colonial context. The idea of "tribal" had an almost exclusively African application, for instance.

As an example of the sorts of distinctions that were drawn consider this statement by the British political thinker James Bryce from his *Impressions of South Africa*, published in 1897 after a tour of that country:

Here in South Africa the native races seem to have made no progress for centuries, if, indeed, they have not actually gone backward; and the feebleness of
savage man intensifies one's sense of the overmastering strength of nature. . . .
When the Portuguese and Dutch first knew the Kafirs, they did not appear to be
making any progress toward a high culture. Human life was held very cheap;
women were in a degraded state, and sexual morality was at a low ebb. Courage,
loyalty to chief and tribe, and hospitality were the three prominent virtues. War
was the only pursuit in which chieftains sought distinction, and war was mere
slaughter and devastation, unaccompanied by any views of policy or plans of
administration. The people were – and indeed still are – passionately attached
to their old customs . . . and it was probably as much the unwillingness to have
their customs disturbed as the apprehension for their land that made many of
the tribes oppose to the advance of the Europeans so obstinate a resistance. . . .
Their minds are mostly too childish to recollect and draw the necessary inferences from previous defeats, and they never realized that the whites possessed
beyond the sea an inexhaustible reservoir of men and weapons. (Quoted in
Thompson, 1995, pp. 93–4)

It is easy to see how colonialism formed essential preconditions for these sorts
of constructions. The imposition of colonial rule was important in a number
of related respects.

First, and most obviously, it was the context for an interaction between culturally different populations so that for the settlers the differences between
themselves and those they were coming into contact with must have appeared
to overwhelm their own internal fields of contrast between manual worker,
administrator, farmer, Protestant, Roman Catholic, lawyer, etc. This is not to
say that the settlers and the colonial administrators had no presuppositions at
all. Rather the constructions they made of the peoples they encountered were
always in terms of conceptual frameworks and distinctions they brought with
them from Europe. We are talking, after all, of a period in which Western
Europe was in the throes of drastic social change. The Industrial Revolution
brought with it science and associated notions of rationality, the rejection of
magic, and the idea of progress. At the same time there was the articulation
of methods of rational administration on the part of the state, the development of bureaucracy, and of statecraft. From that standpoint – and only from
that standpoint – what was encountered in Africa and Australasia in particular must have seemed primitive, backward and irrational.

The definition of difference was associated with attempts to explain, to
understand and make meaningful. These too were in terms of the conceptual
baggage the colonial authorities and the settlers brought with them. An interesting case in point is the equation that was made between civilization and
the European institution of wage labor. It was implied in numerous com-
mentaries that the reason for the supposedly uncivilized nature of the African
was that he or she had never been inducted into the disciplines and rewards
of the wage worker. The civilizing process, therefore, depended on precisely
that. Thus, after visiting South Africa in 1878 the novelist Anthony Trollope
wrote: "Looking as we are bound to look to the good that we can do to these
people, rather than to the extension of our own dominion, we ought to rejoice

greatly at their readiness in adapting themselves to the great European institution of daily work and weekly wages" (Cope, 1990, p. 487). Likewise, the High Commissioner for Natal in 1879:

> The Zulus are, I believe, by nature a light-hearted, thoughtless, very intelligent and very teachable race . . . as easily led to habits of civilization as they can be trained by Chaka or Cetywayo to become a man-slaying human military machine of enormous power. [What little had been done by education and training] . . . has shown that the Zulus are by no means indolent, unimprovable savages, but that they have in a degree far superior to most barbarous races, the power of becoming at once a useful class of native laborers. (Ibid.)

In terms of their experience of "progress" in Western Europe the link between wage labor and "civilization" must have seemed entirely reasonable. After all, the Industrial Revolution and its technical accomplishments had been achieved through the conversion of the majority of the population into precisely that status.

One of the most interesting of the interpretations made, of course, was the racist one: the idea that the native or aborigine was not only backward, irrational, promiscuous, childlike, but that this was a necessary expression of his or her biological makeup. Whites regarded themselves as superior and this was by virtue of their race, as in "The White Man's Burden." Whatever the reason, whatever the relationship to Darwinian theory, there is no doubt that racist thought was rife in Western Europe during the nineteenth century and that, indeed, it only showed serious signs of dissipation in the wake of the Second World War. Heredity and genes were important, and physical appearance was an expression of that genetic makeup. Humanity, or mankind as it was referred to then, could be divided into races akin to different species of animals and it was believed that, in accordance with the idea of evolutionary progress, some of those species were better fitted for survival than others.

But the absolutely crucial and overwhelming point is that these characterizations, these differentiations, suited the (highly practical) purposes of the mine and plantation owners, the settlers and the colonial governments. As Luli Callinicos has written in discussions of the mind sets of the South African mineowners at the turn of the century:

> Many mine-owners and managers liked to think of blacks as backward and lazy, or otherwise as children.
>
> "The position of Kaffirs is in many respects like children," wrote the editor of the mine-owners journal, the *South African Mining Journal* in 1892. Both children and blacks needed "special control and supervision when exposed to temptations." The black worker could not be allowed to "roam unrestricted, not improbably (drunk) at his own sweet will." Blacks needed to be put into compounds[1] for their own sakes, concluded the editor.

---

1 Dormitory-type lodgings, typically attached to the place of work, and constantly under the surveillance of management.

A mine-owner warned: "We should not over-pamper the native and thus weaken his naturally strong constitution."

In these ways mine-owners used racism to justify the treatment of their workers. (Callinicos, 1981)

In the imposition of these identities on the people so identified the colonial authorities, the missionaries, the mining companies, and the settlers had immense advantages. For a start, to the indigenous population the European must have appeared as the vehicle of almost magical powers: powers of medicine, of communication through reading and writing, the telegraph, modern weaponry, and a general ability to mobilize natural forces for human purposes. Their statements and public pronouncements about "civilization" must have appeared convincing merely by their association with the bearers of such seemingly wonderful abilities and powers.

Think and Learn

We have been talking about how the settlers and the colonial administrators construed indigenous peoples, how they defined them and why. But these images were also reflected in thought and education back in the mother countries. Why do you think this type of construction of the native was not simply condoned by the authorities but taken for granted as an essential part of the education of children? What do you think its implications would have been for national identity in countries like Britain, France, the Netherlands, and Belgium? And how has it made it hard to relate to indigenous peoples who now migrate back from the former empire to the (former) mother country?

At the same time the Europeans enjoyed, by virtue of their status as rulers, powers to organize and to name in ways which complemented the categories that had been formulated in the African and Australasian context. Not the least, special places were devised for native and European. Residential segregation, the creation of native locations,[2] for example, along with many other forms of segregation (see figure 7.1) served to validate in a material form the view that settler and native were different. Likewise in much of Southern Africa influx control[3] helped create the idea of the urban as something exclusively European, while the place for the native was the reserve.

Proper names were also important. In the British Empire places were given English-language names, often celebrating their monarchs and the architects of colonial rule and so emphasizing for the native the alien presence and its

---

2   Residential areas specifically set aside for Africans in the colonies of Southern Africa.
3   Limits to the permanent residence of blacks in cities.

Figure 7.1 Scenes from a colonial landscape. Racial segregation in public facilities was an endemic feature of colonial society. These are examples from South Africa under apartheid. South Africa has been independent since 1910 but until the overthrow of apartheid in 1990 it functioned much as it had in colonial days. The top two photos are self-explanatory, though the fact of racial segregation in hospital facilities might surprise. Note also the use of the code words "European" and "Non-European." The reference to "retain your own facilities," bottom left, is about a controversial move to abolish beach segregation.

Source: Photographs by author.

"here-to-stay," "this is now our place, so like it or lump it" character. There are for example, endless "Victorias" around the former empire: a town in British (that's right!) Columbia, a waterfall in what used to be Rhodesia,[4] a state in Australia. "Londons" are found in both Canada and South Africa (East London). Newcastle is another popular place-name with instances in Australia, Canada, and South Africa. Colonial governors and administrators likewise left an imprint as in place-names like Carnarvon, Beaufort West, and Port Shepstone in South Africa.[5]

In numerous ways the colonial authorities organized native populations so as to leave no possible doubt of their difference and of their subordination. The state was organized to take this into account. Local governments had Departments of *Non-European* Affairs, central government had its *Native Affairs* Department, while the *Native* Commissioner was a ubiquitous feature of the reserves. Government censuses collected statistics on a racial basis. Not the least the African was excluded from the vote. The situation in Australia was hardly different. Aborigines did not gain the vote till 1949 and there is still a ministry of the federal government called the Ministry for Aboriginal Affairs.

Think and Learn

National censuses often collect data by race group and these data are presented for each race in turn. In South Africa the state has historically defined four different race groups: blacks, Coloreds, Indians, and whites. In what order would you expect the data for these different race groups to be presented? Which would be first and which last? And what of the United States? What does this tell us about the taken-for-granted nature of racial hierarchies and their resistance to change?

The African, the Indian, the South American Indian, the Oriental, and the Arab were defined as backward, politically inept; the European, the white person, as advanced, politically astute and altogether superior. The all-encompassing nature of this discourse, the ability of the Europeans to define the essential differences in the world, had effects on the way native populations saw themselves. The discourse, in short, had strong political effects of an incapacitating nature. The colonized came to believe in their own ineptitude and their dependence on Europeans and this made it hard for them to organize themselves politically.

This is not to argue that identities in settler societies have remained etched in stone. The ways in which settlers have differentiated themselves from the

---

4 Named after the arch-imperialist Cecil Rhodes.
5 Alongside, it should be pointed out, places named after the leaders of Afrikaner settlers: Pietermaritzburg, Pretoria, Louis Trichardt, Piet Retief. These are poetic to Western ears but hardly likely to resonate with the indigenous populations.

native populations have tended to shift over time. Race has declined as a form of differentiation. This is partly because of the exposure of racist argument as scientifically indefensible and also due to external political pressures which based their position on a rejection of the category of race. But other differentiations have emerged to take their place and to perform similar functions. One such is the identification of the native as "traditional" – a more acceptable designation than one in terms of "race." This in turn came to justify the view that the indigenous population could not be incorporated into the "Western," i.e. white, political system.

The sort of evidence drawn on to justify this "traditional" attribution included: the persistence of (e.g.) lobola (bridewealth), "traditional" healers, initiation rituals, and the authority of chiefs. These were implicitly contrasted with ideas of "reason," "rationality," and "science," "modernity," "cultural dynamism," and "innovativeness." In contrast to "Western individualism" "traditional" native culture was defined as communal, conservative, and backward. This also meant, of course, that their poverty had an explanation. This in turn was other than the differential degrees of privilege legislated by colonial governments and by subsequent settler states and so whites were not to blame for the poverty they saw all around them: no need for guilt feelings there!

Nevertheless, it would be wrong to assume that the identities imposed by the settlers eliminated all possibility of resistance and for all time. For a start some formal education was introduced, if only to staff the lower echelons of the colonial civil service, or to produce new generations of missionaries. This provided access to a world of ideas, particularly those of socialism, which was sceptical and critical of colonialist ideology. This access was further fortified by connections outside the colony. Numerous nationalist leaders – Ho Chi-Minh, Nehru, Kwame Nkrumah – obtained their university education in the imperial capitals of Europe.

There was also the fact that in many ways the settlers and the colonial authorities treated the native populations as all equally backward, in thrall to superstition, racially inferior.[6] This was regardless of many of the differences of language, clan, tribe, and perhaps religion that divided them. There were counter tendencies. The practice of indirect rule often made use of the existing structure of tribal chieftainships and this tended to confirm the fragmentation of native society and encourage the congealing of those differences into ethnic ones. But the unifying consequences of colonial rule should not be forgotten. These were all the more powerful for being unintended.

In addition there were counter-hegemonic movements of a non-national character. The most important of these was black consciousness or what was known in the French colonies as *négritude*. This was a conscious and studied revalorization of black African accomplishment. It was intended to throw off the cultural shackles of white domination and the psychological servitude to which it condemned the black person by demystifying white claims and

6   Much as, and as we saw in chapter 5, the British treated the Jamaicans, the Trinidadians, the Bahamians, etc. as all the same: members of an inferior and unwanted race.

bolstering the sense of pride of native populations in their own achievements and histories: making them feel significant and not the inferior characters of the colonial imagination. It was important in numerous nationalist movements of Africa, particularly that of South Africa, and also had effects among blacks outside Africa, as in the United States.

We should also note, however, that these contacts have produced much more complex geographies of identity than might be presumed from the simple idea of Western–native contact. The view which the colonized, the native populations, had of themselves indubitably changed as a result of the colonial experience. But so too did that of the Europeans and Americans. From now on both would see themselves, situate how they saw themselves, within the context of a new, global field of contrast. On the side of the colonizing societies this laid one of the foundations for the virulent racism that emerged in Europe during the late nineteenth century. More generally it engendered a view among Western peoples of themselves as guardians of civilization and the bearers of that civilization to the rest of the world.[7]

Think and Learn

A common way in which people in North America and Western Europe organize their thinking about the world is in terms of a division into First and Third. How is this consistent with what we saw in chapter 5 regarding the way in which identities are formed in part to justify inequalities? And what do you think the effect is on peoples in the so-called "Third World" to be defined in that way? Does it help them or harm them? And why? Is the notion of "developed" as opposed to "undeveloped" countries also problematic? How about "developing" countries?

The broad patterns of contact which occurred, and regardless of particular imperial countries or colonies, have also fostered new, *international* identities. The most obvious of these is the idea of "Europe" and of being "European." Like other identities this one is not to be taken for granted. Like "English," "American," "Western," it is socially constructed and the colonial/imperial

7   Even so, those on the frontlines of these cultural contacts developed somewhat different ideas of themselves than those back in Europe. For the latter the experience was filtered through the pages of the newspapers and other media and had to compete with much more that was closer to home. This was not so in the colonies. Their life was dominated by the daily experience of "having to deal with the native." In consequence, and particularly in the context of struggles to overthrow colonial rule, it was a common belief of the settlers that the imperial country, the metropolis, didn't and couldn't understand their situation and the conditions they faced. This was the justification given for settler attempts to forestall native-majority rule by overthrowing the imperial relationship and declaring independence on their own terms. This was the goal of the revolt of the generals in Algeria and it was actually realized in Southern Rhodesia. Something very similar happened in Israel when the British withdrew from what was then Palestine.

experience is deeply implicated in that construction. It is true that there is much that Europeans have in common. These include a legal system strongly influenced by the legacy of the Roman Empire. Christianity is another. But contacts with other non-European peoples have also been important; besides which it was those contacts that often underlined just what it was that Europeans had in common. There is, for example, the early instance of the Crusades where the division was defined as primarily religious and Europeans came to see themselves as united in their Christianity. More recently the contrast between European "civilization" and the backwardness of the rest of the world has further served to cement a sense of commonality, though the idea of Europe has also given some ground to that of the West, which, curiously, includes Japan and presumably, now, Russia (on the idea of Europe see Delanty, 1995).

The more contemporary version of this cultural faultline is the division between First and Third Worlds. The very terms "First" and "Third," of course, positively shriek "hierarchy." And indeed the Third World is defined as the First World's inferior: as backward, traditional, given over to superstition, and in its public life corrupt – a set of little better than banana republics. These are places whose facade of modernity is seen as just that: the air conditioning has been installed in the international hotels along with the bathroom fittings but they only work fitfully, if at all. Such are the images projected through television into the living rooms of the First World and which congratulate us on being different: on being modern, scientific, efficient, incorruptible, and so on.

Politically these images matter. Members of "Third World" elites have often studied at universities in the "First World" and/or attended conferences there, and have certainly been treated patronizingly. And more developed countries are indeed impressive in the degree to which they have been able to turn the forces of nature to their own practical purposes. So in the "Third World," "First World" ways of doing things carry the day. There is no need to ask the natives how aid projects should be implemented. The more developed countries have a monopoly of the relevant knowledge because it is "scientific." If the locals aren't convinced, on the other hand, then their governments will help implement the schemes over their heads since they too are impressed with that knowledge and those technologies. But all too often the result is failure. Development does not ensue or it occurs in a highly ambiguous manner so that whether people are better or worse off is debatable. In different social and ecological contexts than those prevailing in the more developed countries, in situations where local knowledge and experience can deliver superior results over methods whose efficacy has been demonstrated in very different conditions, the results can be, and often have been, disastrous.[8] But still the

---

8 Good examples are the various agricultural projects planned by the post-war British Labour government for their colony in what was then called Tanganyika (now the larger part of Tanzania). For an excellent review of the disasters that can attend the imposition of expert thinking in disregard of local, indigenous knowledge see James Scott's discussion (1998, pp. 225–9).

onslaught of inappropriate approaches and technologies continues. For more is at stake than Western pride. Aid projects use capital goods – tractors, dams, irrigation works – and open up markets for more Western-produced products: fertilizers, insecticides, seeds. In other words, while decolonization of a material kind has indeed taken place, what we might call the colonization of the mind continues to be a force moulding change in the less developed countries.

Not that the definition of new identities is all one way. Analogous identities, inventions of cultural unities, have also emerged in the less developed world. One of the more interesting of these is that of "Africa," and particularly that of sub-Saharan Africa. There the influence of Islam was much weaker than in the Sahara and to its north. In addition white racism and colonialism, whether that of the French, the Belgians, the British, or the Portuguese, was a uniting influence: an experience shared by black Africans. To some degree this is being turned to political account by corporations. For to the extent that they can present themselves as "African," they can avoid some of the political problems multinational corporations have faced on that continent, even though they are themselves multinational. One of the most interesting cases is Ashanti Goldfields, based in Ghana, which has a black management which is now active in ten other African countries:

> Ashanti Goldfields usually divests itself of the non-African properties that come with its acquisitions. It sees and presents itself as an African company run by Africans, of all colors, for the benefit of Africa. There is calculation as well as idealism in this. Ashanti Goldfields is well aware that many African governments are leery of foreign multinationals, and especially of multinational mining companies, which they suspect of exploiting African labor and African natural resources for the benefit of already affluent overseas shareholders. (*The Economist*, Sub-Saharan Africa Survey, September 7, 1996, p. 13)

Moreover, in a decolonizing world those white settlers who remained, defending their privileges like erstwhile Canutes demanding that the tide retreat, found themselves on the defensive. Where their social definitions had ruled they now found that they were the ones being defined in derogatory ways, and not just by those they had oppressed. Particular attention focused on whether or not their relation with the indigenous population was a "colonial" one or not. For defining a situation as a colonial one has political intent. It is aimed at galvanizing support for altering it, at throwing the settlers out, bringing moral pressure to bear on the colonial powers to give their colonies independence. Those whose material interests will be affected by decolonization therefore have an incentive to fight back, to resist the identification of them as "colonial," as outsiders and as exploitative.

The postwar travails of South Africa are a case in point. South Africa has been an independent country since 1910 so in formal terms it was not a colony. But by many, if not most, the whites were regarded as settlers who had oppressed and exploited an indigenous population. And to be sure anticolonialism was always part of the rhetoric of the movement for a non-racial franchise. Under apartheid, from 1948 on, the pressures on the white South African government intensified. There was a significant black movement for the vote led by the African National Congress; and in addition the ANC was able to mobilize many other independent states to argue on its behalf in world fora like the United Nations and generally to bring pressure to bear on the South African authorities.

They resisted these pressures in diverse ways. Most of these were material: intensifying surveillance of blacks, removing them from urban areas, beefing up the security services, providing some material concessions to a favored few. But there was also a discursive side to it in which the government tried to convince, primarily the outside world, that the whites could not *possibly* be defined as colonial in their relation with blacks. There were two aspects to this.

First, it was claimed the whites were not colonizers. This was because while they did indeed come from the outside they did not get there after blacks. Rather they both got there at the same time:

> More than 300 years ago, two population groups, equally foreign to South Africa, converged in rather small numbers on what was practically an empty country. Neither group colonized the other's country or robbed him by invasion and oppression. Each settled and gradually extended his settlements, and in the main each sought a different area in which to dwell. . . . The first point, therefore, is that there was no colonialism, only separate settlement by each group, nearly simultaneously. (Information Service of South Africa, *Progress through Separate Development*, 1973, p. 11)

This was backed up by carefully drawn cartographic materials (see figure 7.2) with the inevitable conclusion that the colonizer–colonized distinction was inappropriate. This interpretation of events has always been contested by South African blacks and the historical evidence is probably on their side. But clearly both sides had stakes in promoting their particular versions of "the truth."

The second claim was that the relation between whites and blacks had not been an exploitative one. Blacks had sought out employment with whites because of their own inability to achieve higher material standards on their own. This was despite the fact that "only in South Africa did the white man deliberately reserve land for the Bantu (i.e. black) and endeavoured (mostly in vain) to train him to make the best use of it" (ibid.). The conclusion deriving from this logic: "The white man, therefore, has not only an undoubted stake in, and right to, the land which he developed into a modern industrial state from denuded plains and empty valleys and isolated mountains, but according to all principles of morality it was his, is his and must remain his" (ibid.).

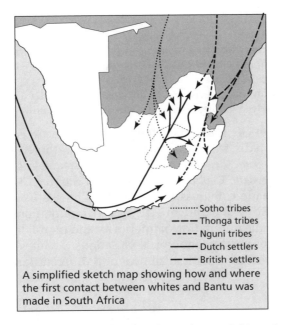

Figure 7.2 Contesting the charge of colonialism: the spatial imaginary of white settlerdom in South Africa.

Source: Progress through Separate Development. New York: Information Office of South Africa, 1973, p. 10.

Racial Encounters

Imperialism not only brought European settlers into contact with indigenous populations. It also generated vast diasporas of those same indigenous peoples of the non-European world: movements over long distances to colonies elsewhere. These transfers were to fulfill the labor needs of the settlers and the colonial authorities. The most obvious case was the movement of *slaves* from Africa to North and South America. Less salient in the public mind was the importation of slaves into South Africa from East and West Africa and from the Dutch East Indies.

Other movements were of *indentured workers*. Indentureship amounted to an agreement on the part of the worker to pay off his or her passage overseas to some colony by working for a given employer for a period that could vary from three to five years. It was attractive to the landless and poor but as they were tied to a given employer and could not seek better conditions elsewhere their lives varied from depressing to awful. A major movement was that out of India. Indians went as indentured workers to Uganda, Kenya, Malaysia, and South Africa as well as to the Caribbean, especially Trinidad and what was known as British Guiana (now Guyana). There was also a substantial movement of Chinese – at least five million – into Pacific coast areas of North America.

Racism

> [T]he racial differences between American whites and Asiatics would never be overcome. The superior whites had to exclude the inferior Asiatics by law, or if necessary, by force of arms.
> The Yellow Man found it natural to lie, cheat and murder and ninetynine out of every one hundred Chinese are gamblers. (Quoted in Hill, 1975)

Thus wrote Samuel Gompers, president of the American Federation of Labor almost uninterruptedly from 1886 to 1920 in a co-authored tract published in 1902 and entitled *Some Reasons for Chinese Exclusion: Meat vs. Rice, American Manhood Against Asiatic Coolieism – Which Shall Survive?* The purpose of the tract was to persuade Congress to renew the Chinese Exclusion Law, introduced in 1882 and due to expire in 1903. The two quotations nicely exemplify the idea of racism: the imputation of inferiority and moral decadence to others, seemingly on the basis of phenotypical variations ("whites" and "The Yellow Man") but often confused with attributes of a different kind (as in "Asiatic" and "Chinese").

Think and Learn

Gompers was speaking as the leader of organized labor. Why do you think the manual workers which provided the majority of labor unionists in those days might have been susceptible to racist argument?

In the second half of the nineteenth century this sort of sentiment was rampant in the Western States of the United States. Chinese had come to California subsequent to the gold rush of 1849 and later moved to neighboring States like Oregon and Utah, partly to work in railroad construction. While anti-oriental feeling was fairly general, the agitation to exclude the Chinese altogether was spearheaded by organized labor. The Workingman's Party of California, formed in 1877, combined, for example, a platform of compassion towards poor white workers with extreme racism towards non-whites. It believed that the rich and the Chinese were engaged in a tacit conspiracy to oppress white workers. This agitation culminated in the Chinese Exclusion Act of 1882 prohibiting further Chinese immigration into the United States but leaving about 100,000 in the country. The latter became a target for white working class hostility in the depression of 1882–6. In numerous places anti-Chinese organizations formed calling for the physical removal of Chinese and their belongings from the area.

Although racism is by no means exclusive to the lower echelons of the working class[9] it seems to reach among them depths of viciousness not

---

9 I qualify "working class" in this way because an objective definition of the working class as those separated from the means of production applies to the vast majority of the population.

apparent elsewhere. This was true also in South Africa where the white labor unions fought in the early 1920s against the attempts of mineowners to replace many of their members with black workers. They used the slogan "Workers of the World Unite, to Keep South Africa White," which is interesting because it recalls the close links between many of the world's labor movements and Marxist thinking. And Marxist thinking has always been totally antithetical to racism, seeing it not merely as an obstacle to effective worker organization but as a denial of a common humanity.

So how do we explain this equation between racism and organized labor? There is an obvious answer: that it is those concentrated in less skilled work, lower-paid jobs who are more at risk from the labor market competition of the new arrivals. This would explain their opposition to immigration or, in the case of South Africa, their support for policies aimed at limiting the numbers of blacks in the cities (so-called "influx control"). But if this is the reason it is not clear why they don't make a clean breast of it and argue simply in terms of their rights as citizens rather than as members of a supposedly superior racial caste. Part of what is going on here, one suspects, is the attempt to build coalitions with other segments of the population not affected materially by the immigrants. In the case of the Chinese, for example, much was made of stories according to which white children were being lured by the Chinese into opium dens.[10] As Gompers asserted in the tract referred to above, "What other crimes were committed in these dark fetid places when these little innocent victims of the Chinaman's wiles were under the influence of the drug, are almost too horrible to imagine. . . . There are hundreds, aye, thousands, of our American girls and boys who have acquired this deathly habit and are doomed, hopelessly doomed, beyond a shadow of redemption" (quoted in Hill, 1975, p. 170).

But there is more to working class racism than material insecurity. It also expresses a pervasive status insecurity. For if you are towards the lower end of the pecking order, the object of middle class contempt and the butt of their humor, then you obtain a sense of significance where you can find it. One of those places, as we will see, is in your masculinity. But historically it has also been as members of a supposedly superior race. You may work long hours at the coal face for money that still leaves you anxious for the next paycheck but at least you can take pride in being white and so express your contempt for those who aren't.

This is not to argue that racist ideas find their origin among manual workers. In the nineteenth century, for reasons discussed in the first section of this chapter, racism was in the air. Racist argument was common throughout Western Europe and North America and was used to subordinate peoples around the world to European rule, as well as to justify their treatment to those more liberal voices who expressed doubt about it. It was, in other words, a common assumption; and as such we should not be surprised if those most anxious about their status, as well as about their jobs, should take it up with vigor.[11]

---

10 The use of the word "den" is interesting because of its non-human associations.
11 Compare Marx in his discussion of the Irish in England: "All English industrial and commercial centers now possess a working class *split* into two *hostile* camps: English proletarians

The social construction of race

As I argued in chapter 5, difference, whether gender, race, or class, is socially constructed. In the case of race the discussion of the colonizer–colonized distinction earlier in this chapter has sensitized us to how this works out in the case of race. I now want to press this point home by contrasting the different ways in which similar spectra of phenotypical variation have been construed in different countries. From this it will be clear that the construction of race does not simply work on differences of physical appearance but incorporates all manner of other differences, including ones of perceived culture. The particular focus for developing these points is the contrast between the US on the one hand and South Africa on the other.

Both in the US and in South Africa there was considerable miscegenation between whites on the one hand and blacks on the other. In South Africa, however, a distinct mixed-race category emerged to be known as the Cape Coloreds or simply "Coloreds." In the US, on the other hand, all the offspring of mixed marriages or miscegenation have been defined as black or, to use the terms common in the earlier twentieth century, "Negro" or "Colored," though clearly "Colored" meant something very different from what it meant in South Africa. The US has toyed with the idea of a mixed-race category and indeed the US Census used to use the term "mulatto" to signify that category but it was dropped in 1920.

The role of material interests in these constructions is also clear. In the case of the Cape Coloreds the context for their emergence was job market competition. Throughout the nineteenth century to be called "Colored" in South Africa was to be black: there was no mixed-race category. It was only in the closing years of the century that the term "Colored" started to be confined to those of mixed-race while blacks became known as Africans or black Africans or simply the derogatory "kaffirs." Labor market competition in the Western Cape and particularly Cape Town, however, led the whites to petition the government to restrict the access of blacks to more skilled jobs: to impose a color bar, in other words. This led to protest on the part of mixed-race people on the grounds that while they were not the same as white people they were sufficiently different from black Africans to warrant less harsh treatment. In the context of this threat to their life chances they came to emphasize what they had in common with whites and how they differed from blacks.

By no means was this simply a matter of differences of physical appearance. They shared with whites, for example, higher levels of education,

and Irish proletarians. The ordinary English worker hates the Irish worker because he sees in him a competitor who lowers his standard of life. Compared with the Irish worker he feels himself a member of the *ruling nation* and for this very reason he makes himself into a tool of the aristocrats and capitalists *against Ireland* and thus strengthens their domination *over himself*. He cherishes religious, social and national prejudices against the Irish worker. His attitude is much the same as that of the 'poor whites' towards the 'niggers' in the former slave states of the American Union" (Letter to Meyer and Vogt (1870), *The First International and After*. Harmondsworth: Penguin, 1974, p. 169).

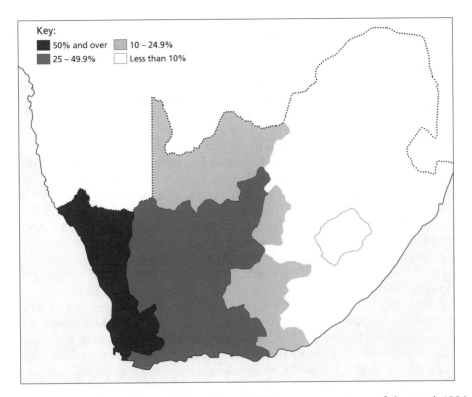

Figure 7.3   The Colored population of South Africa as a percentage of the total, 1904. This map shows the distribution of Coloreds relative to the remainder of the population as it was about the time that they were being differentiated out from blacks. Note the heavy concentration in the southwestern parts of South Africa, the area that is often known now as "the Western Cape."

Source: Tables 2.1 to 2.4, *Urban and Rural Population of South Africa 1904–1960.* Pretoria: Central Statistical Office, 1968.

Christianity, and the same language – Afrikaans – as that spoken by those whites of Dutch descent. They began to see themselves, in other words, as different not just from whites but also from blacks. This resulted in different treatment on the part of the South African government so that to claim a Colored identity came to have certain beneficial material effects: at least to the extent that Coloreds received better treatment from the government than did blacks (see figure 7.3).

But if in the US the tendency has been for blacks to be treated as, and to see themselves as, a homogeneous group, change is in the air, once more emphasizing the social nature of the distinctions made and therefore the social character of race. At least two forces for change can be identified here. In the first place there is a move on the part of blacks for a mixed category on government forms and in the census. The lifting of State bans on interracial marriage has resulted in a significant increase in the number of black–white marriages, from 1.7 percent of all marriages in 1960 to 6 percent in 1990. Yet given current racial classifications the offspring of these unions are automatically defined as black, creating a situation where the child will tend to identify with one parent rather than the other. As we will see below, however, there is also a strong segregationist tendency in black politics today which has lent opposition to this proposal, since the creation of a mixed race category could dilute black numbers. This is important at a time when there are Congressional district reapportionment cases in which race is playing a role and in which reapportionment is seen as an aspect of affirmative action.

A second condition for change has been the arrival in the US of considerable numbers of blacks from the Caribbean. For reasons of employability West Indians have tried to differentiate themselves from native-born blacks and resisted attempts at imposing a shared identity. That this differentiation is beginning to stick is apparent in the employment practices of firms. Apparently all blacks are not seen as alike: there are American blacks and immigrant blacks. New identities are being created.[12]

One of the reasons for this greater perceived employability, however, has to do with the illegal status of many of these immigrants. As illegals they are much less likely to complain about work conditions and employer practices, and much more likely to be "dependable" employees. This is because illegals cannot complain to the authorities unless they want to reveal their illegal status. In addition blacks have the alternative of welfare benefits which, again, illegals do not enjoy. So blacks may indeed be choosy compared with some black immigrants but this has nothing to do with physical appearance and everything to do with their social position as citizens. Once more the racial category has conditions that are social in character.

Think and Learn

The way in which people of color have differentiated themselves has clearly been different in South Africa than it has been in the United States. I have provided information about the South African case but no explanation for why differentiation was so weak in the United States. Why do you think that that would have been the case?

Assimilation versus segregation

As discussed in chapter 5, racism has not gone unanswered. It has been the focus of social struggle and an important element in what has come to be known as the politics of identity. In pressing their claims to inclusion, to

---

12 In Britain something similar has happened, though there it is Indians and Pakistanis that have resisted the appellation "black" while those from the Caribbean have tried to use it for purposes of coalition-building.

recognition, as well as to throw off their own shackles of a sense of inferiority, the oppressed have constructed new narratives about themselves. They have tried, in other words, to rewrite history. This is in recognition that *all* histories – as well as geographies, for that matter[13] – are representations and that all representations are partial; they are written from a particular point of view, *for* particular people and *against* others. Rewriting history involves foregrounding those who have been marginalized, bringing into the open what has been silenced, and with the purpose of raising not just the consciousness of the oppressor, appealing to contradictions between his or her professions of equality and treatment of racial minorities,[14] but also that of the oppressed. So forgotten heroes are remembered and celebrated, the stories of exploitation and violence that are whitewashed in existing accounts are retold in all their awful detail, and the contributions of ancestors in Africa, in the Andes or wherever, to world culture are rescued from the oblivion to which a Eurocentric view of the world has tended to consign them.

These stories also represent claims to space. Historically an important part of racial oppression has been limits to movement. The oppressed races have been segregated in their own neighborhoods and their own schools, with their participation in a common society restricted. In the colonies their movement to the cities has been limited. So in pressing for recognition, for respect, and an undoing of the harm caused them in the past, what they are claiming is an equal right to the space in which they coexist with whites. And not just materially but also symbolically. They want to be recognized, for their presence to be marked and represented in the landscape or, at the very least, for the landscape to be racially neutral. As we will see, one of the effects of the black civil rights movement in the United States was a variety of projects aimed at renaming streets and celebrating its heroes. In post-apartheid South Africa the black majority government has chosen to move more circumspectly. So while, on the one hand, the names given to airports celebrating Afrikaner politicians of the apartheid period have been eliminated – there is no longer a Hendrik Verwoerd Airport at Port Elizabeth or a John Vorster Airport at Durban – on the other hand, white sensitivities to black majority rule, concerns about making the white technocracy on which the country must depend for some time into the future comfortable, mean those names have been replaced by more neutral sounding ones, like "Durban International Airport."[15]

But these drives for integration, for assimilation, for recognition as equals have all too often been only qualified successes. This has been one of the reasons (not the only one) for a resurgence in the United States of segregationist feeling on the part of the oppressed. In the case of both blacks and hispanics we can find this pattern, or rather struggle between the protagonists of assimilation – often those who have most benefited from it – and those who believe salvation lies through claims to a space of their own rather than to one

---

13   As in rhetorics of colonization, for example.
14   And sometimes racial majorities, as in the case of pre-1994 South Africa.
15   "Durban" or "Cape Town" are still colonial, however, if less offensive to black African ears than names which celebrate their Afrikaner oppressors.

shared with whites: their own neighborhoods, their own schools, their own local governments, their own Congressional Districts.

Consider now, and in more detail, the case of American blacks. For many American blacks the fruits of the civil rights revolution, the desegregation of housing and of schools, the elimination of barriers to their exercise of the franchise, have been very qualified. Those with the money to move into the suburbs and send their children to the schools there have generally benefited. Blacks have achieved political office in increasing numbers. Black leaders have been nationally recognized. They have succeeded to some degree in achieving recognition within the same space as whites: in making a claim to that space (but see box 7.1). Advertising and the media have changed though tokenism still abounds and TV programs are either all-white or all-black. And it is rare to see marriages across the racial line depicted in the media; if dating and marriage enter into the story lines of the soaps then they will be invariably within rather than between the races. Government programs aimed at uplifting blacks have been a disappointment. Busing for racial balance rarely had positive effects. Rather, and as we saw in chapter 4, the introduction of busing for racial balance in central city school systems was typically followed by massive white relocation to suburban school districts where busing was not being applied. This meant that schools again became segregated, but this time on an inter-school district rather than inter-school basis.[16]

The most serious qualification to the effects of the civil rights revolution, however, is that most blacks still exist in serious poverty. Almost half of black children are brought up in single-parent (usually female) headed households. Unemployment rates run far ahead of those for whites and even for those of similar educational levels there are differences.[17]

The failure of civil rights legislation to make a difference for the vast majority of blacks has led in turn to an increase in separatist sentiment: pressures for all-black schools, for the construction of new public housing in all-black neighborhoods, and, in some instances, as in Boston, for the establishment of black areas as independent municipalities (see box 7.2). The news magazine

16   Attempts to introduce busing on a metropolitan scale between school districts were rebuffed in the courts. In addition segregation asserted itself not just at larger geographic scales but at smaller ones too: in schools themselves. School integration in Charlotte, North Carolina, is often hailed as a success since white flight has been so limited; not surprising when it is realized that the school system there is county-wide so that the possibilities of escape are restricted. But within the schools there is tracking and there is a marked tendency for blacks to be in slower classes: "Until they take a core course that is required of everyone, many whites have little contact with blacks in high-school classrooms" (*Wall Street Journal*, May 8, 1991, p. A6). So busing for racial balance, from diverse standpoints, has done little for black self-esteem, let alone for their education.

17   Some have argued, however, that the achievement of full political and civil rights by American blacks just happened to coincide with important changes in the global economy which worked against them. It is precisely the sorts of lower skill, historically well paid jobs in the US for which they might qualify that have been under most pressure from the newly industrializing countries. Rather, to the extent that these activities remain in the US they are likely to be looking for ultra-cheap, often illegal, labor which will allow them to compete with the foreign producer.

Box 7.1   *Claiming Space through Symbols*

Claiming space is in part material and in part symbolic. Materially it requires people taking some sort of material control or at least sharing in the control of a space: the transfer of power in some cities from whites to blacks as a result of voting in city council elections is an example of how claims are advanced materially.

But symbolic practices are also important. What are the symbols that are selected to represent those who live in a particular geographic area? In few areas of the United States have these issues been so contested as in the American South. One example is the Confederate flag which many blacks see as an affront to them as blacks. This is because the Confederacy fought the American Civil War to retain slavery in the South; something that blacks had always opposed. In 2000 a battle over precisely this issue raged in South Carolina: should the flag be flown over public buildings? Black groups, like the NAACP, were arguing that to fly the flag was racist. They were also bringing the power of money to bear on resolution of the issue, threatening to relocate conventions and sports meetings held in the State to other States. A year later the issue had shifted to Georgia which, in 1956, in an act of in-your-face defiance against a nascent civil rights movement had adopted the rebel flag.

Another area of contestation has been in the naming of places. Place names constitute a form of recognition; they celebrate the contributions to a common life of individuals who represent, symbolize through their actions, a particular group. Since the civil rights revolution in the South black councillors have been able to form majorities in many towns and have proceeded to rename streets after the civil rights leader Martin Luther King. In other towns there has been strong white opposition, however, and the renaming process has been limited to streets in black neighborhoods, side streets, or even dead ends: so the symbol is not just a question of *what*? It is also and equally a question of *where*?

See Derek Alderman (1996) "Creating a New Geography of Memory in the South: (Re)Naming of Streets in Honor of Martin Luther King, Jr." *Southeastern Geographer*, 36(1), 51–69.

The Economist (July 10, 1993, p. 18) has described it thus, though perhaps a little too definitively:

In black America the long-standing argument between separatists and integrationists is being resolved on the side of the separatists. Educationists point out that black children learn better with black teachers and black students have happier and more successful careers in all-black colleges. School systems in several cities, including Milwaukee, Detroit and New York, are trying to set up blacks-only schools.

There has always been a tension between those advocating assimilation to white society as the vehicle for black salvation and those opting for segregation. The black civil rights movement represented an attempt at assimilation and there are many blacks, particularly middle class ones, who have gained from civil rights legislation, and who continue to support that viewpoint. But even in the heyday of the civil rights movement during the sixties there were always black voices articulating a separatist argument: the belief was that only by their own efforts, and without white cooperation, could blacks be emancipated. They had, in other words, to emancipate themselves. To rely on the help of whites was to rely on those who had other agendas and who did not feel black pain, and who therefore would be unreliable.

Part of the separatist program has been that of controlling their own spaces: a reasonable enough goal given the high level of residential segregation to which blacks have been subjected. Blacks have defined themselves as a colonized population and their residential areas as the colonies in question to emphasize the way in which they have not had control of those spaces. They have been governed from outside, even to the point of the race of those who police them and teach their children. And urban renewal programs in the fifties and sixties showed how little they were regarded by the white authorities. Black areas were so disproportionately affected by the clearance of housing entailed by urban renewal that the programs earnt the sobriquet "negro removal."

Claiming their own space is also facilitated by residential segregation. Much depends, however, on the form of electoral organization. Where this is territorially based, as in wards, then the fact that, as a result of residential segregation, blacks can claim a majority of the population in particular wards means that they can return to city council, or whatever, candidates that will represent their, and only their, views. Indeed, for this very reason the form of electoral organization has become an important political stake for blacks.

An alternative to election by wards is what is known as election "at large." In "at large" elections electors vote for people from a list of candidates each of whom will, if elected, represent the whole city rather than some subsection of it. If there are ten city council seats then voters may choose ten names from a list of, say, twenty. Under these circumstances, where blacks are a minority of the city's population, and assuming that people vote along racial lines, then even though blacks are, say, 40 percent of the electorate no black councillors may be elected. In consequence in some cities blacks have pressed for the adoption of a ward system or some mix of the two which would facilitate the election of black representatives.

Think and Learn

In discussions of the advantages to blacks of ward rather than at large elections we have abstracted from the question of whether or not blacks are in a minority in the city as a whole. What difference do you think that would make as to whether or not blacks would demand a change in electoral organization from at large to wards? Why would that be?

Similar logics apply to the election of judges and school board members. In fact in a number of cases black groups have brought law suits charging that it was only subsequent to the civil rights movement that a change from a ward system to an at large one was engineered by white political elites. In yet other cities, however, segregation can provide a precondition for secession and the creation of black-dominated municipalities, as in the case of Mandela (see box 7.2).

In some cases, as a result of white flight, blacks are now in an electoral majority in the central city. The realities of tax base, however, have made it difficult to implement programs that might allow the realization of black power ideals. The relocation of business and more affluent residents to the independent suburbs means that, more often than not, black city administrations do not have the revenues out of which to facilitate the collective economic uplift of their constituencies. Even so, there are other benefits. In predominantly black cities juries tend to have black majorities. There is an increasing tendency for race to affect black jury persons.[18] In acquitting black defendants they typically draw on their own life experiences of police harassment and so are suspicious of police evidence. In the New York City borough of the Bronx juries are more than 80 percent black and hispanic. Black defendants there are acquitted in felony cases 47.6 percent of the time – nearly three times the national acquittal rate of 17 percent for all races (*Wall Street Journal*, October 4, 1995, p. A1).

Black nationalism has been turned against other minorities, and not just against the white authorities. The presence of immigrant-owned businesses,

---

Box 7.2 *Mandela*

Black power has a variety of different expressions depending on local circumstances. In Boston one of its expressions has been the call for the secession from the City of Boston of black and hispanic areas to form a new city of Mandela. It is argued that the area has been neglected by the white power structure of Boston. However, there may also have been concern over redevelopment and possible residential dislocation subsequent to the construction of a new elevated rail line through the area. In the event a referendum polled only 25 percent support for secession. Interestingly, black ministers – a conservative element historically associated with the civil rights movement – came out against the idea. The very threat of secession, however, does seem to have sparked a flurry of activity in the area on the part of the City of Boston: activity designed to defuse black concern. The issue of Mandela also needs to be seen against the broader backdrop of intense racism in Boston; by reputation, at least, one of the most racist cities in the US.

---

18   Note that this is no different from the way white juries behaved in the South. Whites were usually acquitted in lynching cases, while blacks were almost invariably convicted in cases involving whites.

particularly retail outlets in black ghettoes, has been a particular irritant. In Los Angeles it is Koreans who more often than not own the corner groceries. In Detroit it is Arabs. The goods they stock, the people they hire, their attitude to black customers have all become issues. In Chicago, for example, a group calling itself Black Women for Economic Parity boycotted Korean merchants to force them to stock more black-made goods and hire more black employees.

Questions of Gender

ED.   I must go. I'll have a light meal. Take a couple of nembutal and then bed. I shall be out of town tomorrow.
KATH. Where?
ED.   In Aylesbury. I shall dress in a quiet suit. Drive up in the motor. The Commissionaire will spring forward. There in that miracle of glass and concrete my colleagues and me will have a quiet drink before the business of the day.
KATH. Are your friends nice?
ED.   Mature men.
KATH. No ladies?
           Pause
ED.   What are you talking about? I live in a world of top decisions. We've not time for ladies.
KATH. Ladies are nice at a gathering.
ED.   We don't want a lot of half-witted tarts.
KATH. They add colour and gaiety.
ED.   Frightening everyone with their clothes.

(From Act I of Joe Orton's *Entertaining Mr Sloane*)

These few lines of conversation from Orton's (1964) black comedy *Entertaining Mr Sloane* between a middle aged gay and his nymphomaniac sister dramatize the exclusion of women from particular spaces: from top decisions in this instance, though Ed also seems to have doubts about including women in social gatherings. The obvious fact is that there is very serious spatial separation along gender lines, and this is a separation from which women emerge at great disadvantage. What is known as the gender division of labor sees women confined to lives of domestic labor while men work outside the home for a wage with all the subsequent advantages that the social power of money conveys. Increasingly, of course, women are found in the paid workforce, though this does little to reduce their domestic responsibilities, and they tend to be confined to lower-wage positions. This applies to the state as much as it does to the private economy. The glass ceiling that is referred to in the context of the promotion of women to higher levels in the corporate hierarchy is also a feature of the state: not just the civil service but also elected office.

There *are* women in politics but as one ascends from local through state to national politics so their presence diminishes.

Accordingly, since the sixties gender has become an important focus of identity politics in North America and in Western Europe. The feminist movement has been to the fore in pressing for equality between men and women in a whole variety of arenas: not just the workplace, important as that is, but also in running for political office, in how women are represented in the media, in the admissions policies of universities.[19] This is not to marginalize the earlier political achievements of women. In particular we should recall the suffragette movement which led to the extension of the franchise. But the depth of the contemporary movement in terms of the variety of issues it has dealt with and in its ability to change the way people think about gender – rebutting, for example, Ed's claim that "ladies" are "a lot of half-witted tarts" – makes it qualitatively different. And at the same time it has served to draw attention to gender inequalities, gender identities, and gender struggles worldwide.

As with other cases of identity politics, this is again a matter of claiming space, which is what gives it a resonance for the political geographer: being, for example, part of Ed's "world of top decisions" and not just on Kath's terms ("Ladies are nice at a gathering"). In gender politics issues of movement, the control of movement and of particular spaces is paramount. There is an old saying that an English*man's* house is *his* castle, implying that within those four walls patriarchs can do much as they please. But one of the objectives of the women's movement has been to bring the state into those four walls in order to protect women from the spousal abuse which, apparently, is extraordinarily common, and among all classes.[20] Part of the way in which the women's movement has tried to combat this problem has been to establish shelters for battered women: to give them a space in which they can be protected from their husbands, and receive counselling on how they can get out of the abusive relationships and achieve some independence; spaces of their own.

Quite what accounts for this gender inequality, these exclusions, has been the object of intense debate and the arguments are many:

1   One of the most common claims has been that men sought, through agreements with employers, through the control of apprenticeship schemes and through legislation, to exclude women from the workforce or to confine them to lower paying jobs. In this way men would not only obtain material advantages for themselves; they would also secure their authority within the household.

19   It seems hard to imagine, for instance, that those staunch bastions of masculinity, the Oxford and Cambridge colleges, would have admitted women, particularly on the scale they have, without the change in climate of opinion inaugurated by the modern feminist movement.

20   Some statistics in the form of the percentage of adult women ever physically assaulted by an intimate partner: 29 percent in Canada, 35 percent in New Zealand, 30 percent in the United Kingdom and 22 percent in the United States (Source: *Population Reports*, Series L, No. 11, p. 4).

2   Another argument is that gender inequality has had to do with the inter-
    ests and beliefs of employers. The fact of women receiving lower pay than
    men was well established before industrialization and employers took
    advantage of this. On the other hand, men were also thought to be more
    productive and so merited higher levels of remuneration.

But what these arguments tend to overlook is the remarkable degree of
variation in gender inequality, in the confinement of married women to the
home. In the British coalfields the male breadwinner was indeed the norm.
But in the textile weaving areas of Northeastern Lancashire the proportions
of married women employed in wage work were quite high. In some instances
the fact of women's employment might allow them to live independently of
men. This at least seems to have been the case in Dundee, Scotland, where
women dominated employment in the jam factories.
    An equally informative case is South Africa. There the participation rates
of white women in the paid labor force were substantial well before the same
could be said of women in Britain or the US. A major reason for this had to
do with the cheapness of domestic help, which in turn was a function of South
Africa's historic racial order. This latter meant that black women would be
available as general purpose maids, to child-mind as well as to cook and clean,
at very low wages indeed.
    But much of this discussion abstracts from the question of class and the role
of the specifically working class family. Accordingly another line of argument
bases itself on the idea of the working class family as, essentially, a survival
unit: a form of association whose *raison d'être* was material survival through
cooperation. There are a number of different variants of this approach. One is
that married women supported legislation and union activity aimed at limiting their participation in the waged labor force since this would induce a
scarcity in the supply of labor and so serve to drive up the wage that the,
henceforth male, breadwinner would bring home. Another variant has to do
with the interests of working class families in reproduction: in producing children who would then, at an appropriately early age, go out to waged employment while continuing to live at home and so swell the family's income. But
this, given the conflict between pregnancy, childbirth, and lactation on the one
hand, and factory work on the other, meant the characteristic gender division
of labor (Brenner and Ramas, 1984).
    Nevertheless, whatever the original reason for this gender division of labor
it is clear that men benefited from it. By virtue of earning a wage they had
powers to control what went on in *their* families that would otherwise have
been denied them. The wage was the working class family's passport to means
of subsistence. As a result the wage earner could command his wife, subordinate her to his will, in a way that would not have been possible if the wife
too had had a wage income; and indeed this has been the lesson of history as
women *have* acquired wages and been able to resist the exercise of patriarchal
power.
    But given the advantages it provided men, we should not be surprised
to see the elaboration of a particular interpretive framework, a discursive

construction which, in essence, justified the position men enjoyed not just as wage earners but also as politicians or states*men*, and which at the same time explained why women were – and *should* be – consigned to lives of domestic labor. We can grasp this interpretive framework through a set of contrasts which, even now, given all that has happened in the past fifty years, should still resonate:

male/female	work/home	public/private
strong/delicate	active/passive	providing/nurturing
controlled/emotional	discipline/sympathy	force/persuasion
cerebral/physical	quantitative/verbal	rational/irrational
science/arts	dominant/subordinate	protector/protected

These are contrasts into which we have all been socialized. In consequence gender inequalities have tended to be reproduced. Dolls and helping mummy around the house are *still* for little girls just as toy guns and soldiers are for little boys. Likewise it is still the case in school that needlework and cookery lessons are for girls while carpentry and the metal shop tend to be monopolized by the boys, though this may now be breaking down. Assumptions also affect the way teachers deal with the two sexes. The assumption that girls are more verbal in their aptitudes and less quantitative often means that they get less attention in math and science: they just aren't expected to do as well. So it is little wonder that the world of hi-tech is predominantly a male world (Massey, 1995).

Think and Learn

We are talking here about the role of discourse in creating social order. But for discourse to have these effects the definitions have to stick. Why do you think that men have been able to make their particular definitions of women, their social roles, prevail? To what degree and how do you see these sorts of definitions continuing to prevail?

To the extent that women have entered public life in either the labor market or that of the state, and this has happened increasingly over recent years, then the old stereotypes come into play and channel them into certain positions in respective divisions of labor. Women are supposed to be good at dealing with people so they make good salespeople in the retail trade, good nurses, good flight attendants, good receptionists. On the other hand their "shortcomings" in terms of rationality, discipline, their tendencies to emotional reactions[21] make them poor leaders, not good when it comes to making quick decisions.

---

21   Often justified in terms of the biology of the female reproductive system.

So the doctors are predominantly male, as are airline pilots and anyone in a position of leadership. To the extent that women do accede to positions of control, then it is always qualified: more often city council than a Congressperson or MP, more likely mayor than Prime Minister (of which there has been only one instance so far) or President (of which there have, of course, been no instances as yet); and in the world of work, more likely to be principal of a grade school than that of a high school, more likely to be principal than superintendent in charge of the school district as a whole.

Even in the world of leisure women are marginalized. There is women's sport but it is taken far less seriously than men's, as any perusal of the sports pages of a daily newspaper would quickly affirm. The major sporting events in the annual calendar are either all male (the various national soccer championships, the Superbowl, the World Series, the Master's Tournament, the British Open, the Ryder Cup, the Stanley Cup, the Tour de France) or ones in which the male winners are taken more seriously than the female (Wimbledon), as is evident in the disparities between prizes. Likewise one expects men to climb Mount Everest but not women (though some have), and likewise for men to write travel books, though again there are exceptions. The widespread belief is still that these are not female activities and, moreover, shouldn't be.

Even so, there has clearly been change. Women now have the vote. Provision of day care, sexual harassment, and the problem of domestic violence have become political issues. There is an active women's movement. The hiring of women has become a measure of the acceptability of an organization's hiring practices. Most importantly, and underlying most of these changes, since the Second World War women have entered the job market in increasing numbers in both North America and Western Europe. This has been decisive because money in our societies is social power. This has worked in a variety of different ways.

In the first place it has facilitated a shift in the balance of power within the home. With an independent income there can be a limit to what women are prepared to take from husbands or live-in boyfriends. The burgeoning divorce rate is depressing news to many but the other side of the coin is that it expresses, in many cases, a new willingness on the part of women to assert themselves against the demands of men and, often, their violence: a willingness that would be greatly reduced if their physical subsistence depended on access to a husband's wage.

The changes are still more subtle. The fact of a dual income in many households has meant being able to afford a bigger house, more vacations, more eating out, more consumption in general. This has resulted in an enhanced appreciation on the part of men for the wage earning of their wives. A wife's employment is now something that many men encourage and this too has, in many instances, resulted in a reduction of women's domestic labor as many men share, at least some of, the chores.

The conditions for the rise of female employment are complicated. In part it has been due to declines in demands on a woman's time as wife and mother. The widespread adoption of birth control has meant smaller families to care for. The mechanization of the home, the vacuum sweeper, the dishwasher, the refrigerator, the microwave, etc., along with the rise of packaged foods and fast food, have all served to reduce the need for domestic labor and created time for wage work. The provision of day care, if still inadequate, has worked in the same direction.

These changes have meant that women have been able to go out and work for a wage instead of staying at home: that there was, in other words, a potential *supply* of female wage workers. But at the same time there has been a *demand* for their labor. Women have been discovered by employers. Much of this is due to shifts in occupational composition, shifts that have worked in favor of women's supposed gender-specific skills and shortcomings. So the rise of office employment at the expense of blue collar work has reduced the premium on the physical strength that women are supposed to lack. The expansion of the so-called service industries – in particular that of the health care field, the demand for customer service representatives, travel agents, social workers, drug counsellors, the continuing expansion of retailing – has opened up opportunities for which women, by virtue of their supposed skills in dealing with people, are most fitted.

To a degree, of course, this occupational channeling simply reproduces the old stereotypes. There is also the fact of vertical as well as horizontal gender stratification in the job market. Just as blacks still for the most part take orders from whites, so women are confined to those jobs where they are under the control and supervision of men. Those who run the bureaucracies are more likely to be men while the people behind the VDUs on a continual basis are more likely to be women, etc. And just as many whites still have trouble taking orders from blacks, as if it is some violation of the natural order, so too do many men have trouble taking orders from women.

So the struggle is far from over. The authority of men in "their" households has been diminished. Even women who do not work for a wage have found their ability to resist enhanced, not just because of any widening of legal rights but also because of a sense of empowerment coming from the women's movement and the publicity it has obtained. The response of men has varied. In some cases it has assumed the form of resistance to the hiring practices of employers, particularly if they are biased towards the employment of women rather than of men. A striking instance of this comes not from a more developed country but from South Africa, but it is so rich in its implications that it is retold in box 7.3.

In North America and Western Europe, on the other hand, the crisis of male authority in the household has been most evident in the phenomenon of domestic violence. No one knows the extent of this prior to the recent past. But as Colleen McGrath (1979, p. 16) has put it:

> Interestingly, battering has emerged as a recognized social problem in a period when the traditional nuclear family is declining as the major instrument of male dominance in capitalist society; when women's power relative to men is increasing in the marketplace, the legal sphere and the cultural arena; when "the family" as an institution seems threatened by feminists and "modern life."

Box 7.3 *"Angry Men and Working Women"*

Under apartheid in South Africa the government tried to relocate most blacks into so-called homelands where they would enjoy political rights and eventually become independent from South African rule. The whole project was a gigantic ruse by which the whites of the country hoped to finesse demands from blacks for the vote. In an attempt to give the homelands some degree of economic development the South African government also introduced policies designed to persuade firms to set up plants there. Many of these employed large numbers of women. The article on which this box is based (Bank, 1994) describes the events surrounding some of these factories in the capital city of the homeland of QwaQwa, Phuthaditjhaba.

In October of 1984, over 400 unemployed black men marched through the city and made for the industrial parks. They stoned the factories, assaulted women workers, and chased them away from their jobs. They then demanded of the employers that female employees be dismissed and their jobs be given to men. They argued that factory work was not for women and that a woman's place was at home looking after their children.

The immediate trigger for this incident was the failure of a contract for (male) migrant workers to arrive in the capital. Large numbers of unemployed men had come to the city that morning from the surrounding area expecting to obtain jobs as migrant workers. But it also needs to be placed in a broader context. For it was not just a matter of jobs. It was also about the self respect of frustrated patriarchs. Among South African blacks patriarchy, particularly in the rural areas, has always been strong. If money is to be earnt then it is men who should be earning it. For a man to live off his wife's wages results in a loss of self respect. This helps account for the violent nature of the reaction in Phuthaditjhaba: an expression not just of the frustration of unemployment but of a masculinity denied.

A particular flashpoint according to McGrath is the performance of the domestic role. As women go out to work so they find this more and more onerous. But for men, doing the cooking, the laundry, making the beds, cleaning the bath is not male work and to see it not done or done poorly is an affront to their authority within the household.

Summary

If nations are divided one from another, they are also, as we have just seen, internally divided. Gender and race have recently been, and continue to be, the basis of a politics of identity in Western nations. Much of our thinking about race dates from just after the Second World War. Prior to that racist thinking was taken for granted. The retreat from racism, prompted in part by the policies of Nazi Germany, posed a challenge to discourses of colonialism as much as it did to racialized inequalities in North America and Western Europe.

Western colonialism was racist from the start. The natives were defined as inferior along a number of different axes, the most prominent of which was that of "civilization." Europeans were "civilized." Indigenous populations in contrast were "barbaric." This was explained in significant part in terms of race. There was also, however, a view that cultural backwardness was a result of the absence of wage labor. This is an important observation since it points to the hidden agenda behind this "othering" that the colonial authorities, the missionaries, the settlers engaged in. This was one of development on Western terms for which a supply of wage labor was needed. In turn this called for the establishment of colonial governments, the government *of* the native rather than *by* the native. For only in that way could the native be coerced into working on the plantations and in the mines. These acts were in turn justified in terms of the betterment of the native: bringing civilization to him – "him" because the native was almost always seen in male terms – making him a better person. It should be noted, however, that racism always stood in a somewhat contradictory relation to this rhetoric. This is because while racism could be called on to explain the backwardness of the native it lost its force if one assumed that the native was capable of improvement, even of becoming "civilized."

What magnified the importance of these discursive strategies was the fact that the colonial authorities disposed of very limited manpower to enforce their rule. In consequence it was important that the natives accept the way in which they were being differentiated. And to a considerable degree they did. The white man did indeed seem to be the agent of superior powers, though in addition we should note the way in which he organized colonial society so as to leave the natives in no doubt as to their inferior status.

Even so, resistance movements did emerge. The manpower needs of the colonial state required the creation of at least a thin stratum of educated natives and they were often the catalyst for embryonic nationalist movements. To some degree the necessary discursive constructions were provided from outside the colonies; Garveyism and communism provided understandings of the predicament of the colonized which, accurate or not, fortified their opposition. Discourses of black consciousness were also important: narratives explaining to the black person why he or she could have pride in being black, what it was in their history that justified that pride, and how their treatment at the hands of whites was one of exploitation and oppression.

The colonial experience was immensely important in producing other forms of identity. It is not for nothing that under apartheid in South Africa, park benches reserved for whites bore the words "Europeans only." The colonial experience tended to cement a sense of the differences not just between whites and blacks (and browns too) but also between Europe and Africa, Europe and the Orient, and so forth. This is a sense of identity that has carried over into the present day, albeit recast as one of the First World versus the Third World or more developed and less developed countries.

Decolonization led to some deracialization of social relations in what had been the European Empires: "some" because the colonization of the mind continues to be important. But Europeans and non-Europeans had also been brought into contact with one another in the white-dominated, settler societies of North America and South Africa. The slave trade was immensely important in this process. In addition, and later, indentured workers were brought from India to work in South Africa, and Chinese and Japanese crossed the Pacific to work in the western regions of both Canada and the United States. In all these instances racism was a common response, particularly among those who felt most threatened both economically and culturally by the new arrivals – the white working class.

Moreover, just as the concept of the native can be seen as a social construction, and one nicely calibrated with the needs of whites to subordinate indigenous populations to their own, highly material, ends, so too was it the case in those destinations where the vast tide of Africans, Indians, and Chinese eventually came to rest. These constructions were by no means bipolar. The case of the Coloreds of South Africa should alert us to the ways in which constructions are undertaken by anyone wanting to advance a claim for material privilege. Differences are drawn from those who are a challenge, even though the resultant racial constructions leave one in a position of inferiority with respect to some others.

In the United States the struggle against these demeaning categorizations got under way with vigor in the sixties. A similar movement arose among hispanics, though this was partly in response to the black civil rights movement and the fear that blacks might come out of it relatively privileged *vis-à-vis* hispanics. In all cases these movements have involved not only a rewriting of history in order to rectify the omissions and biases of mainstream accounts but also in a very real sense a rewriting of geography: a laying claim to space. This has had both assimilationist and separatist streams. Blacks, for example, have drawn on rhetorics of colonialism in struggling to control their own lives within the spaces they dominate numerically.

In discussions of the so-called "new social movements" the feminist movement has, along with the civil rights movement, been one of the principal instances drawn on. It has been prominent in pressing for gender equality in a whole variety of areas; and while the purpose has been in part material it has also been about recognition in place of marginalization. Space is once more central to the struggle: should women be confined to the home or, for example, pink collar ghettos or should they share a common space with men as equals? There is considerable variation here of course. Restrictions on a woman's movement are far more draconian in Islamic societies. Initially in the North American and West European cases there emerged a "male breadwinner/female domestic worker" form of household arrangement. This in turn was bolstered by a rhetoric of difference which confirmed the restriction of women to the role of mother and housewife (or more accurately, unpaid housekeeper) in terms of "natural" dispositions. But more recently changes in the technology of the home, changes in the demand for labor, have resulted in a vast increase in the numbers of women in wage work.

In turn, this enhanced access to money as a vital social power has laid the foundations for change in gender relations; though as the recent emergence of wife battering as a serious social issue testifies this change still has a considerable way to go.

REFERENCES

Bank, L. (1994) "Angry Men and Working Women: Gender, Violence and Economic Change in QwaQwa in the 1980s." *African Studies*, 53(1), 89–113.

Brenner, J. and Ramas, M. (1984) "Rethinking Women's Oppression." *New Left Review*, 144, 33–71.

Callinicos, L. (1981) *Gold and Workers 1886–1924*. Johannesburg: Ravan Press.

Cope, R. L. (1990) "C. W. de Kiewiet, the Imperial Factor and South African 'Native Policy'." *Journal of Southern African Studies*, 15(3), 486–505.

Delanty, G. (1995) *Inventing Europe: Idea, Identity, Reality*. London: Macmillan.

Hill, H. (1975) "Anti-oriental Agitation and the Rise of Working Class Racism." In H. I. Safa and G. Levitas (eds), *Social Problems in Corporate America*. New York: Harper and Row, pp. 162–72.

McGrath, C. (1979) "The Crisis of Domestic Order." *Socialist Review*, 9(1), 11–30.

Massey, D. (1995) "Masculinity, Dualisms and High Technology." *Transactions of the Institute of British Geographers Series*, 20(4), 487–99.

Scott, J. C. (1998) *Seeing Like a State*. New Haven, CT, and London: Yale University Press.

Thompson, L. (1985) *The Political Mythology of Apartheid*. New Haven, CT, and London: Yale University Press.

FURTHER READING

An excellent survey of the issues discused in the early part of the chapter on colonizers and colonized can be found in Stuart Hall (1996) "The West and the Rest: Discourse and Power," in S. Hall et al. (eds), *Modernity* (Cambridge: Polity Press). For discussion of the black civil rights movement an excellent introduction is still parts 3 and 4 of the book by Richard Cloward and Frances Fox Piven (1975): *The Politics of Turmoil* (New York: Vintage Books). Likewise a good discussion of black political struggles in an urban context can be found in Susan Anderson's (1996) discussion of the Los Angeles case: "A City Called Heaven: Black Enchantment and Despair in Los Angeles," chapter 11 in A. J. Scott and E. W. Soja (eds), *The City* (Berkeley and Los Angeles: University of California Press).

Alistair Rogers (1990) "Towards a Geography of the Rainbow Coalition, 1983–89." *Environment and Planning D: Society and Space*, 8(4), 409–26 is good on the attempt of those prominent in the black civil rights movement to move beyond a politics of race to one of class, embracing African-Americans, labor unions, and hard-pressed farmers in the election of 1988.

On more specific issues: on the politics of ward versus at large electoral institutions see Lee Sloan (1969) "'Good Government' and the Politics of Race." *Social Problems*,

17(2), 161–75. On the attempt to create a new municipality of Mandela within the city limits of Boston: M. Kennedy, M. Gastón, and C. Tilly (1990) "Roxbury: Capitalist Investment or Community Development?," chapter 3 in M. Davis et al. (eds), *Fire in the Hearth* (London: Verso Books). Finally, on geographic variations in the gender division of labor in contemporary Britain see S. S. Duncan (1991) "The Geography of Gender Divisions of Labor in Britain." *Transactions, Institute of British Geographers* NS, 16(4), 420–39.

Part III

Territory and the State

Chapter 8

The State in Geographic Context

Context

Hovering in the background of virtually all our discussions so far has been the state. It is now time to confront it directly, examine its rationale, its relationship to geography and in particular to territory, and also its characteristic tensions and transformations. In this regard we will be speaking of the specifically *modern* state. This is the centralized state that emerged alongside capitalist development and which has had a symbiotic relationship with the third element in that troika: national identity and its Janus-faced offspring, nationalism.

The state is a very peculiar institution. It separates government from society and then subjects both of them to itself through the rule of law. Law displaces custom. Government regulates the relation between employer and employee, between husbands and wives, parents and children, slowly substituting itself for employers, parents, husbands. But as the saying goes, the government, at least in democratic countries, is not beyond the law and the state can take steps through the judiciary to ensure that justice is served. All this is done in the name of the people. Governments govern on behalf of the people and the state sees to it that they do. This begs the question, of course, of what constitutes "the people" and this is where the state connects with nationalism: typically the people has been defined as the nation.

But if the relation of the state and national identity is symbiotic, so too is that between the state and capitalist development. As we will see, capital needs the state and vice versa. This is where we commence our discussion. What this section will bracket, however, is the way in which the state is a territorial organization; it has sovereignty with respect to a particular bounded space. It is divided internally into the jurisdictions of its various local branches. Furthermore, there are, of course, many, many states in the world. Just how are we to understand this territoriality? This is the question we will address in the chapter's second major section. Finally we will explore some of the points raised through some case studies.

Capitalism and the Modern State

The state is an essential part of the social division of labor in a capitalist society. It performs functions, carries out activities, that capitalism could not possibly carry out, but which are essential to its reproduction and development. This does not mean to say that what the state does is not contested; that, for example, particular groups of firms may not argue that such-and-such could just as well be performed by the so-called private sector. And the boundary between provision by the state and provision by the market does indeed fluctuate over time. During the past fifteen to twenty years there has been a strong movement for transferring some state functions to the market: what has come to be known as the marketization or privatization of the state. States have sold assets, industries have been denationalized, there have been experiments designed to introduce market principles into public provision as in the case of school vouchers in the US. Likewise there are few local governments in the US or in the United Kingdom which are not required to put out requests for services to competitive tender; no longer is it inevitable that the state agency set up to perform that function gets the job. At the time of writing this is a transfer that has yet to completely run its course. But having said that, it still remains the case that there is an irreducible core of activity that the state *must* perform; activities that are beyond the market because the principles according to which markets operate make it impossible.

> **Think and Learn**
>
> Privatization of activities formerly carried out by the state, the attempt to mobilize the disciplines of the market in public provision, as in the case of school vouchers, have been dominant themes in state–society relations over the past twenty years. Thinking of the various ways in which this has occurred, the varying forms of public provision which have been the object of privatization and marketization, how would you explain them?

The general conditions of production

One way of looking at this is in terms of what one might call "the general conditions of production." Production under capitalism depends on commodity exchange. But the actual fact of commodity exchange cannot be taken for granted. There are all sorts of barriers in the way; barriers which the play of market forces cannot resolve and which therefore require something above the market, like the state, if they are to be eliminated. Some examples follow.

The money supply

For the commodity exchange on which capitalism is based to occur smoothly, there has to be a means of exchange which people trust; which they believe will be accepted as means of exchange, and which will act as a store of value. It used to be the case that banks issued their own monies. The money supply, in other words, was left up to the market. But the competition between banks tended to undermine the confidence of the public in their respective monies. Banks make money out of the interest on loans and this brings them into competition with other banks. Competition, however, can result in lending to relatively high credit risks. If the loan was not paid back the bank would be short of money to cover its liabilities to depositors. But since it controlled its own money supply this concern was not always strong enough to counter the temptation to get business and steal a march on rival banks. As a result the value of the bank's currency could be debased, to the obvious disadvantage of anyone using its money as a means of exchange. Control of the money supply, therefore, had to be taken out of the hands of the banks and given to an organization not subject to competition. That organization was the state.

Transportation infrastructure

Exchange requires transportation. Coal has to be moved from the mine to the blast furnace, cement from the factory to the building site, wheat from the fields to the grain elevator, workers from their homes to the factory and office, and so on. Accordingly we live in a world criss-crossed by rail and highway networks and, in an earlier age, by canals. But the actual provision of these conditions of exchange and therefore of production is far from unproblematic. Under capitalism rights in private property are precisely that – private. Property can be transferred from one person to another in exchange for money but it requires the consent of the property owner. This is an obstacle to the provision of transportation networks. The laying down of a highway or railroad requires land assembly along the line of the proposed route. This, however, gives inordinate power to the owners of property who will be asked to sell some of their land to whoever is building the highway or railroad. They can hold out for a huge price knowing that the success of the whole project depends on their cooperation. This dilemma is the origin of a right that the state invariably assumes: the right of eminent domain or the right to purchase land for a public purpose at what is determined to be a market price, and if the property owner still refuses to sell, to impose such a price and so avoid the risk of monopolistic extortion that would otherwise be incurred. All the networks that are necessary to an advanced industrial society – gas lines, electricity lines, water lines – benefit from the use of that power.

Think and Learn

States use the right of eminent domain to facilitate the assembly of land for the construction of transportation infrastructures. What other land assembly projects does the state get involved in for which it requires the use of the right of eminent domain? Hint: in what other construction projects is the state involved?

The definition and protection of private property rights

Again this has to do with exchange and therefore, given the centrality of exchange to production under capitalism, with production as well. Clearly exchange would come to a complete standstill if a purchaser could not be sure that having paid money for something it could be taken away without compensation (as in theft) or that someone else might have a claim on it. In consequence, an important function of the state is the definition and protection of private property rights. There is a police force to protect private property and law courts to settle disputes over ownership. Of course the buyer could rely on the seller to say that there are no other claims on the property. But the desire of the seller to sell and at the best possible price – a result of the forces of competition – could get in the way of full disclosure. This is a problem that the state does not face.

Regulating the capital–labor relation

A second broad function of the state, apart from providing what I am calling general conditions of production, is regulating the capital–labor relation. This is perhaps the most significant of exchange relations of all. Again, it is something into which the state intervenes because business itself is structurally incapable of responding, at least on a level that would not be threatening to the ability of firms to reproduce themselves. There are, in other words, practices that are in capital's interest but which it would not be able to enact itself. Examples follow.

Think and Learn

If the state is responsible for the definition and enforcement of private property rights how do you explain the existence of private security services? Is it not possible that they could substitute for the state in these matters?

The reproduction of labor power

Business depends on the qualities of its workers. It needs workers who are appropriately educated, workers who are healthy, and also workers who will stay around and wait for the next business expansion even though at present they are unemployed. None of these functions can be performed adequately without state intervention. Capital therefore needs the help of the state.

Education. In the advanced capitalist societies education has become immensely important. Basic skills of numeracy and literacy have become fundamental. Instructions have to be read, calculations done in the head, time and space rationed, all of which demand some form of education. On the top of that, of course, there is the education required of the technical strata: the engineers, the lawyers, the scientists, the accountants, the doctors; and without some form of elementary and secondary education they would never have reached the point at which they could even apply for university or professional school.

Education is not something that can be provided adequately by employers. Workers are free to move around from one employer to another. This means that investment in worker skills that are generally in demand may not yield an adequate return on their money for the firm so investing. It is true, of course, that parents have a motivation to see that their children are educated. A more educated person can gain a higher paying job and that is one of the reasons that formal education has, among other things, become such a significant part of the residential choice calculus of households, as we discussed in chapter 4. In itself, however, this does not mean that the state has to provide education or to insist that people make provisions for the education of their children. Education can be provided by the market and to some degree it is. There are private schools and there are private universities. But if *all* education was private then many children would not go to school. And for those who did there would be a tendency for their parents to underinvest. The reason for this is that education is said to be characterized by a large degree of publicness in the benefits it provides. There are benefits that the individual is not able to internalize. Among other things firms are interested in more educated workers because their working together, their cooperation, can produce useful effects not replicable by uneducated persons working together. There are collaborative effects, for example, which will be reflected in higher productivity and which the firm can appropriate in the form of profits and the state in the form of taxes. The fact that an educated person can be so productive to the advantage not of himself or herself but to that of someone else is hardly likely to encourage an investment in education that is optimal from the standpoint of business. As a result the state steps in, mandating a period of formal education for all children and typically either making it free or heavily subsidizing it.

> ## Think and Learn
>
> How might the same logic that we have developed for the case of public regulation and financing of education apply to public health? Is there a similar logic in the state mandating immunizations against certain infectious diseases and in regulating the provision of water and sewage disposal in cities? To what extent was urbanization a necessary condition for a health crisis in earlier stages of capitalist development and why?

Unemployment compensation. As we saw earlier capitalism is always accompanied by a degree of unemployment. As a term "unemployment" only comes into being with the development of capitalism (Williams, 1976, pp. 273–4). From the standpoint of business firms, unemployment is a problem because although they may not need the workers now, they may have reduced production in order to cope with a decline in orders, they may need the workers in the future. But if they have no incentive to stick around then there will be a labor shortage when business takes an upturn. On the other hand, any business firm itself is unwilling to pay for the subsistence and shelter of former workers and their families as an insurance against future labor shortages since this would put it at a competitive disadvantage. If other firms want to, fine, since it can take advantage of the presence of the unemployed so supported when it wants to expand. Since, however, other firms have the same calculus no benefits will be forthcoming. The state, on the other hand, not subject to the same competitive logic, can provide for unemployment compensation by forcing firms to pay into an insurance fund for precisely that purpose.

> ## Think and Learn
>
> If it is in business's interest to provide income supplements for the unemployed why do you think they rail against "welfare bums" and constantly strive to reduce the magnitude of those supplements? What is it about programs of support for the unemployed that arouses their concern and why is it in their interest that they should be concerned? And what does the resultant stigmatization of the unemployed have to do with the questions of identity that we discussed in chapter 5?

Workers as consumers

Employers relate to their employees precisely as that: as workers who must be paid a wage. On the other hand, we know that workers are also consumers and that if firms are to find markets for their products, in most cases it will be among workers.[1] As an employer, however, engaged in competitive markets, the firm feels the necessity to keep wages down as much as possible. Rather it looks to the employees of other firms as the consumers of its products or services. Unfortunately, and obviously enough, their employers too are subject to the same sort of logic. This means that there is a risk of a shortfall in purchasing power.

Whether intendedly or not a diversity of state practices have the effect of putting purchasing power into the economy: effective demand that would not otherwise exist. One example is unemployment compensation. This puts a floor under demand, as indeed does the institution of minimum wages. Furthermore, one of the reasons states have supported the drive for unionization and mandatory collective bargaining is the recognition that this has expansionary effects on the market and therefore on the profitability of business, even though for individual firms to undertake such action unilaterally would be ruinous for them.

Reducing class tensions

Finally, but by no means least, there is the fact that capitalist societies are subject to class tensions of no inconsiderable magnitude. As I have pointed out earlier, *the* social cleavage in the polities of North America and Western Europe is that of class. Inequality, disparities in wealth, in the opportunities that are in consequence open to different people are all sources of worker discontent: discontent that can break out into labor unrest, demands for the extension of the welfare state and generally for an expansion of the workers' share of the product at the expense of business. These are demands which it is also in the state's interest to resist. This is because they could threaten the ability of businesses to invest and hence the source of the state's own revenues.

The state tries to alleviate these tensions through practices of both a material and a discursive kind. As such it acts as an important condition of social cohesion. Materially it reduces some of the insecurity to which workers are subject through unemployment compensation and income supplements like food stamps. The creation of a social safety net in the form of national health services[2] and old age pensions have served the same purpose.[3] The state also

---

1   Not in all cases, however. The state has become a major consumer of the products of private firms. Consider all the expenditures that are made on behalf of education: textbooks, school desks, the schools themselves. Since the Second World War, and particularly during the Cold War, defense has been a major source of demand. Likewise firms producing machine tools have other firms as their consumers; though if the firms that need the machine tools are producing consumption goods then worker consumption again enters the picture.

2   The United States is unique among advanced capitalist societies in *not* having a system of universal and public health insurance. In consequence large numbers run the risk on a day to day basis of an incapacity that would quickly drain their savings.

3   This is a social safety net, moreover, that offers little threat to capital's investment fund and hence its ability to invest and operate profitably. This is because the welfare state is funded largely by workers themselves. Old age pensions, for example, come out of the pocket of those who are still working; and when they retire they will benefit from the same redistributional scheme.

promulgates discourses of unity. Much is made of the fact that the law is neutral between those of wealth and those without. Likewise universal schooling is touted as leveling the playing field in the competition for better paying positions, even though its effects in that direction are very qualified ones.[4] And the state always presents itself as representative of the nation as a whole, as acting in the public interest, though clearly in so acting it cannot undermine the conditions for capital accumulation since to do so would threaten its own sources of revenue. But to argue it is acting on behalf of the nation requires that there be a nation; and as we saw in chapter 7 the state takes its own steps to ensure that there will be.

The State in Geographic Context

The territorial organization of the state

When we turn and try to put the state in a geographic context perhaps the first thing we note is its territorial character. The state enjoys the right to define what is legal – in legally defined ways – and to enforce the law within a bounded space: its geographically defined jurisdiction. This territorial character has many aspects. In the first place there are many states. The surface of the earth, and much of that of its oceans, is partitioned among them. Within their jurisdictions states are sovereign; their power cannot be challenged. Citizens, once their constitutional and statutory rights are exhausted, cannot appeal for justice to other states. This does not mean that the relations between states are unregulated. There are numerous supranational bodies. These include: common markets in which states agree to eliminate the barriers to trade between each other; consultative organizations like the United Nations or the Group of Seven (G-7); and organizations that have come into being to facilitate trade, like the General Agreement on Tariffs and Trade (GATT) and its successor, the World Trade Organization (WTO). What all these have in common, however, is that they do not operate by majority rule. Decisions have to be reached through negotiation and, in effect, any member has a veto; though if they want to continue to enjoy the benefits of the organization they cannot veto too many times or their own demands will count for nothing.

The second point to notice is that states have territorial structures internal to themselves. Legislators commonly represent territorially defined popula-

---

4   To the extent that per pupil expenditures make a difference in the United States there are often huge disparities between one district and another. Children in tax-base rich suburban school districts are taught by more qualified and experienced teachers (since they can be bid away from districts that cannot afford to pay them so much) and have access to a wider variety of physical equipment, like language laboratories, more PCs, and so on. There is some doubt among educationists, however, that these things make a tremendous difference. More significant is what is referred to as "social background." This means that children from more affluent homes have advantages by virtue of their exposure to a particular cultural milieu, including parental encouragement, an expectation of success rather than failure, as well as exposure to books, control of TV watching, and so on.

tions: parliamentary constituencies, Congressional Districts, and State-elected US Senators. Many cities elect councillors by ward, something we discussed in chapter 7 in the context of black claims to space. But that example also alerts us to the fact that not all cities have such a territorial form of electoral organization. So too is it the case with national elections. Proportional representation does not need districts: legislative seats are awarded to parties on the basis of the proportion of the national vote that they polled respectively – 60 percent of the vote translates into 60 percent of the legislative seats.[5]

Think and Learn

Think about the mechanism by which councillors are elected in your city, suburb, district; or how (in the US) county commissioners are elected. Is it by ward, at large, or is there some mix of the two approaches? And has the mechanism been at all controversial? If so, why, and what are the alternatives that have been suggested?

The state itself has a division of labor which is in part territorial. There are central branches that have powers and responsibilities with respect to the entire territory of the state and local, in some cases regional, branches which have authority within their own, smaller jurisdictions and with respect to a different set of functions. We should all be familiar with this sort of thing. Central branches the world over manage the money supply, look after defense and regulate trade. Local governments are more likely to be involved in things like land use regulation, schools, sewage disposal, and some aspects at least of the police.

Finally, it is also the case that many of the state's interventions into social life have a territorial form. For land use regulation purposes the state divides land up into zones, each with a prescribed use. Recreational needs are taken care of through a patchwork of national, and local, parks. There are geographically defined Enterprise Zones designed as a palliative to localized

---

5 Though there are few instances – Israel is one of the few exceptions – in which there is not some element of territorial representation. This can be in the form of multi-member districts where a district is represented by several legislators. In a two-party system with one party gaining 60 percent of the vote, for a five-member district three would be from the majority party and two from the minority one. The new Scottish Parliament is elected on a variant of proportional representation. People vote for candidates in electoral districts or constituencies and the one receiving more votes than other candidates gets elected. But additional seats are awarded on the basis of the proportions polled by the different parties in the country as a whole. So if there were thirty constituencies but forty seats and party X won in fifteen of the constituencies but polled 60 percent of the vote in Scotland as a whole then its number of representatives would be 15 + 9 (which equals 60 percent of the assembly's members). On the other hand, if it only polled 50 percent in the country as a whole it would only get 15 + 5 (which equals 50 percent of the assembly's members).

unemployment. As we discussed in chapter 3 many of the European countries have designated areas of high unemployment as ones that are eligible for various forms of aid to new employers, and so on.

Having made these general points it remains that there is tremendous variety in the territorial structure of the modern state. In most countries legislators are elected to serve geographically defined populations, but not in all cases. Likewise there is considerable difference in the state's scale division of labor betweeen unitary states on the one hand, like the United Kingdom or France, and federal ones like the United States and Canada. But the United Kingdom is not France, and neither is the United States Canada. Even before the recent creation of representative bodies for Wales and Scotland, both areas had their own branches at the executive level in the form of the Welsh and Scottish Offices respectively; something for which there is no equivalent in France, despite histories of separatist sentiment in areas like Brittany and Corsica.[6] Likewise Canadian federalism has a vigorous program of revenue equalization designed to ensure that the constituent provinces, despite their geographically uneven development, can all provide similar levels of public services to their populations; something which does not exist in the United States, which has been strongly resisted but which, it has been argued, results in a creeping centralization as many States lack the resources they need to provide for their citizens (see Théret, 1999).

Understanding the state's territorial organization

Let us now turn and look at this geography, this territorial organization of the state, from the standpoint of those with various place-dependent interests, and with place-specific identities – national identities, regional identities like Poppie's "East London people" (chapter 5). In this way we can hope to grasp something of the forces behind the patchwork of states and their various territorial subdivisions.

Clearly, as discussed earlier, the state is a necessary complement to the activities of market agents; to those engaged in commodity exchange. To be active they need the state's regulatory interventions, and its various financial outlays. This point was argued primarily from capital's standpoint: issues of assembling land for the geographically extended infrastructures required if business is to access markets and sources of inputs; of maintaining effective demand; and of regulating the currency so that values remain stable. But to the extent that labor sees capital as an immovable horizon for its own activity, so the argument applies to it as well. Again, as with capital, labor wants things from the state that market transactions simply cannot provide. Business is interested in unemployment compensation because it wants to main-

---

6   This may be about to change. In 2000 the French prime minister proposed some limited devolution of legislative power to Corsica. The proposals, however, have been highly controversial, particularly among the French right, which fears for the unity of a state that, since Napoleon, has always been highly centralized.

tain a labor reserve for future business upturns; but for many of the unemployed it is a case of sheer survival. So too is it with public health. From capital's standpoint it is a good investment, if not its own, since it reduces labor turnover and work absences, as well as providing a market for new sewer systems and domestic plumbing! From the worker's standpoint, of course, it can be literally a matter of life or death.

So the realization of interests, whether on the part of capital or labor, is dependent on state mediation. But, and here is where geography enters in with a vengeance, those interests have a place-dependent character. In short, it is not just a matter of securing state action; rather it is a question of securing a state intervention that will have mitigating, supportive effects *in particular places*. This is where the state's territorial character, its territorial organization, becomes so germane. For there is obviously going to be some relation between these interests, given their place-dependent character, and the state's own territorial structure.

Consider the possibilities here. For a start, locally dependent business interests may want the state to protect them from low-cost competitors from elsewhere or provide the regulatory relief, or the money out of which to finance restructurings, necessary to counter that competition. But if those low-cost competitors are within the state's jurisdictional limits, then securing that particular end becomes that much more difficult. The same applies to labor in *its* respective markets, as the continuing furore over immigrants willing to work for lower wages affirms.

Think and Learn

Think of the various ways in which place-dependent interests play a role in debates about the state's territorial organization. Is it just a matter of excluding competition? Recalling arguments in chapter 4, what role might resident concern about property tax levels play in the creation of new local governments? And what difference would it make if those local governments lacked the power to control land use?

In consequence, we should not be surprised at some congruence between patterns of geographically uneven development on the one hand and the territorial structure of the state on the other. This is not always about protection from low-cost competitors, by any means, or at least not directly. Rather it is enough to recall the Padania case, the drive to divide Italy into two separate states, a more developed Padania and a much less developed South, that we reviewed right at the beginning of this book. In that instance an important issue was the diversion of taxpayer money to support projects in the South rather than invest in the infrastructure of the more developed North so as to secure *its* economic future and those of the various companies embedded in that region.

Separatist movements aside, similar logics apply within the state's own boundaries. Here it is again, at least in part, a question of inclusion and exclusion, of the drawing of territorial boundaries in order to secure the effective representation and realization of place-dependent interests. In the US there is a history of proposals for partitioning various of the States or for attaching segments to other States typically based on a belief that place-dependent interests will be better served by the change (see box 8.1). We should also recall in this context the discussion from chapter 7 regarding the geography of representation in US central cities. The belief of many blacks, particularly where they are in a minority, is that their interests, those of black businesses as well as residents, would be better served by an electoral organization through which councillors were elected to represent wards rather than the city as a whole, as in at large elections. In ward elections candidates contest particular subdivisions of the city, or wards, and this, in conjunction with racially segregated neighborhoods, enhances the likelihood that blacks can elect councillors more in tune with their (place-specific) interests.

But these conflicts around territorial organization are not just a matter of how space should be divided up. There is also the question of what powers and responsibilities should reside at different levels of the state's scalar divi-

---

Box 8.1   *Two Californias?*

In California there is a history of calls for partitioning the State into two, one in the north and one in the south. A ballot in 1915 to do this failed. In 1941 several counties in northern California along with some adjacent ones in southern Oregon started a movement on behalf of a new State of "Jefferson" but the Second World War intervened before it could gather steam. And again in 1992 a majority of voters in 27 of 31 northern counties called for a north–south division that would exclude from the proposed Northern California all the major urban areas, including San Francisco and Sacramento.

Partition sentiment, therefore, is currently concentrated in the northern part of the State, though it should be pointed out that the intensity of feeling waxes and wanes over time. A major issue at the present time is a feeling of minority status as the growth of population in more southern areas, including San Francisco and the major urbanized area that includes Los Angeles and San Diego, results in a redistricting of the State legislature. This perhaps would be of less concern if it were not for laws that are passed which mandate action on the part of local governments, often expensive action, but which, it is believed, reflect conditions in the urbanized areas more than those in the rural areas. This is particularly the case with costly environmental regulations. Examples: (a) a State law requiring gas stations to install pollution-reducing benzine vapor-recovery systems; these cost $3,000 per pump and have led to the closure of gas stations in areas where there is no serious air pollution problem; (b) a State law forbidding the incineration of refuse, which adds to the cost of garbage disposal, again seemingly inappropriate in areas where air pollution is very, very limited.

sion of labor. These are equally revealing of the forces that drive, transform, the state's territorial organization. Consider here two ideal-type, fairly common, problem situations that exemplify the logics involved.

(1) The first concerns the division of welfare state responsibilities between central and more local branches of the state. Typically what one finds is that welfare state provision, to the extent that it is vested in more local branches, as in the US where the States are important, marches in step with incomes: higher-income areas have more generous income support provisions, for example. But in the past thirty years it has been common for firms to relocate some of their operations out of high-wage into low-wage areas. This has led to demands from various place-dependent interests in the high-wage area that are dependent on local economies – retailers, banks, some fractions of labor perhaps – to lobby for a shift in responsibility for income supports to a more central level. At the same time, the request has been for a redefinition upwards of benefit levels to those prevailing in the high-wage areas. The logic of this is that to the extent that income supports in low-wage areas increase, then wages there will have to increase. The result of that will be a reduction in the hemorrhage of employment from high-wage areas. This has been an issue both in the United States and in the EU: in the latter instance, the degree to which the EU should legislate union-wide welfare standards.

(2) The second applies to metropolitan areas in the US. As we have observed earlier, these are typically divided into many local governments each with responsibility for land use control and for raising much of their own revenue. They therefore have stakes in embellishing their own *net* revenues through attracting in industrial and commercial land uses, which bring in money without requiring large expenditures, as opposed to housing, the residents of which will entail local government expenditures for education, parks, libraries, and so on. These local governments also, by virtue of their electoral dependence, have to be sensitive to the wishes of their residents regarding land use: another form of place-dependent interest.

There are often, however, place-dependent interests at larger geographical scales in the metropolitan area: in fact, interests dependent on the metropolitan area, its labor markets, etc., as a whole. A feature of many metropolitan areas is the close relations between different firms: one's output is the input for another, for example, and they may all depend on the same sorts of labor skills, the same local knowledges on the part of the banks which lend them money. These close relations make it difficult for them to relocate to some other metropolitan area. They are indifferent between one local government and another in their metropolitan area but not between different metropolitan areas. But in order to keep on expanding they need certain things to happen and these bring them into conflict with the local governments.

For a start, if they are to attract more workers at a wage they can afford housing costs have to be kept down. But overzoning for industrial and commercial and underzoning for housing is likely to produce the perverse result of increased demand for workers without the housing to accommodate them.

Furthermore, as the urban area expands in population so its built environment has to be reorganized: the airport expanded, new freeways inserted, space for new housing and industrial estates found. All these have the potential for resistance from residents and the various local governments that they invariably appeal to.

In this context, major industrial employers, those dependent on a local knowledge, a network of relations that can only be replicated elsewhere with great difficulty, have looked to the creation of new metropolitan planning authorities, or to a revivified county as the answer: as a means, that is, of asserting their needs over those of people in the different neighborhoods and constituent local government jurisdictions, though this often also requires some help from the State in shifting some of the responsibilities for land use planning away from local governments to this new layer of government at the level of the metropolitan area as a whole.

In short, these two ideal-type cases illustrate not just the way in which territorial organization is related to place-dependent interests but also how it becomes an object of struggle. What is an appropriate territorial organization of the state for one set of interests is not necessarily appropriate for others: what local governments and their respective residents want may not fit in with the plans and projects of those firms that comprise the area's economic base and whose competitiveness depends on policies that will upset people in particular neighborhoods, threaten their amenities and their property values.

Think and Learn

Thinking about struggles around land use in cities, what about proposals in some cities and/or suburbs to devolve control of land use rezoning decisions to neighborhood boards? Who might support such an initiative and why? And who might oppose it and why? In what ways would such a move be analogous in its effects to the present devolution of power over zoning to local governments?

This is not to say that we can make sense of these struggles purely in terms of place-dependent interests, in terms of local dependence. By no means are these the only forces in play in contesting the territorial organization of the state. As we saw in the previous three chapters, identities often have a place-dependent character. National identity is a case in point. The nation always has a geographic expression: it refers to a country, the "motherland" or "fatherland." As such it plays a very active part in many separatist movements like those of Quebec and Slovakia, though not in all, since a sense of common identity was very weak in the Padania case. And clearly the relation between national identity and a particular place is more complex in the case of Afrikaner nationalism for reasons that we discussed in chapter 6.

In yet other instances space is a weapon in the tool box of social movements as they press their cases in struggles over recognition. The segregationist arm of the black civil rights movement has, as we saw in chapter 7, a history of pressing for ward-based elections and, in some cases, political separation and the establishment of largely black local governments. Nationalist movements in the European colonies likewise pressed for spaces of their own, and the expulsion of the settlers. Yet in talking about interests and identities we should be careful to avoid a simple opposition between them. Identities are closely related to struggles in the economic sphere. This was plain in our discussions of Quebecois and Afrikaner nationalism respectively as well as in black struggles for recognition.

The upshot of these logics is that every state, every branch of the state, central or local, has its own cluster of supporting, place-dependent, interests, typically a mix of business, government officials and at least some popular fractions that are workplace- or living place-based, which then promulgate a particular conception of place-based identity in order to mobilize a wider support base. But the fact of place or local dependence returns us to a theme that we focused on in part I of the book. There it was the conflict between local dependence or place dependence on the one hand and the mobility of a fraction of capital that attracted our attention. We used that as our fulcrum for an understanding of the politics of local economic development as well as neighborhood change, antagonisms between central cities and suburbs. This is a theme argued out by the geographer David Harvey (1985) in terms of what he has called the geopolitics of capitalism. It is to that which we now turn.

The Geopolitics of Capitalism

The mobilization of branches of the state, central, local, subnational, by place-dependent interests is typically with respect to a wider flow of value. Capitalist economic geography, we should recall from chapter 3, is subject to a high degree of flux over time. For sure, there is some degree of constancy from one year to the next. Broad patterns tend, by and large, to get reproduced so the major industrial regions this year will still be major ones five years hence. But there are also changes. Plant closures impact particular places and growth is faster in others. New industrial regions, even countries, as in the Far Eastern NICs, emerge to challenge the economic bases of particular urban areas, regions, or even, again, countries. Shifting flows of investment and declining foreign markets can therefore mean that value no longer flows as it once did through the firms of a particular area. They have trouble paying their bills, they start releasing workers, their workers have trouble paying their mortgages, firms dependent on the market provided by this now declining economic base find that they too are facing mounting business difficulties. As we saw in our discussion of growth coalitions in chapter 3, it is in this context that the firms affected, to the degree that they are place-dependent, will organize in an attempt to recapitalize the local economy, replace the declining sectors with newer, more promising ones. In this they may seek to build alliances with workers, and workers, with stakes in the values of their housing, and limited in their knowledge of job opportunities by local social networks, may see this as providing their only glimmer of hope. Cross-class alliances, therefore, form, though clearly always subject to pressures of a dis-integrating kind. This is because there is a continual struggle with business over who should bear the costs of the necessary restructuring. So ideologies of community become important as the cement through which the alliance can be held together: defending, therefore, not just material circumstance but also the conditions for the realization of particular, place-specific, identities – a French-speaking culture in the case of the Quebecois, or, in more local instances, status as "a major league city."

Accordingly it is in the context of defending a particular "home" or "geo-graphic base" that state agencies are mobilized, expand (or contract) in terms of what they do, emerge via partition of some pre-existing state, or find their scale division of labor, the territorial character of their systems of representa-tion, being reworked.[7] But this matter of defending a "home base" has a more concrete expression. Rather the contradiction between fixity and flux, immo-bility and mobility, between place-dependent interests and those that are relatively place-*in*dependent, at whatever geographical scale, is expressed in a struggle to defend positions in wider geographic divisions of labor or consumption.

Think and Learn

Recalling chapter 3, how does this claim jibe with the arguments there? To what extent was economic development policy focused on moving up a hierarchy of positions in the geographic division of labor?

Agricultural countries want to be industrial countries. Countries produc-ing low-cost, low-skill, labor-intensive products want to "graduate" to the pro-duction of goods which have a higher "value added" content: products, or services for that matter, which are more skill- or information-intensive. This is the story of the Far Eastern NICs, and let us not forget, of an earlier vintage of "newly industrializing countries": the McKinley tariff in the United States and Germany's industrial tariffs of the late nineteenth century – tariffs in the latter case which their apologist Friedrich List argued were essential for the protection of "infant industries."

And so it is repeated at other geographic scales. Cities want corporate head-quarters, R&D, high-paid service industries in place of branch plants. Low-tech cities want hi-tech. Still others, with more limited prospects of so graduating, struggle to defend what they have: to preserve their sunset indus-

---

7   For example, and with respect to systems of representation, wards versus at large in the case of city elections, or first-past-the-post versus proportional representation in national elections.

tries in the face of Chinese competition, for example. In metropolitan areas, on the other hand, attention tends to shift from production to consumption: defending or achieving a more favored position in the geographic division of consumption, whether it be a matter of a central city pushing for increased gentrification or some decaying rural service center on the urban periphery which dreams of becoming the nucleus for an upmarket suburb.

It is altogether reasonable, therefore, to anticipate attempts to restructure economic bases through a restructuring of the conditions of production. This has been particularly clear in the case of the Far Eastern NICs. There, as we described in chapter 3, states have used their powers over the allocation of credit to purposely shift production, as industrial experience accrued, from lower-skill, lower-value-added lines to higher-skill, more profitable ones. But even in countries like the United States which, historically, have been unwilling to exercise this sort of top-down power, local and State governments have been active in retraining initiatives, providing supports to growth industries, attempting to slow down the decline of those in the sunset stage, and altering the employment relation – labor law, workers' compensation – so as to appeal to new investors.

As a productive force, space has also received state attention. In downtowns local governments, often with central government support, have attempted to create through urban renewal policies patterns of land use there that can facilitate a transformation from a former retail function to one that is more focused on office employment and its support services. Likewise, as observed above, metropolitan planning has been seen as a means of avoiding the land use bottlenecks that can impede the ability of a city's economic base to expand and so defend its position in wider geographic divisions of labor.

But state agencies have also been pressured to defend and enhance those positions through intervention into the world outside. In the late nineteenth century and at the national level states sought to underpin access to raw materials and exclusive control of markets through policies of imperialism. This is a theme we will take up in more detail in the chapter to follow. The challenge of defending and achieving more advantageous positions in a division of labor that is, to some degree at least, international has also led to attempts to reconstruct the state. It is believed, for example, that new supranational forms of state organization like the EU can confer competitive advantages that are denied the nation state, unless, that is, it has its own massive internal market, as is the case with the US.[8] And certainly, to take an example that is very *au courant*, it seems unlikely that the pre-eminence of Boeing in international airplane markets could have been challenged by Airbus Industrie without the ability of the latter to bring together, unimpeded by trade restraints, different suppliers in different member countries of the EU; and also for the EU to provide the heft in international trade forums against American claims that Airbus has been the recipient of "unfair" subsidies.

---

8   Though the formation of the North American Free Trade Area would suggest that the US is concerned about the potential competitive advantages that the common market provided by the EU could confer on European firms relative to American ones.

This reaching out into wider political and economic spheres, this development of "foreign policies" to defend particular, geographically defined, economic bases, also extends to the local and other subnational scales. Local growth coalitions look to national governments for regulatory relief or for the subsidies that will allow them to restructure or attract in, say, the assembly plant of a Japanese transnational. In many instances pressure is exerted through forming coalitions with other localities that would similarly benefit from the central government policies in question. Often these are localities with similar positions in the division of labor to defend and they are not always purely business-led coalitions either (see box 8.2). What happens at one scale therefore is not divorced from what is happening at others. Central governments may adopt policies because of bottom-up pressure, perhaps exerted by some cross-locality coalition, though not necessarily. And to be sure, what is vital to a local economic base may also be regarded as nationally important, and local pressure groups may package their demands and policy proposals to take advantage of precisely that.

Divisions of labor, conceived as geographic, link up different places. But the pressure to maintain a favored position or achieve an improved position, to define new niches in that division, imposes a geographic dynamic on the process. This is one of reaching out to new markets, new, cheaper sources of raw materials at greater distances, and in some instances, labor. There is, in other words, an impulsion to geographic expansion of a firm's markets, the areas from which it obtains raw materials, perhaps even from where it obtains its loans. This does not mean to say that the more distant always has the advantage over the closer. The oil consumed now in Britain is much more likely to come from the North Sea rather than from the older source region in the Middle East. But judged against the historical geography of capitalism, the tendency has been inexorable and has been one of the forces behind the instability of capital's space economy. And the globalization of production that is given so much attention in the media is, of course, a particular expression of this geographically expansionary dynamic.

The struggle for improved, advantageous positions in geographic divisions of labor generates tensions between places. This is because its outcome affects the business prospects of firms there, unemployment levels, wage levels, the stability of the state's revenue streams. The ability of one place to maintain an exalted position depends on other places occupying subordinate positions. So it is inevitable that the various cross-class alliances of the place-specific that tend to lose out, backed up by respective state agents, will challenge the subsequent inequalities. And equally inevitable is the fact that those which benefit from it will not be bound by any supposedly neutral rules of the market but will do all in their power, including resort to monopoly[9] and even to force, to impose a solution to their continuing advantage.

---

9   One example: the history of the jet airplane industry might have been different. The first intercontinental jet plane was the product of a British firm; this was the Comet. The US government, however, citing safety reasons, but knowing full well that Boeing was shortly to unveil its, ultimately highly successful, 707, withheld landing rights. There *were* safety issues, but clearly more was at stake than that.

Box 8.2   *Inter-locality Coalitions*

The idea of inter-locality coalitions can be exemplified by two cases. The first is one that brought various local governments in Britain together around questions raised by their common situation of hosting auto assembly plants. This was MILAN or the Motor Industry Local Authority Network. In its own words: "Until recently the voice of communities dependent on the motor industry has been muted when compared with those of the manufacturers, trade unions and government. The basic aim of MILAN is therefore to work together on behalf of those communities which depend on the motor industry, and to support the further development of motor and components manufacturing in the United Kingdom."* Towards this end MILAN saw itself, among other things, working with its member local authorities to develop automobile industry training initiatives, to develop auto-related research in local institutions of higher education, and to assist the automobile companies in local development issues. But in addition it also saw itself as mediating between member authorities on the one hand and the national government and the EU on the other ("MILAN will seek to become an effective lobbying organization at the national and international levels to make the strongest possible case for the UK motor industry and the communities which depend on it"*).

The second example† is that of EUR-ACOM (loosely translated, "European action for mining communities"). This is an association of local and regional authorities in the European Union's coalfields and comprises seven national associations of present and former coalmining areas. Coal is a rapidly diminishing presence in the West European division of labor, along with the communities dependent on it, and there are many highly distressed localities formerly reliant on the coal industry. Accordingly EUR-ACOM has had two goals: help secure a future for what remains of the coal mining industry in Western Europe; and promote the economic, social, and environmental renewal of coal areas. Its major successes have come with respect to the second objective. In particular it has been able to influence EU regional policy, their lobbying leading to the introduction of a new program, RECHAR, specifically for economic regeneration schemes in the coalfields.

*Motor Industry Local Authority Network (1987) *Local Government and the Chal-
  lenge of the Motor Industry: Report Summary*. Birmingham: Institute of Local Government Studies, Birmingham University.
†S. Fothergill (1994) "The Impact of Regional Alliances: The Case of the EU Coalfields." *European Urban and Regional Planning Studies*, 1(1), 177–80.

There are, in short, tensions surrounding that geographically uneven development that is the outcome of the struggle for positions in the geographic division of labor and we are going to address these at length in the next chapter. But in addition, and as we might surmise from what we know of the politics of globalization, tension also surrounds the tendency to geographic expansion. It is a case of the more local being threatened by the more

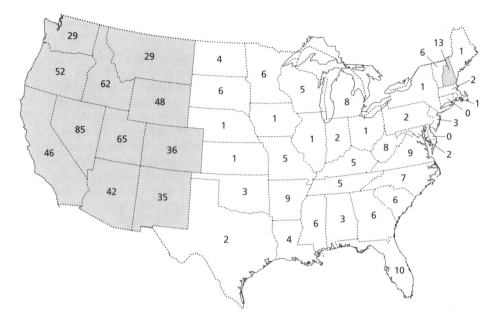

Figure 8.1 Federally owned land in the US, 1992. Numbers refer to percentage of land federally owned.

Source: US Statistical Abstract. Washington, DC: US Government Printing Office, 1993, table 365, p. 229.

global: a tension resulting from the expansion of the geographic scale on which the division of labor is being progressively reorganized. This will be the focus of the final chapter of the book. But for the time being we should be careful not to separate in too radical a fashion these two types of tension. It is, rather, the struggle for positions in the geographic division of labor which drives the expansion of the geographic scale on which that struggle occurs. As a result the tension between the local and the global is often misinterpreted, as indeed is the case in contemporary arguments about the politics of globalization.

Case Studies

The Sagebrush Rebellion

In the Western States of the US the federal government owns very large proportions of all the land: in the States of Nevada, Utah, and Idaho over 60 percent of the land is so owned (see figure 8.1). The use of this land is subject to federal regulation yet at the same time the use, and hence that regulation, is important to local and regional business interests, particularly ranchers who lease land and mining companies developing mineral bodies. In consequence federal regulation has periodically become an issue in the West. In the 1970s it took the form of the so-called Sagebrush Rebellion: a variety of State initiatives, with substantial popular backing, to have the federal government relinquish control of this land to the States and so reassert States' rights in land use regulation.

Much of the concern at that time seems to have come from a surge of environmentalist sentiment on the part of the federal government as to how federal lands should be managed. There were concerns for land conservation and for the preservation of wildlife which ran counter to the interests of ranchers; and limits on mineral exploration and the imposition of land rehabilitation standards which were contrary to the interests of mining companies. Much of this was at the prompting of various environmental lobbies.

The latter were also strongly opposed to federal government relinquishment of its land in the West. This is because State law requires that State-owned land (which is what the Sagebrush Rebellion wanted federally owned land to become) be managed on a revenue-maximizing basis; federal law, on the other hand, requires that *its* land be managed on a multiple-use basis, according to which all uses, including non-profitable ones such as some recreational uses, must be weighted on an equal basis. The Sagebrush Rebellion was, therefore, seen as a clear threat to environmental interests. However, the major reason for its collapse appears to have been the relaxation of environmental standards in regulation of land use on federal lands by the Reagan Administration.

Fifteen or so years later, with a Democratic administration in office, as it was in the seventies, there was a renewed outbreak of revolt in the West. A major issue was the attempt of the federal government to charge ranchers higher rents for the use of public land. These are commonly regarded as very low and even with the increases would have remained below those charged for private land.[10] Even so, this – albeit modest – increase was opposed strenuously on the grounds that it would force many ranchers out of business.

Environmental issues also continue to be a concern. The damage to public lands by grazing cattle has been considerable. They chew off the natural bunchgrass, for example, and this is replaced by fire-hazardous cheat grass. They also erode stream banks where they congregate. This in turn has led to attempts on the part of the Bureau of Land Management and the Forestry Service (who administer grazing on national forest lands) to limit the number of cattle allowed on federal land.

The major polarizations have been much as they were before: ranchers versus the federal government and its environmentalist supporters. In addition a number of other local economic interests in the Western States have been involved. This is because a threat to the economic viability of ranching in the areas is also a threat to them: local suppliers and local banks with loans extended to ranchers, for example. The revolt has also resonated more widely still, in large part because it has coincided with a marked and more general

10 It should be pointed out, though, that unlike on private lands those renting public lands have to put up fences and make their own improvements; in addition public land tends to be less valuable as grazing land.

rise in the political temperature around the issues of State regulation, citizen rights, the right-wing militia movement, and the bombing of the federal building in Oklahoma City in 1995.

Apart from agitation at the State level, as in the past, a distinctive feature of the current Sagebrush Rebellion is what has come to be known as the County Supremacy Movement. The distinctive mark of this is the passage of ordinances by counties repudiating federal control of public lands. Between 1992 and 1994 there were over one hundred counties in the Western States passing such ordinances. In Catron County, New Mexico, for instance, local ordinances were passed making it illegal for the Forestry Service to regulate grazing. There was also a proposal that would require environmentalists to register with the county and procure a license or risk arrest (*Wall Street Journal*, January 3, 1995, p. A1). Apart from these formal measures threats of violence against Bureau of Land Management and Forestry Service personnel are far from uncommon. The overall mood has become thoroughly unpleasant as the locals, ranchers and those dependent on them, dig in their heels.[11]

The overriding character of the Sagebrush Rebellion, however, is the way it has brought together sets of locally dependent interests in different places – ranchers, those with stakes in the local economies for which the ranchers constitute the economic base – around a common agenda. The fact that it is the federal government from which redress is sought has dictated the national (or more accurately regional) nature of the alliance.[12]

The focus of the County Supremacy Movement on property rights, however, is significant because it has allowed further coalitions to be forged with more nationally, as opposed to regionally, based movements. Two in particular are the property rights movement asserting the rights of private property owners against so-called State and federal takings and the anti-gun control lobby. Interestingly in the Catron County instance referred to above the county has passed a measure requiring heads of households – presumably males! – to own firearms to "protect citizens' rights" (*Wall Street Journal*, January 3, 1995, p. A1).

This case study illuminates a number of the themes developed in this chapter. It is, in the first place, a good instance of struggles over the scale division of labor of the state and how those struggles are typically generated by interests of a place-dependent nature: those of the ranchers and the commu-

11 The major arguments of the County Supremacy Movement are nothing if not radical. Rather than simply opposing federal policy it is federal ownership of the land itself which is called into question. Arguments vary. One is that since the US constitution says nothing about the federal government owning land it can't. Another is based on the so-called equal footing doctrine. According to this the federal government's acquisition of land in Western States as part of the bargain for the States being admitted into the union was illegal since the original thirteen States did not have to turn over any of their land as a condition of Statehood. In other words, the federal government stole the land. Neither of these arguments is given much credence by legal scholars.

12 We should also note the irony at the heart of this revolt. The federal government has become the object of loathing as a threat to the rights of individual citizens but the ranchers have been taking federal subsidies for years in the form of below market rents.

nities which depend on their profitability are clearly to the fore in this instance. And the ultimate stake is who should have control over the vast swathes of federally owned land in the West: the federal government or the States themselves.

Less evident from this case study is the way in which the battle is also being joined at the level of identity: in particular, different visions of "the West." The vision dear to the environmental movement is of the West as wilderness. This plays on values important to the American sense of national identity since the wilderness experience and the exploration by whites of the "wild, untamed West" have been elaborated into such an important theme in American history. This has obviously been repackaged into a sense of the West as a consumption experience – something to gaze at and admire – and has been shorn of the notion of the *conquest* of the wilderness as another, related theme in that same American history. This, however, is the theme emphasized by the protagonists of the Sagebrush Rebellion: a different sort of wilderness experience in the form of the rugged, individualist American still imbued with the values of frontier law (or so it would seem).

Internal restructuring

This is a topic that we have already broached. In chapter 3 reference was made in particular to the attempts of governments in Western Europe to fashion spatial arrangements for productive purposes; we talked, for example, about new town policies and those designed to help industries relocate from areas where diseconomies of agglomeration seemed to be in evidence. In this case study I want to take these arguments further.

Spatial arrangement is a productive force. It can facilitate productivity and hence the ability to compete in wider markets. It has accordingly been a focus of the state and of various territorial coalitions anxious to defend investments in particular places from devaluation and to use them to enhance their own prospects for accumulation. These processes, both the input of territorial coalitions and the actions of state agencies, can be observed at a variety of geographical scales.

In cities these arguments can be illustrated to some degree with the case of urban revitalization. In the United States, from the 1930s onwards, central business districts found their prospects faltering in the face of an accelerating suburbanization. Retailing interests saw their market being siphoned off in favor of new retail establishments further out, something that was to reach its climax in the post-war period. There was also some tendency towards the suburbanization of employment which gave further impetus to the suburbanization of the retail market. In many cities this led to the emergence of coalitions of forces in search of remedial action, though it was unclear whether this should be aimed at preserving a position in the metropolitan division of labor/consumption or searching for a new niche. Typically these coalitions consisted of downtown department stores, owners of downtown real estate and city governments anxious about their tax base (Weiss, 1980). The remedies they sought crystallized after the Second World War around two particular courses of action: (a) urban renewal; (b) radially focused freeways.

Urban renewal was legislated into being by the 1949 Housing Act. It gave cities the right to acquire through eminent domain land on which structures were designated as "deteriorated," to demolish the structures, insert new highways and utility lines and sell the land to real estate developers. Two-thirds of the difference between the price the city paid for the land and the lower price at which it was able to sell it to developers was paid for by the federal government: potentially, therefore, a very significant subsidy to private developers. Moreover, it created perverse incentives in that it reduced the pressure for local government to drive a hard bargain, in either purchasing the property to be demolished or selling the land so cleared.

Most of the structures that cities sought to clear were in the downtown area and the goal was a spatial restructuring so as to enhance new investment and channel value once more through downtown real estate. Radially focused freeways, designed to improve the downtown's access to suburbanizing populations, had to wait a little longer. But the 1956 legislation bringing into being a program of constructing Interstate, limited access, highways provided the opportunity. Local governments sought input on network design and one of the consequences was a repetitive pattern of directing Interstates through downtowns.[13]

To a significant degree these two programs achieved their goals, though it is easy to see how they might just as well have failed. For urban renewal and related highway construction programs created the conditions for the downtown location of major offices, hotels, and, later, new enclosed shopping malls designed to bring back the shopper: some shift, in other words, in the roles that the downtown served in wider geographic divisions of labor. In the postwar period the demand for office space in downtown locations soared. This was driven in part by the hiving off of the headquarter functions of firms from their production functions so that they *could* be located downtown. It was also affected by the desire to take advantage of co-location with other firms where face-to-face communication was important: proximity to major banks, to insurance companies, to lawyers and accountants and marketing firms, to the offices of Congresspersons, to municipal and county government. At the same time the construction of the freeways placed the downtown at an advantageous position *vis-à-vis* the metropolitan professional and managerial labor market.

This is an instance in which the initiative was clearly bottom-up: it was a matter of urban-based, more specifically downtown, territorial coalitions lobbying central government for assistance. But there are cases in which the initiative is in the other direction: where central government intervenes in the

---

13 Interestingly this was *not* the policy in the case of the analogous motorway construction program in Britain. Also interesting is the fact that the interventions of local interests in network design placed at serious risk the justification given for the Interstate program when it was legislated into being: to facilitate the evacuation of urban populations in case of nuclear attack. The sense of evacuating through other urban areas, however, strains credulity.

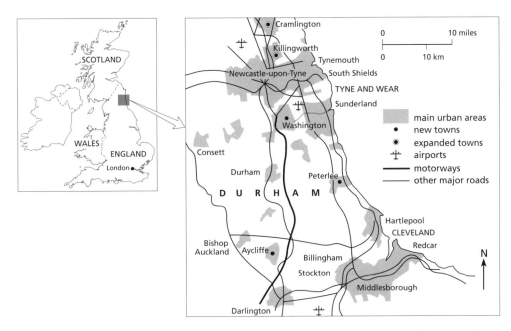

Figure 8.2 Washington new town in regional context. The northeast of England has been a relatively depressed area for some time. One of the strategies for making the area more attractive to major corporations looking for sites has been to concentrate population in new towns, three of which are indicated on this map. There has also been investment in other items of physical infrastructure like the motorway that runs through the heart of the area connecting with other four lane, limited access highways. Note also the category "expanded towns": these were town expansion schemes which, like new towns, were intended to take populations relocated from elsewhere displaced by slum clearance programs, and motivated by the desire to geographically concentrate populations and to relieve pressure on housing in major cities.

Source: After R. Hudson (1982) "Accumulation, Spatial Policies, and the Production of Regional Labor Reserves: A Study of Washington New Town." *Environment and Planning A,* 14, figure 1, p. 666.

restructuring of particular urban, or indeed regional, spaces in order to achieve some goal defined as in the national interest. As we remarked earlier one of the aims of British new town policy was to create labor market conditions appropriate to the needs of inward investors. Through the concentration of new housing development and the provision of updated highway access firms would be able to rely on a steady supply of workers. One such case comes from County Durham in the United Kingdom and the new town of Washington, though similar conclusions apply to other new towns (see figure 8.2). As Hudson (1982, p. 688) has written: "the pattern of expenditure by Washington Development Corporation can be seen as part of a deliberate policy to assemble work forces – both by providing housing within the town and (in collaboration with other parts of the State apparatus) by improving road links between the town and its surrounding subregion." As Hudson goes on to show, the state concentrated new public housing in Washington relative to the surrounding area, one that included substantial urban areas like Gateshead, Chester-le-Street, and Sunderland (see figure 8.2). So while at the start of development in 1964 just 1 percent of all public housing completions were in Washington new town, some seven years later this amounted to almost half of all completions. This bias in state housing location policy meant that in order to obtain public housing[14] people from the surrounding area had to move to Washington. Indeed of the various reasons people gave for moving to Washington new town, housing applied to just less than a half as compared with fewer than 30 percent who gave employment as the reason (Hudson, 1982, p. 673).

This should also be seen in the light of policies of settlement planning in the surrounding county of Durham. This is an area largely of small mining towns and villages, many of which in the 1960s were losing their colliery employment. This of course was one of the reasons the government was anxious to promote new employment in the area. But population concentration was seen as part of the answer to attracting that employment. Accordingly, and in addition to the creation of new towns at Washington and Peterlee, the council identified a set of villages and small towns which should be allowed to decline. No further housing would be allowed there, either public or private, and investments in supporting physical infrastructure should be renewed sparingly.[15] This, however, proved thoroughly controversial among the people directly affected (figure 8.3) (Barr, 1969).

Several further comments on these materials are in order. Note first in the case of urban renewal in the US the formation of an inter-locality coalition. Downtown business associations from cities across the country came together in order to lobby the federal government for the funds necessary to realizing their vision. This recalls other instances of cross-locality coalitions that we have discussed in this chapter.

The second point to note is the role that serendipity plays in these instances. It is clearly a mistake to impute to restructuring projects, spatial restructuring in these instances, a clear vision of the position a place will ultimately occupy in the geographic division of labor. When urban renewal legislation was introduced there was little or no sense of the opportunities that would eventually appear in the form of corporate headquarter location, convention centers, and hotels. Indeed, the original impetus behind urban renewal was to revive downtown retailing. Likewise in the case of Washington new town. Washington new town is now synonymous in Britain with the location of a major Japanese auto plant: Nissan. But the concentration of population

14   It bears noting that at that time public housing was a very important source of all housing: just less than a third in Britain as a whole and in the region under discussion almost certainly considerably higher.
15   To quote from the relevant planning document: "any expenditure on houses, facilities and services to those communities which would involve public money should be limited to conform to what appears to be the possible life of existing property in the community" (Blowers, 1972, p. 147).

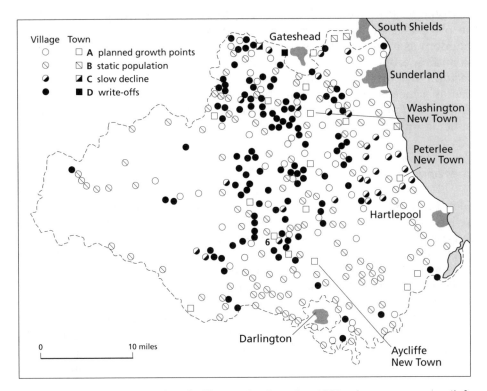

Figure 8.3 Durham's murdered villages. During the 1960s the county council for Durham had an ambitious policy of settlement consolidation. Some villages were to be allowed to die. No expansion would be allowed in them. In other cases, however, growth was to be encouraged in so-called "planned growth points." These included the new towns of Aycliffe, Peterlee, and Washington.

Source: After J. Barr (1969) "Durham's Murdered Villages." *New Society*, April, pp. 523–5.

there was long in the making and the policy did not foresee the particular role that Washington would come to occupy in wider geographic divisions of labor.

Finally there is an interesting contrast between the urban renewal and the Washington new town cases. In the urban renewal one the initiative is clearly bottom-up: the drive for legislation comes from the cities, or rather the downtown businesspeople, who see themselves benefiting. In the Washington new town case the initiative is top-down from the British government in Whitehall. So too is the action of Durham County Council to limit the growth of a large number of villages in the county. This has to do with the nature of the place-dependent interests in play in the two instances. In the urban renewal one there was a cluster of downtown property owners whose rents were dependent on the revitalization of the downtown; they were joined by local government concerned about its tax revenues – another case of place dependence. In the Washington new town cases the conditions of place dependence are different. Primary is the place dependence of the British government and the place it is dependent on is not Washington new town but the United Kingdom as a whole. This is where the government must establish a viable, expanding economy if it is to enjoy growing tax revenues as well as healthy electoral support. Washington new town, like other new towns in Britain, was seen as a mechanism for taking the pressure off big city housing and labor markets and so smoothing out the course of national economic growth by reducing inflationary pressures.

Cities and countries

An important concept in our discussion in this chapter has been that of scale. We have made reference to the state's scale division of labor, for example: the division of responsibilities between central and local government. We have drawn attention to the fact that the various coalitions of private and public interests that come together as a result of their interests in the future of particular places occur at a variety of geographic scales: the places in which they have their interests may be more local, a city perhaps, a region, or a nation. The phenomenon, in other words, is scale-independent. Likewise the geographic division of labor itself has a scale: it can be organized more locally, nationally, or internationally, or businesses in a particular place can function at all three scales, selling to their own workers, to the domestic market, and to the international market.

This latter point means that the different scales are not mutually exclusive. And indeed countries can be as interested in the future of particular cities as are interests located in those cities themselves. Alternatively, cities may compete for a country's economic policy, marshalling arguments as to their importance to the economy of the country as a whole and to its position in the international economy. Historically, for example, cities have sometimes been to the fore in pushing for policies of an imperialist character at the national level. This has obviously been to the degree to which the interests of local government, businesses, even workers there could be made to coincide with the realization of some national objective.

A case in point is the history in Britain of what has come to be known as "social imperialism."[16] This was a set of policy agendas, of some diversity, which emerged around about 1900 in the context of increasing competitive pressures on British industry, particularly from Germany and the US, the rise of organized labor, and a questioning of free trade as the national policy. The dominant form of social imperialism was linked to Joseph Chamberlain, who was closely associated with the city of Birmingham. He had been the mayor of the city as well as a local MP. Among other things he was noted for what came to be known as municipal socialism.[17]

---

16   In what follows I have drawn heavily on Semmel (1968) and Foster (1976).
17   "Municipal socialist" in the sense that it involved taking into municipal ownership private utilities like gas and water in order to generate revenues (in the case of gas) that could be used

Chamberlain's version of social imperialism was closely tied up with the changing fortunes of the metal industries dominant in the Birmingham area. The centerpiece of his proposals and those who supported him was twofold: (a) a reversion from Britain's long-term commitment to free trade in favor of tariff protection; (b) imperial preference according to which markets in the colonies would be reserved for British manufacturers. In these ways he hoped to secure markets for the metal trades of the Birmingham district and bring business and labor together around a policy from which both would gain: a cross-class alliance, in other words. He also believed that tariff protection would make Britain attractive to foreign investment as overseas manufacturers sought to retain access to British and imperial markets. This would further serve to cement an alliance between British industrial employers and their workers.

Chamberlain's proposals did not go unchallenged. There was another faction of British capital, located largely in London, that liked things as they were. Britain's early industrial supremacy, based on cities like Birmingham, had made London a major financial center. The country's huge ability to export relative to its demand for imports made the pound sterling as desirable then as the dollar is today. It was an attractive currency to hold and as a result the London banks attracted deposits from around the world, and owing to the desirability of sterling deposits, at low rates of interest. This money could then be lent out to finance development projects in other countries, and to finance the huge trade that made its way through British ports.

From the middle to later years of the nineteenth century this stream of money was augmented by the savings of smaller British businesses, the least competitive firms, which were being squeezed by foreign competition and so seeking alternative outlets for their capital. In effect this money was used to finance the economic development of the rest of the world: railroads in the US and Argentina, gold mines in South Africa, guano fields in Peru, and, just after the turn of the century, oil in the Middle East. The status of Britain as a major trading country – up to 40 percent of the world's trade passed through Britain in the mid-nineteenth century, much of it coming from Europe for shipment elsewhere – also created business for the city in the form of a demand for bills of credit. Trade had to be insured as well, which was the origins of Lloyds of London.

To some degree this financial activity complemented Britain's industry. Constructing railroads overseas created markets for British steel (though by the end of the nineteenth century this was being supplanted by German steel). Ocean-going trade stimulated British shipbuilding. But clearly there were also tensions. In effect the City was financing the development of competitors for British industry. The tariff protection and imperial preference demanded by Chamberlain and his supporters, acting on behalf of the metal goods industry of the Birmingham area, was seen as limiting the trade that the City could

to upgrade the sewage system, and improve operations (in the case of water), all with a view to improving the health of the people and particularly that of the most vulnerable, the poor. See Briggs (1963, chapter 5).

finance. Likewise, it was believed that imperial preference, by channeling trade in favor of Britain, would limit the trade of other countries in which the City had an interest. The issue was: would wider flows of value be channeled to the benefit of industrial interests in cities like Birmingham, or to the advantage of the City based in London?

In short there are connections between the imperial ambitions of cities, or more accurately the interests located there, and the imperial policies of nation states. Chamberlain wanted a different form of imperialism; a neo-mercantilist one that would protect British, and particularly Birmingham, manufacturing. The City wanted a *laissez-faire* imperialism of economic penetration and the establishment of informal political controls through the British Foreign Office[18] through which to safeguard its investments. The tension between the City and British manufacturing, moreover, is one that is ongoing, as we will learn in chapter 10; and this is despite the liquidation of formal empire.

In addition to the major point of this case study, set out in its final paragraph, it serves to illustrate other points. The first is the way in which the coalition that Chamberlain sought to forge in order to save the metal working industries of the Birmingham area was a cross-class coalition. This is a common source of the appeal of protectionist policies.

The other point is how the study dramatizes the significance of global connections for more local conflicts. It was, after all, the rise of Germany and the US which was so threatening to the metal industries. On the other hand, the growth of industry elsewhere in the world and the trade it was generating was a major boost for the banks and the insurance businesses in London. In short globalization had both a downside and an upside for one particular national economy. This is worth noting, not least because today the politics of globalization is often treated as a new stage in the economy history of capitalism. Clearly that assumption needs to be questioned. What was being fought out in Britain a century ago was part of the politics of globalization.

Summary

Capitalist development is a tension-ridden process. The tensions occur between the different firms as they compete one with another, and between the two major classes as they each struggle to increase their respective shares of the total product. These tensions, moreover, mean that there are certain preconditions for capitalist development which the market agents through which it works cannot provide. These include the general conditions of production, the reproduction of labor power, and the provision of some degree of cohesion between antagonistic forces. It is into this void that the state steps.

At the same time, firms and workers, while they have respective interests in profits, rents, and wages, find that their interests are also in particular places. It is a profit or a wage in a particular place that is at issue as a result

---

18   Parodied by the "send in a gunboat" ethos.

of diverse forms of place dependence. Place-specific identities also enter into play since these may provide the means through which cross-class alliances can be cemented. As an organization that can respond to these needs and constraints, the state therefore has to have a territorial expression. Agents want states that will respond to their place-specific needs and that means that there must be many states, each with a degree of territorial coherence; and in turn those many states must have their own internal territorial breakdown into subnational jurisdictions.

State agencies, therefore, central governments, local governments, subnational governments, emerge as the vehicles through which firms in particular places pursue their goals of accumulation. But these goals always assume a concrete form, i.e. enhanced positions in a geographic division of labor. To some degree this may proceed through alliances with workers, though tensions always threaten to break out into open conflict (see chapter 2).

Now, however, the tensions can assume geographic forms, and two of these have been identified. The first originates in the geographical unevenness of development and the attempts of some to contest the rules and conventions on the basis of which the struggle is being played out. We will take up this theme in the next chapter. The second tension is exemplified by the dilemmas of the Birmingham metal industries at the beginning of the twentieth century. This is the tension between the local and the global, which we will examine at length in the final chapter. But note here already how the story of Chamberlain's social imperialism demonstrates how difficult it will be to separate the politics of globalization from that of uneven development; in that instance the struggle between Birmingham, on a downward trajectory, and London, on an upward one.

REFERENCES

Barr, J. (1969) "Durham's Murdered Villages." *New Society*, April 3, 523–5.

Blowers, A. (1972) "The Declining Villages of County Durham." In *Social Geography*. Milton Keynes: Open University Press.

Briggs, A. (1963) *Victorian Cities*. New York: Harper and Row.

Foster, J. (1976) "British Imperialism and the Labour Aristocracy." In J. Skelley (ed.), *The General Strike 1926*. London: Lawrence and Wishart, chapter 1.

Harvey, D. (1989) "The Geopolitics of Capitalism." In D. Gregory and J. Urry (eds), *Social Relations and Spatial Structures*. London: Macmillan, chapter 7.

Hudson, R. (1982) "Accumulation, Spatial Policies, and the Production of Regional Labour Reserves: A Study of Washington New Town." *Environment and Planning A*, 14, 665–80.

Semmel, B. (1968) *Imperialism and Social Reform: English Social Imperialist Thought 1895–1914*. Garden City, NJ: Doubleday.

Théret, B. (1999) "Regionalism and Federalism: A Comparative Analysis of the Regulation of Economic Tensions between Regions by Canadian and American Federal Intergovernmental Transfer Programmes." *International Journal of Urban and Regional Research*, 23(3), 479–512.

Weiss, M. (1980) "The Origins and Legacy of Urban Renewal." In P. Clavel, J. Forester and W. W. Goldsmith (eds), *Urban and Regional Planning in an Age of Austerity*. New York: Pergamon Press, chapter 4.

Williams, R. (1976) *Keywords*. London: Fontana.

FURTHER READING

Despite its mainstream character an article that students find useful in understanding the issues raised by the state in capitalist societies is one by Otto Davis and Andrew Whinston (1965). Its focus is urban renewal but its applicability is much broader than that: "Economic Problems in Urban Renewal," pp. 140–53 in E. S. Phelps (ed.), *Private Wants and Public Needs* (New York: W. W. Norton). More advanced but introducing the question of territory is the paper by Michael Mann (1984): "The Autonomous Power of the State: Its Origins, Mechanisms and Results." *Archives Européennes de Sociologie*, 25, 185–213.

The paper on the geopolitics of capitalism is one with the same title by David Harvey (1985), chapter 7 in D. Gregory and J. Urry (eds), *Social Relations and Spatial Structures* (London: Macmillan).

On more specific issues: for an excellent study of the politics of shifting to new positions in the geographic division of labor to replace ones that have been lost see Mike Davis's "Junkyard of Dreams" on the changing fortunes of the town of Fontana in Southern California, in M. Davis (1992) *City of Quartz* (New York: Vintage Books, chapter 7). The paper by Marc Weiss listed in the references is excellent on the historical origins of urban renewal legislation in the US and the link to those with stakes in downtown fortunes.

Chapter 9

The Politics of Geographically Uneven Development

Context

"Development" when applied to people in places is a highly complex category. As lay people we use it without giving careful thought to precisely what it means, or even if it might mean several things rather than just one. Indeed, the reader may already have noticed how at least two different notions of development have crept into the discussion in this book. There is the sense elaborated in chapter 2. This is development as development of the productive forces. According to this meaning development corresponds to the progressive elaboration of labor processes which facilitate the productivity of the worker. A somewhat different meaning emerged in the course of chapter 8. In that instance development corresponds to positions in the division of labor, and sure enough we do indeed think of countries that are still agricultural as less developed than those that are industrial; and countries still focused on manufacturing as less developed than those where services have become a more important element in the occupational composition. Nor should we neglect another frequently encountered, perhaps the most common, meaning, at least for the lay person. This is the notion of development, at least as it applies to places, as having to do with income, wealth, the ability to command the labor of others through the purchase of consumption goods.

In sorting out this apparent confusion, there are several things that we should bear in mind. Historically, as development has occurred in the sense of increased labor productivity, so the division of labor has been transformed. This is commonly recognized in the way in which occupational compositions by country change with increasing labor productivity. An initial concentration in agriculture and possibly mining gives way to greater proportions employed in manufacturing. Later the rise of service industries serves to displace both manufacturing and agriculture and mining in terms of the proportion of the labor force accounted for. This is the "primary versus secondary versus tertiary" conception of the division of labor. But the logic also applies to finer categorizations. There is no way, for instance, in which public forms of transportation – the streetcar and the railroad – could have given way to the more privatized form of the automobile and the rise of automobile production as a major sector in the economy without an increase in real disposable incomes: an increase that was predicated in turn upon the increasing productivity of workers and therefore on the downward trajectory of the real prices of those goods that had been subjected to that logic of developing the productive forces. The rise of privatized entertainment and with it the TV and radio industries can be accounted for in similar terms, as can the growth of white collar employment alongside the earlier preponderance of blue collar.[1]

But in reconciling these different concepts of development one with another, more is involved than this: more, that is, than the way in which the division of labor between firms and people, and therefore the division of labor between places, gets transformed as a result of the development of the productive forces. Part of the reason for the development of the productive forces is, of course, as Adam Smith recognized long ago, the development of the division of labor and the virtues of the specialization, in both worker and means of production, that it allows. However, the division of labor not only facilitates production. It also imposes its own logic of a redistributional sort. It helps in production and mediates the distribution of what has been produced. And this is why local growth coalitions want to see "their localities" graduate from "lower" functions in that division to "higher" ones.

As we discussed in chapter 2, the division of labor can be considered from two standpoints: as a social division of labor between firms in terms of what it is that they produce; and as a technical division of labor between different employees in terms of their contribution to the (collective) labor process of the firm. The first yields a geographic division of labor in the form of (e.g.) textile towns, insurance centers, and also less developed and more developed regions; the latter, one in terms of (e.g.) branch plant towns and corporate headquarter cities, blue collar and white collar towns.

As far as the social division of labor is concerned, firms developing or occupying new niches in it typically have advantages. Their market is expanding rapidly yet the skills and understandings that would allow competitors to set up in business are as yet scarce. This means that the firms in question, and their employees, will have some degree of market power that is not available to longer established branches: those, that is, that may become the next round of sunset industries. Typically, therefore, the terms of trade work to the advantage of the newer sectors and all the more so to the extent that what they are producing results in productivity revolutions in the older sectors and drives

1 Obviously, in understanding the emergence of new branches of production more is involved than simply increasing disposable incomes. The branches that come before often provide preconditions for those that come later. The bicycle was important for the automobile because of the way it provided scope for some development of the components of the automobile: brakes and pneumatic tires in particular. Early models of the automobile also used chains as a means of transmitting power from the engine to the rear wheels.

real prices there down at the same time as it drives up demand for the product or service coming out of the new sector.

Market power also counts in the technical division of labor. To the degree that (e.g.) white collar workers, like programmers, or maintenance engineers, are in short supply they may be able to demand a premium wage over less skilled assembly line workers. This also helps to explain the difference between blue collar towns and research and development centers. But there is also an important role here for administrative fiat: for the decision of the management as to how the firm's product should be divided, who should get what. This is particularly the case for firms that enjoy strong market positions, creating some space for granting lavish salaries and expenses to higher management.

The upshot of these considerations is that the geographic division of labor, constituted by firms occupying different positions in the social division of labor and by workers performing different roles in the technical division of labor, is something important to workers, businesses, and state agencies in particular places.[2] Moreover, as was argued at length in chapter 8 the resultant struggle among places and over space is one in which the activities of the state are central. There is in short an extremely lively, sometimes brutal, struggle over the geographic division of labor.

In this chapter I am going to exemplify and explore this through two separate sections. The first concentrates largely on the sorts of conflicts that have emerged *within* states. The second focuses primarily on inter-state relations; on imperial, colonial, and neo-colonial relations. Nevertheless, there is a degree of artificiality about this separation as I will demonstrate in a brief third section to this chapter. For charges of colonialism are not unknown in the contest among regions within a country; and the sorts of appeals to social justice that are the predominant rhetorics in the politics of geographically uneven development within countries often resonate in struggles between them as well.

Territorial Justice

Development within countries is invariably geographically uneven. The processes producing this unevenness are certainly complex. As we have seen in chapter 8 they include the activities of various local and regional growth coalitions. More central branches of the state are also implicated. Through its urban renewal policies the federal government helped central cities and especially downtown property owners resist the implications for their tax bases and property values respectively of suburbanization. Likewise in our discussion of settlement policy in County Durham, England, the county administration was quite clearly picking winners and losers.

---

2   We should be careful, however, not to reduce wage levels to positions in the division of labor since labor processes in the same branch of production can vary considerably in terms of their productivity.

In fact it seems hard to imagine a situation in which the activities of more central branches of the state did *not* play a role. Almost every policy that they implement has unequal regional effects. Their expenditures channel money through some places and away from others. Their regulatory policies facilitate private investment in some but not in others. And much of this is quite unintended. The government has to locate military bases *somewhere* but its intention is not to pump money into local economies, though that is one of its effects. Interstates likewise: and again, the way they alter the accessibility relations of places, making some more attractive to investment than others, is, at least on a broad regional scale, inadvertent.[3]

Think and Learn

What other central government policies (in the US, either State or federal) have clear, geographically differentiating effects, even if inadvertent? Try to identify some policy that differentiates through its expenditures and one which is regulatory in nature and which has effects through that on private investment decisions.

So some lose out in the battle for more desirable positions in the geographic division of labor and it seems as if central branches of the state are likely to be implicated, if only accidentally. In that context it would be surprising if arguments were not elaborated as to the injustice of it all. And that is precisely what happens. There is, in short, an issue of territorial justice, or more accurately *issues*, since it assumes different forms. The disadvantaged press their case not simply through material interventions designed to change it, but discursively, attempting to mobilize popular opinion, create larger alliances by pointing to the unfairness of it all.

A major form this assumes, for example, is that of requests to "level the playing field." This is particularly the case in federal or quasi-federal systems where subnational levels of government exercise considerable power to change, through their competition, the geography of the division of labor; but where, equally, there will be losers. This was broached in chapter 8 when I discussed the tensions that arise in the context of a division of welfare state responsibilities between central and more local branches of the state. Differences in labor law can also be the occasion for "leveling the playing field," particularly where they are associated with the relocation of employment.

A case in point has been the debate about the right-to-work clause in the 1947 Taft–Hartley Act in the US. This piece of federal legislation gave States the option of banning the closed shop and declaring themselves "open shop

---

3 I insert the caveat "on a broad regional scale" because there is evidence that at more local levels the routing of Interstates – and indeed the placing of military bases – *is* subject to interference from local growth lobbies.

Figure 9.1   "Right-to-work" States in the US.

Source: National Right to Work Foundation (after http://www.nrtw.org/rtws.htm).

States." This meant that in those States employers could not legally come to an agreement with labor unions in which employment would be conditional upon joining the union. This weakened labor unions and the ability of workers to press claims on employers, which was what the States introducing such laws aimed for. For it was their belief that such a provision would make them attractive to inward investment. But only in States where labor unions were already weak did it prove possible to obtain the necessary public support for such legislation and these States have tended to be the lower-wage States of the South and West (figure 9.1). These are the ones, moreover, which have shown quite large increases in manufacturing employment, especially over the past thirty years or so, while at the same time manufacturing employment in those States dominated by the union shop have tended to lose it (compare figure 9.1 with figure 3.1). As a result, and in the context of some widely publicized cases of plant relocation, such as the shift of Mack Truck from Allentown, Pennsylvania to North Carolina,[4] growth lobbies in the Midwest and Northeast, the so-called Coldbelt, argued for "a leveling of the playing field" in the competition for investment by eliminating the right-to-work clause.

A second type of territorial justice argument rests less on degrees of geographically uneven development and more on the fiscal relations between different regions and the central state. Fiscal balance refers to the difference

4   The State with the dubious distinction of the lowest rate of unionization in the US.

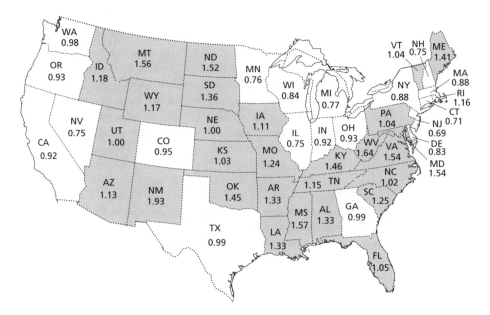

Figure 9.2 Fiscal balances of the States with the US federal government, 1998. These figures indicate "returns on the federal tax dollar." In other words, states with a ratio greater than 1 received more than they sent in federal tax dollars. Those with ratios less than 1 received less than was sent.

Source: After table 15, p. 61, in P. Duggan with G. Starnes (1999) *Flow of Federal Funds to the States.* Washington, DC: Northeast-Midwest Institute.

between what a particular region sends to the central government in the form of taxes and what it gets back in the form of central government spending. It is not that difficult, for example, to calculate the amount of taxes paid by the residents and firms located in a given region. Using some not-too-heroic assumptions it is also possible to calculate how much is returned to the region in the form of (e.g.) subsidies to local governments there, pension payments, research grants awarded to universities, government orders for goods and services, and central government employees (civil servants, military personnel) living there. Interest then focuses on the difference or balance; and in particular how it varies from one region to another, supposedly shedding light on the degree to which some regions are benefiting at the expense of others. Obviously this bears on local economic development, since spending by the central government affords a boost to local economies.

In the United States the issue of unevenness in fiscal balance between the States first attracted attention in the context of the Coldbelt–Sunbelt divide: the fact that Sunbelt States had, since the early seventies, shown considerably greater rates of economic growth than the Midwestern and Northeastern States of the so-called Coldbelt. Statistics and maps were prepared (see figure 9.2) purportedly showing serious inequalities. Some States tended – and still tend, since the controversy is still alive – to send more in taxes to the federal government than they got in return in the form of federal expenditures: those

States were primarily in the Midwest and the Middle Atlantic region. Other States, on the other hand, tended to get more back than they sent: these States were primarily in the South and West.

Congresspersons from the Northeast and Midwest organized themselves into a coalition (the Northeast–Midwest Congressional Coalition) in order to seek redress for what they regarded as a violation of territorial justice. In particular they have pushed for legislative acts which would favor the Coldbelt rather than the Sunbelt. One way in which the geography of fiscal imbalance with Washington shows up is in federal infrastructural investments. The bias of federal legislation towards money for *new* waterworks or sewers tends to work, for instance, against the interests of the Coldbelt where the main problem is the rehabilitation of existing infrastructure. A major focus of the Northeast–Midwest Congressional Coalition, therefore, has been a rewriting of the legislation in question in order to provide monies for the modernization of facilities already in existence. More generally, and significantly, a major thrust of the Coalition has been greater federal spending on the rehabilitation of physical infrastructure, an area in which the Northeast and Midwest – as opposed to the Sunbelt – have a considerable backlog of needs. Another concern has been the disproportionate degree to which military orders go to Sunbelt firms, particularly those in California. As a result there have been attempts to legislate preferential bidding from suppliers located in areas of unusually high unemployment.

This in turn has stimulated the emergence of a coalition acting on behalf of the Sunbelt: the Sunbelt Council. But they have not challenged the data on fiscal balance. By and large these seem to be accepted. Rather the response, particularly from the Southern States, is to resort to geographical equality arguments: that for many, many years, and still today, much of the Sunbelt was, and remains, relatively deprived compared with the States of the US industrial heartland in the Midwest and the Northeast. The significance of this is that regional lobbies select the particular concept of territorial justice they will draw on according to circumstances. In Britain, for example, candidates in the recent election (2000) for Mayor of London, part of the favored South, campaigned on the need to redress the city's unfavorable fiscal balance with the rest of the country. If you can't campaign on the need for regional equality, or if those who do so are a threat, then find your own, more appropriate concept of territorial justice.

Think and Learn

How valid do you think the claims of those regarding the unevenness of fiscal balances are? Is it fair to say that a particular State actually *does* benefit from a favorable balance of incomes of federal spending to outgoes of taxes? If (e.g.) Boeing Aircraft Corporation, headquartered in Seattle, Washington, receives large orders from the federal government for the provision of military aircraft does all the money stay in the State of Washington so that it can give a boost to local economic growth? What do you think and why?

> **Think and Learn (Again)**
>
> What do *you* think about these various criteria of territorial justice? Is there one criterion broadly acceptable to everyone? And if territorial justice were achieved, would that necessarily mean that a socially just society had been attained as well?

This is clearly an issue that has cropped up in other countries. In Britain the issue has been one of the relation of London to the rest of the United Kingdom, the various candidates for the recent mayoral election in that city each arguing that it subsidizes the remainder of the country. But as an instance of purported territorial injustice it pales into insignificance when compared with the emergence of what has been called a North–South divide in that country. Moreover, this has resulted in a territorialization of politics in a country where territory has historically been rather subdued as a form of political cleavage. It is to the politics of the North–South divide that I wish to devote the remainder of this section.

Geographically uneven development is no stranger to the United Kingdom. Unemployment and average incomes have always varied geographically. The existence of persistent pockets of unemployment was the focus of a succession of government policies that came into being from the 1930s on aimed at inducing industry to relocate there (see chapter 3). Indeed it had seemed that they were having the intended effect. As figure 9.3 shows, until about 1976 interregional inequality in the country had been on the decline. But from then on things changed quite sharply and only after 1990 did regional inequality start declining once again. Furthermore, the geography of these changes was far from random. Figure 9.4 shows that there was a quite marked divergence between the Southeast, Southwest, and East Anglia, on the one hand, and the rest of the country, on the other. These changes were closely matched by interregional variations in employment growth (see figure 9.5). The major growth centers over the period were predominantly in an arc around London and included places like Crawley, Reading, Swindon, High Wycombe, Aylesbury, Cambridge, and Colchester. London itself, on the other hand, was the site of considerable job loss, which perhaps accounts for the concern we noted above regarding its supposed subsidization of the rest of the country.[5]

These changes had a great deal to do with government policy, though to what extent it is true that, as Mrs Thatcher famously said, "There is no alternative" is debatable. Furthermore, one should be careful not to accord *all* responsibility to the Conservative Party-led governments presided over by Mrs Thatcher. Between 1974 and 1979 there was a Labour Party government.

---

5 The major exception to this pattern is Aberdeen where employment grew on the basis of North Sea oil services.

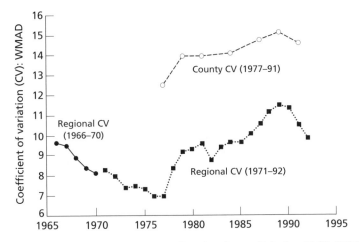

Figure 9.3 Trends in regional inequality in Great Britain 1966–1992. Inequality has been measured by the coefficient of variation for gross domestic product per inhabitant. The coefficient of variation is calculated by dividing the mean by the standard deviation for the various regional GDPs per inhabitant. Note that until 1976 there was a strong trend towards greater interregional inequality. Since that time this has been reversed, though with some return to equalizing tendencies since 1990.

Source: M. Dunford (1997) "Divergence, Instability and Exclusion: Regional Dynamics in Great Britain." In R. Lee and J. Wills (eds), *Geographies of Economies*. London: Edward Arnold, p. 261.

As the increasing levels of interregional inequality from the mid-seventies on suggest, policy changes had already been made,[6] though Thatcherite initiatives greatly deepened and extended these and gave them a distinct ideological inflection.

What Mrs Thatcher and her governments aimed for was a drastic pullback of the state. State involvement in the economy, its regulation of private business, state ownership of business were to be dramatically scaled down. Privatization was a watchword. All forms of economic activity, whether private or public, were to be opened up to the keen breeze of competition. State agencies would have to put all requests for services out to competitive tender. Nationalized industries would be held to strict profit and loss criteria prior to their sale to private business. What remained of central government policy would itself be subject to the disciplines of competition but on an international scale. By making sterling fully convertible the Thatcherites aimed to bring

6 Compare Hudson and Williams (1986, p. 26): "To some extent the [Thatcher] government simply intensified tendencies visible in the policies of its Labour predecessor: generally restricting public expenditure but selectively increasing funding (notably for defense and law and order) while reducing the Public Sector Borrowing Requirement (PSBR); increasing the scope of the private sector at the expense of the public sector as companies such as British Aerospace, British Telecom and Jaguar have been privatized; and reaffirming the primacy attached to reducing inflation."

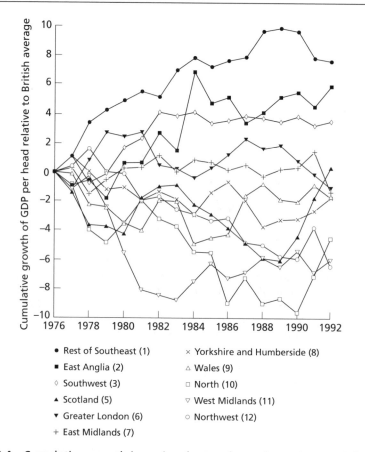

Figure 9.4 Cumulative growth in regional gross domestic product per inhabitant relative to the British national average, 1976–1992.

Source: M. Dunford (1997) "Divergence, Instability and Exclusion: Regional Dynamics in Great Britain." In R. Lee and J. Wills (eds), *Geographies of Economies*. London: Edward Arnold, p. 262.

inflation under control by subordinating British monetary policy to the judgment of international investors.

The Thatcherite revolution was also a remaking of class relations. One of the diagnoses of the United Kingdom as "the sick man of Europe" in the sixties and seventies was that it was owing to a labor union movement that had too much power. By rewriting labor law the Thatcherites hoped to change that and at the same time help refloat the British economy by making the country attractive to foreign investors. The turn to strict monetarism would wring demand out of it and discipline labor through the threat of unemployment. The working class was also seduced. The acquisitive goals of many of its members were flattered by the sale of council houses[7] to their occupants.

7   Council housing is the term used in Britain for public housing.

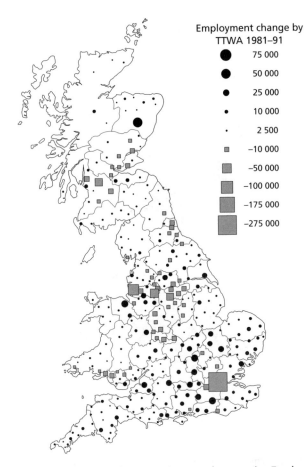

Figure 9.5 Employment change by travel-to-work area in England, Scotland and Wales, 1981–1991. Note the strong tendency for employment growth to be concentrated in the southern parts and particularly in the Southeast, though London does not appear to have shared in this, and actually experienced a decline. Employment growth in the northeast of Scotland is associated with the growth of the oil services industry there, particularly in Aberdeen.

Source: M. Dunford (1997) "Divergence, Instability and Exclusion: Regional Dynamics in Great Britain." In R. Lee and J. Wills (eds), *Geographies of Economies*. London: Edward Arnold, p. 268.

These policies, however, had uneven geographic effects, and by no means is it the case that they were always intended. Thatcherite geography had a good deal of the inadvertent about it. Most notably, perhaps, the policy of holding nationalized industries to market criteria with a view to making them attractive to private buyers had disastrous consequences for those areas of the country that, as we have seen in figures 9.4 and 9.5, were the ones to show slower rates of employment and income growth under Thatcher. For the fact is, employment in the nationalized industries was geographically highly

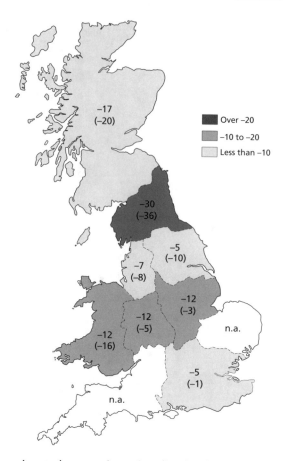

Figure 9.6 Changes in employment in nationalized industries in the United Kingdom, 1973–1981. I: The case of coal. Figures are regional changes as percentages of national decline. The top figure in each area refers to 1978–81 while the figure underneath in parentheses is for 1973–8. The north and northwest clearly bore the brunt of these changes, which is not surprising given the fact that most coal mining was in those areas. Note, however, that the figures in parentheses indicate that these changes were already under way prior to the election of Thatcherite governments starting in 1979.

Source: After table 5 in R. Hudson (1986) "Nationalized Industry Policies and Regional Policies." *Environment and Planning D: Society and Space*, 4(1), 7–28.

uneven. The major industries affected in this way were coalmining, iron and steel, and shipbuilding, all, incidentally, "sunset" industries. As figures 9.6 and 9.7 show, the North and West, including Scotland and Wales, were highly vulnerable to these changes since the nationalized industries represented such a high percentage of total industrial employment there.

Accordingly they were the regions most affected by the rundown in employment in the nationalized industries. The Southeast, East Anglia, and the Southwest, which we have seen were to prosper under Thatcher (figure

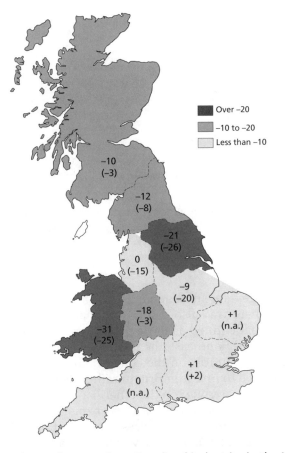

Figure 9.7 Changes in employment in nationalized industries in the United Kingdom, 1973–1981. II: The case of iron and steel. Similar patterns to those in figure 9.6 are observable. Again, note that the top figure in each area refers to 1978–81 while the figure underneath is for 1973–8.

Source: After table 7 in R. Hudson (1986) "Nationalized Industry Policies and Regional Policies." *Environment and Planning D: Society and Space*, 4(1), 7–28.

9.4), were hardly affected at all. On the other hand, it is also important to note that the rundowns were already in place before the accession of the Conservative Party to power in 1979. Indeed under Thatcher the decline in coalmining employment was not much in excess of that experienced over the previous five years.[8] In iron and steel, however, layoffs increased dramatically, with the burden being disparately borne by Wales, Yorkshire and Humberside, and the West Midlands. More generally manufacturing in traditional assembly line industries like automobiles and textiles was squeezed by the high

8   However, employment in coal declined dramatically subsequent to the coal miners' strike of 1984–5.

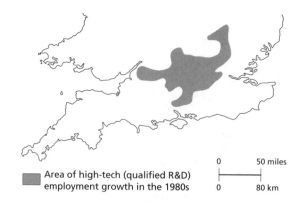

Figure 9.8   Areas of hi-tech employment growth in the United Kingdom in the 1980s.
Source: J. Allen, D. Massey, and A. Cochrane (1998) *Rethinking the Region*. London: Routledge, p. 44.

interest and exchange rates subsequent to the Thatcherite turn to monetarism. And employment here was also disproportionately concentrated outside of the regions we have identified as those particularly benefiting from Conservative Party policies: the Southeast, the Southwest, and East Anglia. Moreover, with the lifting of all controls on the convertibility of sterling investors had less and less incentive to invest in Britain rather than overseas.

The convertibility of sterling, on the other hand, had effects that were positive for London and the Southeast. As was explained in chapter 8, London had long been a major financial services center in the global economy and convertibility accelerated this tendency since it made it so easy for dealers in international currencies to operate from a base in London. These tendencies were accelerated by policies to deregulate money and security markets: in particular the Financial Services Deregulation Act of 1986 which heralded the so-called "Big Bang." This led to a massive boom in financial services employment in London and also in the Southeast in general as firms sought out locations in the hinterland for their back office operations.

At the same time, and for reasons that have little to do with Thatcherite policies, the new hi-tech sectors of industry that emerged in the advanced industrial societies in the seventies took root in an arc of towns to the west, northwest, and north of London: from Bristol in the west to Cambridge in the north. These were the biotech and pharmaceutical industries, computing services, and the independent research and development sector (see figure 9.8). This reindustrialization with a regional slant was reinforced by the effect of increased defense expenditures by Thatcher's governments. Of the major capitalist powers, with the exception of the United States, throughout the 1970s the United Kingdom spent the highest proportion of gross domestic product on defense of all and this increased under Thatcher, giving a further boost to employment in that sector. These layouts also tended to benefit disproportionately the Southeast.

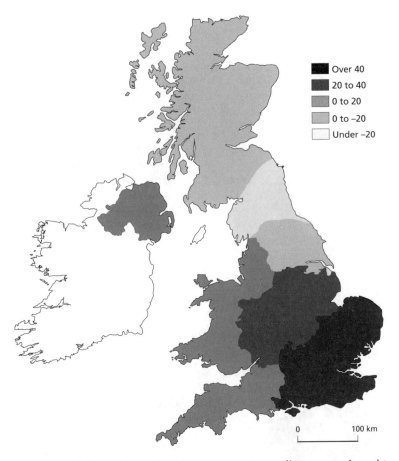

Figure 9.9   Regional change in national government spending on roads and transport in the United Kingdom, 1987–1991.

Source: J. Allen, D. Massey, and A. Cochrane (1998) *Rethinking the Region*. London: Routledge, p. 45.

So for a variety of reasons, and not all related to specifically Thatcherite policies under the Conservative governments, London and the Southeast and adjacent areas in the Southwest (Bristol, Cheltenham, Swindon in particular) and East Anglia (cities like Cambridge, Peterborough, and Ipswich) boomed, while much of the rest of the country lagged behind. This difference in turn was reflected in major government capital projects (figure 9.9), with large amounts of government money finding their way into an orbital freeway for London, a new international airport at Stansted about forty miles north of the capital, and investments in motorways to connect London to its immediate hinterland.

This has been contrasted with the patterns of government expenditure that had prevailed earlier, particularly in the sixties and early seventies when the government provided major subsidies to firms locating in the depressed areas

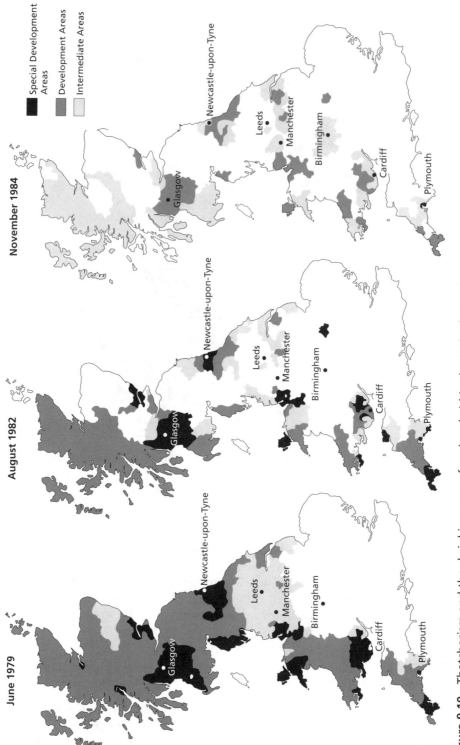

Figure 9.10 Thatcherism and the shrinking map of regional aid in the United Kingdom. Development Areas in England, Scotland and Wales. Under the Conservative governments that came into being from 1979 on there was a dramatic shrinkage in the area covered by the regions qualifying for special government aid to attract new employment.

Source: Department of Trade and Industry, UK.

of the North, Scotland, and Wales, along with money for new towns like Peterlee, Cumbernauld, Washington, and Cwmbran (see figure 3.7). In fact under the Thatcherites, there was a considerable pullback in regional aid programs (see figure 9.10). But it is not entirely fair to say that the governments of the eighties and early nineties lacked a regional program. Their belief was that by making Britain friendly to the multinationals through a reduction in the power of the labor unions, through taming inflation and so reducing currency risk, the relatively low wages prevailing in the depressed areas would do the trick. To some degree this happened. There have been, for example, much publicized Japanese investments in automobile manufacturing in Washington new town (Nissan) and in consumer electronics in South Wales as well as elsewhere in the North (figure 9.11) (see also box 3.5).

Even so, the perception of a North–South divide, or rather, and more accurately, a division between London and the Southeast on the one hand and the rest of the country on the other, has bitten deeply into the national consciousness and continues to be represented by real material differences. As such it has provided the basis for a significant territorialization of British politics. The Conservative Party has become *the* party of those parts of the country that have prospered under Conservative Party rule; though, as I pointed out earlier, not necessarily directly as a result of Conservative Party policies. The Labour Party, on the other hand, has found *its* heartlands outside of the Southeast and particularly in the old manufacturing and mining areas of the Northwest, the North, Yorkshire and Humberside, Wales, and Scotland.[9] This is in some contrast to the situation that prevailed from about 1945 to the mid-seventies when territorial issues were subdued and both Conservative and Labour Parties were seen as predominantly class-based parties: *the* political cleavage in Britain was one of social class and territory was seen as having very little significance.

Some indication of this territorialization is provided in figure 9.12. This shows the percentages of those who identified themselves as working class who voted for the Labour Party in 1983. If there was no territorialization of party voting then the percentages across regions should show little variation. But this is clearly not the case. The degree to which working class identifiers vote Labour is quite a bit higher in the North and West Midlands. Interestingly Greater London shows considerably higher levels of working class voting for the Labour Party than one might have anticipated, suggesting that there are some flaws in the way in which the North–South divide has been constructed. On the other hand, to talk about territorializations assumes that people have commitments to particular places. Conservative commentators and neo-classical economists are fond of arguing that if people don't like it they can always move. But evidently, and as I have argued throughout this book, this is not always that easy. So just what are the embeddednesses at work here? Why are people resistant to moving?

---

9 Significantly, following the 1997 general election the Conservative Party had no MPs from Scottish or Welsh constituencies. In the 2001 general election, the party gained one seat in Scotland, but none in Wales.

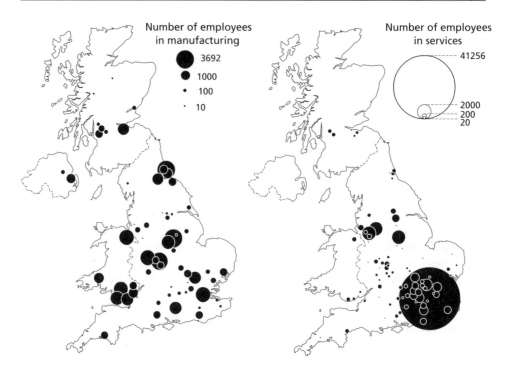

Figure 9.11 Contrasting patterns of Japanese MNC investment in the United Kingdom: employment in 1991. Note the fairly evenly spread character of manufacturing employment, with a strong cluster in South Wales. The concentration of service employment in London and the Southeast, however, is a function of the area's strength in financial services and the desire of the Japanese corporations to take advantage of the synergies available there.

Source: P. Dicken, A. Tickell, and H. Yeung (1997) "Putting Japanese Investment in Its Place." *Area*, 29(3), 208. Copyright: Royal Geographical Society (with the Institute of British Geographers).

In part it is the fact that for most people a move in search of a job elsewhere is like a leap into the dark; great risks are attached. Most job-related movement, particularly for those with families, is sponsored by the future employer. The employee will have been invited for an interview, had his or her expenses paid, and then been assisted financially with the move. Alternatively the employee will be relocated by the firm to another of its branches; but again, that is something likely to apply only to key workers. Most people do not qualify for that sort of treatment and that tends to confine them to very local labor markets (Gordon, 1995), unless, that is, they have relatives who can recommend them to their employer.[10]

10 In less skilled lines of work this is an extremely important avenue of labor recruitment. As a result one often finds that migrants to a particular place come from a select few areas of origin.

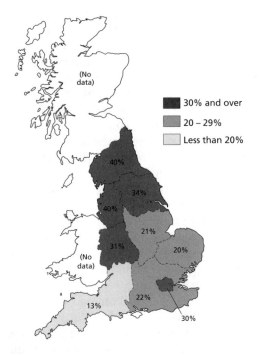

Figure 9.12 Interregional differences in class voting, 1983. Percentage of working class self identifiers voting Labour. Clearly, working class identifiers are much more likely to vote Labour in the less economically dynamic North and Northwest than they are in the South and Southeast.

Source: After R. J. Johnston (1987) "The Geography of the Working Class and the Geography of the Labour Vote in England 1983; A Prefatory Note to a Research Agenda." *Political Geography Quarterly*, 6(1), table 5, p. 13.

There is also the matter of home ownership. It has been suggested (Savage, 1987), for example, that tendencies to voting Labour were particularly strong among homeowners in depressed job markets (found mostly in the North), while among homeowners in booming job markets the reverse tendency applied: a much stronger impulse to vote Conservative. In depressed job markets demand for housing was stagnant. That made it hard for homeowners to realize any capital gain on their property. This in turn made it harder for them to raise the assets they would require if they were to sell up, move to more buoyant labor markets, and purchase a house there.[11] They therefore pinned their hopes on a general election victory by the Labour Party and a return to a macroeconomic policy that was less deflationary in character and more stimulating to investment in the economy (Savage, 1987).

11 This reflects an important feature of British housing markets: if you are going to move then your best chances of securing housing are in the private owner-occupancy market. This is for two reasons: (a) the private rental sector is not well developed in Britain; (b) in order to secure public housing you have to be on the waiting list, and in order to be on the waiting list you have to already live in the area.

The obverse of the homeowner in a depressed job market was the home-owner in a booming job market, often in the Southeast. They prospered under Thatcher as a result of the highly localized boom that occurred during her governments, but which was only in part a function of her policies. Even so there was a strong "feel good" factor at work. But this also raises the question of precisely what territorial coalition it was that the Thatcherites responded to *originally*. For the fact is British capital is unusually multinational.

Thatcher's territorial coalition was constructed as she went along. It consisted of all those interests that were embedded for various reasons in the Southeast and which came to the conclusion that they owed their growth and prosperity to her policies. These included, in particular, the financial services industry of London and the surrounding area, and the homeowners who benefited from the boom in real estate values in the Southeast. A return to Labour policies, certainly "old" Labour, threatened to shift the center of gravity of the national economy away from the Southeast and so work to their disadvantage.

The Conservatives were elected in 1979 because under Thatcher's radical program they promised to break a stalemate in British politics. All the various approaches that had been tried in an attempt to revive the British economy, whose history since the Second World War had been one of relative decline, had failed. Both Labour Party and Conservative Party governments had pursued some variant of managed capitalism, only the degree of management varying between the two. Thatcher promised something different, and so appealed to a sense of national frustration. But because her policies would lead to a cleansing of the Augean stables, labor militancy and inept management both, it wasn't clear at the outset just what the concrete outlines of a reconstructed British economy would look like.

It certainly was not clear that the City would benefit so much from her policies or that old line manufacturing would suffer so much. That had not been the original intent of making sterling completely convertible. Rather the goal had been to subject state regulation of the economy and therefore the economy to the disciplines of an international marketplace. Likewise the role of multinational corporations in British revival was not perceived at the beginning. But the rollback of union power and the stabilization of the currency did make the country attractive as a site for American and Japanese corporations looking for a location within the boundaries of the EU. Once it became clear that the Japanese had very different approaches to labor relations that jibed with the Thatcherite project of reducing union power they were courted even more fervently. And as far as the Americans were concerned Britain had the additional attraction of a shared language.

Think and Learn

Consider this politics in the light of the interregional struggle for control of British economic policy that we discussed in the third case study in chapter 8. Is the politics of the North–South divide any more than a case of *déjà vu*?

Postscript: Despite the fact that the United Kingdom now has a Labour Party government, talk of the North–South divide remains undiminished and there are continuing signs that the Labour Party is becoming (or remains) divided along territorial lines. The fact is, the Labour Party was elected on a program that rejected much of its past policy commitments and has continued for the most part along lines that are clearly Thatcherite. Commitment to the market remains almost undiminished. There is the same concern about not interfering with incentives to invest that were common currency among Conservative Party politicos during the Thatcherite era; the same sensitivity to the constraints of living in an age of "globalization."

There is evidence[12] that the honeymoon between Labour backbenchers and the party leadership is now coming under some strain, however. There is concern that the government is insufficiently sensitive to the needs of their supporters in the Labour Party heartlands that are, for the most part, outside the Southeast: in the North, Wales, and Scotland in particular. This was highlighted by two events in early 2000.

The first was the resignation of the Defence Minister, Peter Kilfoyle, over concerns that Labour was neglecting its traditional support base. To quote the London *Times*: "His decision comes amid growing anger from Labour MPs and activists that the Government's 'southern middle class bias' threatens to destroy the coalition of voters that won it the last election" ("Defence Minister Quits," *The Times*, January 31, 2000). Picking up on some of the issues referred to above, the same article goes on to pinpoint Kilfoyle's frustration that even though his own constituency (in Liverpool) shows no sign of housing price inflation, people have to pay the same high interest rates imposed to cool off the housing market in the Southeast.

The second event which gave the North–South divide a further jolt into renewed prominence was a highly controversial government decision about the location of a major research establishment: the synchrotron facility or so-called Diamond Project. Worth hundreds and possibly thousands of jobs, the original hope was that it would go to a location just to the west of Manchester in the Northwest region. But instead it went to one close to Oxford in the Southeast where it was felt the research synergies were higher. But it was also revealed that pressure in favor of the site close to Oxford had come from the Wellcome Trust, which had contributed a fifth of the cost of the approximately $800 million project. This allowed the event to be constructed not just as a territorial matter but as illustrating the government's subordination to private interests (not exactly the Labour Party's traditional position). As the person who led the effort to locate the Diamond Project in the Northwest commented: "The Government has effectively been held to ransom by the Wellcome Trust. Many people see central government as divorced from them – running the country from the South, for the South" ("Anger as Jobs Jewel Given to the South," *The Times*, March 14, 2000).

12   At the time of writing in 2001.

The Politics of Colonialism and Dependency

A policy through which the major capitalist economies of the mid to late nineteenth century attempted to secure their positions as dominant parts of the expanding global economy was through empire: the acquisition of colonies. Through various means and stratagems all the major European countries acquired colonies, as did Japan and to a lesser extent the United States. Governments and publicists might huff and puff about the white man's burden or *la mission civilisatrice*, and missionaries were always well in attendance suggesting that these slogans did achieve a wider resonance. But the main purpose always was to advance the interests of a national capitalism, of national firms against those of other countries: to carve out geographic areas which national firms could monopolize as fields for profitable investment in mines and land development projects and in trade. If other motivations cannot be totally excluded the primary one was to defend the home base and create a field of opportunity through which it might grow further.

More specifically the colonies provided fields of investment, primarily in raw materials like minerals, foodstuffs, and timber. If they were to function effectively in this regard then the extension of political control was necessary. With fixed investments in the form of mines, plantations, and railroads, civil order had to be created and a wage labor force formed. As we will see colonial governments were vital to both of these projects (see box 9.1).

Think and Learn

In order for capitalist development to occur, and as was set forth in chapter 2, a wage labor force has to be created. Why do you think wage labor might not have been forthcoming spontaneously, without colonial intervention, in the areas over which the European powers claimed imperial jurisdiction?

The colonies also provided protected markets for the products of the imperial country or, more accurately, of the firms headquartered there. Initially trade did not require the creation of colonies. Traders could meet with the indigenous population, exchange their goods, and leave. As the expansion of trade became more important to the trading companies, however, as it became something that was a specialized operation with its own highly distinctive knowledges and further expansion became significant for the ability of the trading companies to raise loan finance, so obstacles to expansion became significant. These obstacles included disruption of trade as a result of warfare between local potentates; and the raising of prices of native products as a result of the activities of those same potentates or native merchants. As Bill Freund (1984, p. 88) has written with respect to conditions just prior to the so-called "scramble for Africa" in the late nineteenth century: "African

---

Box 9.1   *Colonies and Raw Materials*

The search for raw materials as a stimulus to the extension of European rule in the form of colonies is highlighted by a number of important cases. One is the aftermath of the disintegration of the Ottoman Empire in the Middle East after the First World War. In Europe the disintegration of the Austro-Hungarian, Russian, and German Empires had been followed by the granting of independence to successor states, notably Poland, Czechoslovakia, and Yugoslavia. In the Middle East there were voices which argued for the same outcome both from the peoples of the region as well as from outsiders like Lawrence, of Lawrence of Arabia fame. But antedating the war by a few years had been the development of the internal combustion engine and during the war oil was discovered in important quantities in what was to become Saudi Arabia, Iraq, and the various Gulf states like Kuwait. As a result independence was pushed aside in favor of incorporation into the British Empire, albeit as adminstrator on behalf of the League of Nations.

Another case in which raw materials so clearly played a significant role in imperial expansion was the incorporation of that part of South Africa known as the Transvaal into the British Empire in 1902. This was in the aftermath of the Boer War. As we discussed in chapter 7, the Boer War was precipitated by a crisis in the operation of the gold mines in the Transvaal. Gold had been discovered there in 1886. The mining companies were all British but the Transvaal was an independent Boer or Afrikaner state. The Transvaal government had its own views as to how the economy should be administered and these were not to the liking of the gold mining companies. The government operated a monopoly in the sale of dynamite which increased operating costs. There was a tariff designed to stimulate industrialization which increased the cost of manufactured goods for the mines. There was also a tax on the mines which the government planned to use to stimulate agricultural development. It was for these reasons that the British government went to war with the Transvaal in the Boer or South African War of 1899–1902.

---

society badly needed reform in order for the ambitions of Western commerce to be realized. Governments were required which could smash the power of ruling classes, construct telegraph lines and railways, impose uniformly peaceful conditions and permit coastal traders direct access to free peasant producers."

In order that empire function in these ways, however, two preconditions had to be met: creating civil order; and creating a wage labor force. With respect to the first of these, the imperial powers were, more often than not, resisted violently by the indigenous populations: the native Americans in North America, the Xhosa and the Zulus in South Africa, the Maoris in New Zealand, the Berbers in North Africa, etc. Settlers were attacked to the point at which appropriation of the land and other natural resources became hazardous. The establishment of colonial governments with clearly defined territorial jurisdictions was the necessary precondition for imposing social order; albeit social order from the standpoint of the imperial powers.

Colonial governments were also important in creating a labor force for the mines, the plantations, the railroad companies, the ports, and the like. The problem was that although there were often indigenous populations in areas where the Western mining, trading, and land colonization companies wanted to do business they did not constitute pools of easily recruitable workers. Native populations enjoyed rights in land and were self sufficient. They had no need to work for a wage. Their needs could be met through working the land, perhaps through hunting and fishing, and through traditional crafts. The problem was, therefore: how to turn indigenes into wage workers.

The approach finally hit on[13] was the forcible appropriation of land for white settlers and the confinement of the native population to reserves. This seriously reduced their access to land and so forced large numbers to find work in the mines, on the plantations, and, in addition, for the settlers who were now occupying their ancestral lands. This was a common feature of British colonial rule in those parts of Africa where white settlement was possible, in particular South Africa, Kenya, and what is now Zimbabwe (formerly Southern Rhodesia) (see figure 9.13). In short the colonial state presided over that separation of immediate producers from the means of production which, as we saw in chapter 2, is the indispensable condition for converting labor power into a commodity and so unleashing the forces of capitalist development.

As we discussed in chapter 7, however, colonialism ultimately generated its own characteristic responses on the part of the indigenous populations. The primary counter-action, of course, was (eventually) resistance and demands for independence. The nationalist movements, or more accurately anti-colonial movements, emerging in the European colonies from the 1930s on[14] saw the colonial governments as the agents of their respective oppressions. Depending on the particular case in question the colonial government was regarded as:

- limiting the development possibilities of the colony through its monopoly of trade;

13 Not necessarily immediately, therefore. A common initial strategy was to impose a tax payable only in money. Frequently this took the form of a tax levied on each native hut and payable only in money: a hut tax. This meant that for at least part of the year a household member would have to work on a plantation or in a mine or, possibly, on the docks, for a money wage with which to pay the tax. On the other hand there was no assurance that this would work. Natives might choose to earn the money with which to pay the tax in other ways which did not mesh quite so well with the aims of the colonial governments. They might divert some of their land to the production of foodstuffs to be sold to Western traders or those mines and plantations that already existed. They might hunt down elephants and rhinos for the burgeoning trade in ivory.

14 "Anti-colonial" rather than "national" because the national element was so weak. Forging a common nationhood out of diverse language and tribal groups has proven a major challenge in many of the formerly colonial states.

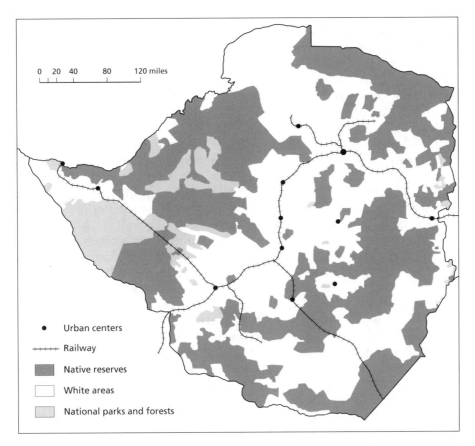

Figure 9.13 White appropriation of land in the former Southern Rhodesia. Note: the extensive areas appropriated for white ownership even though whites amounted to less than 5 percent of the total population; the way in which white areas enjoyed economic advantages of access to the railroads.

Source: After W. Roder (1964) "The Division of Land Resources in Southern Rhodesia." *Annals, Association of American Geographers*, 54(1), 47. Copyright: Association of America Geographers.

- guardian of the interests of European settlers;
- guardian of the interests of those elements of the native population who had gained from colonial rule.

The idea that imperial powers rigged trade regulations so as to impose a division of labor with the colonies is well known. A classic case is that of British tariffs on imported Indian cotton goods which were supposed to have sounded the death knell for the (dominantly craft) industry in that country. But in fact once a certain level of capitalist development had been attained in Western Europe the imperial powers could, as indeed they did, dispense with those sorts of regulations. It was only in the twentieth century that apparently there was some resurgence of restrictions.[15]

Settlers were by no means ubiquitous. But where they were present they were an important factor, as we will see. In many cases there were virtually no settlers at all. This was the case in much of equatorial Africa: Nigeria, Ghana (the former Gold Coast), Uganda, Tanzania (the union of the former Tanganyika and Zanzibar). The colonies in the Caribbean – Jamaica, Barbados, Trinidad – had very few, as did the Middle Eastern oil states, India, Morocco, Egypt, Lebanon, and Syria. In another important category are those cases in which the settlers came to outnumber the native populations. This was the case in North America where the native American population is now only a very small percentage of the total population. In Australia there are very few aborigines left. In New Zealand the Maoris are a minority but numerically a very significant minority. In these instances, and the American one apart, independence was typically granted without any serious anti-colonial movement; government would be in white, "civilized" hands, there was already substantial capitalist development, and they could be relied on to continue as part of a more informal empire, as in the case of the British Commonwealth.

The situations that were exceedingly fraught, however, were those where, again, there was a settler presence but the settlers were in a minority. The extent of this minority varied a great deal. It was larger in the cases of South Africa (approximately 20 percent at one time though now closer to 13 percent) and Algeria than it was in Kenya or Southern Rhodesia. There were settlers in Namibia (the former Southwest Africa), Angola, and Zaire (the former Belgian Congo). In Palestine the Jewish settlers were always outnumbered by the native Arab population and a majority was gained only by the flight of Arabs subsequent to the war in 1948 which brought the state of Israel into existence.

In all these cases, with the possible exception of Palestine which was altogether more complicated, the settlers were a privileged caste and their privileges were clearly state-mediated. They were the beneficiaries of imperial land policies. These dispossessed the natives in order to provide the settlers with land and, since the natives had been deprived of access to the land as a result of state expropriations, with labor as well. They also benefited from government discrimination in agricultural subsidies. And of course, the railroads were located strategically in order to provide access to market for the white settlers. On top of that they enjoyed the vote and had a say in how the colonial government conducted itself. That they should be further favored as,

---

15 Compare Thomas Balogh: "The free play of the price mechanism (as in the case of the "independent" countries of Latin America and the Caribbean) was quite sufficient to restrict the less developed countries to a status of permanent economic inferiority. The implicit preference of the colonial administrations for the metropolitan products did the rest. Their orders on public and private account – and these represented a large portion of the total money demand of the colonial area – in the main flowed toward the metropolis" ("The Mechanism of Neo-Imperialism." *Bulletin of the Oxford University Institute of Statistics*, 24, 1962, pp. 331–46. Reprinted in K. W. Rothschild (ed.), *Power in Economics*. Harmondsworth: Penguin, 1971, p. 324).

for example, in terms of welfare state legislation,[16] should occasion no surprise.

However, and even where there were settlers, the forces the colonial governments could call on were typically rather thin on the ground. What is remarkable in the circumstances is the way in which the imperial powers were able to govern the vast territories and alien peoples they incorporated into their empires. The population of Britain was vastly outnumbered by the non-British populations of its empire. Accordingly, one of the techniques employed in order to foster a social order[17] conducive to continued imperial rule was that of *indirect rule*. Authority would be delegated to traditional leaders: tribal chieftains in Africa, maharajas in India, sultans in Malaya. Traditional leaders became in many respects the instruments of the colonial state. They collected taxes, perhaps supervised recruitment of workers for the mines and plantations or for the construction of public works like highways, and reported on seditious activity, all in exchange for the retention of some of their traditional authority. This latter included the authority to (e.g.) impose customary law on their peoples and distribute land. Again, as with the settlers, their privileges depended on a continuation of colonial rule. They could therefore be the object of animus for the colonized as much as were the settlers.

In the event the settlers proved a major obstacle to the process of decolonization. Initial attempts by the imperial power to suppress an anti-colonial movement would be supported by the settlers. The financial strain of fighting a war of liberation, however, along with loss of personnel, perhaps with a concern for playing into the hands of communist elements in the resistance movement, would bring the imperial power to the peace table. Their conclusion would be that in the circumstances the most rational course of action was to deal with the native nationalist movement and negotiate independence on their terms.

It would be at this point that the settlers, who were committed to staying in the colony, would oppose metropolitan policy. The concern was that giving the nationalist government what they wanted – the right to form a government and rule in the context of an independent state – would be the end of their privileges. This might be expressed through pressure on the metropolitan country not to "sell them down the river." But in at least two instances – Algeria and Southern Rhodesia – it led to attempts to declare the independence of the colony from the imperial power but under settler tutelage, i.e. so as to deal with the natives on settler terms. In the Palestine case, of course, the announced desire of the British to withdraw from the country led the Jewish settlers to a drive to seize power and deal with the Palestinians on their terms rather than wait for the Palestinian majority to deal with them on *their*, presumably very different, terms.

Elsewhere the classic forms of colonial domination and exploitation mediated by political control have given way to independent states. But

16   More money for the education of their children, substantially higher pensions than those for which the natives might qualify, for example.
17   Apart, that is, from the discursive ones which were discussed in chapter 7.

colonialism did not end. Rather relations with the major powers of the Northern hemisphere, including the US, have been ones that could be plausibly constructed as "colonial." There have been interventions designed to protect Western interests. Political elites in the erstwhile colonies have been nurtured in various ways as protectors of Western interests. It was in this context that the idea of "neo-colonialism" emerged.

The substance of classic colonialism[18]

The case of Angola

In a number of states in Africa the struggle for independence is closely bound up with the land question: a question in which, in the African context, land had been appropriated from the native population by European settlers, so reducing its availability for subsistence purposes. In some instances it also led to Africans having to seek wage work which also caused resentment, though depending on the conditions encountered there. This was the situation in erstwhile Rhodesia, now Zimbabwe. But always the appropriation of land by the settlers meant attempts to also appropriate their labor. Depending on conditions on the settler estates this could also generate hostility. So while in Angola the appropriation of land by settlers never occurred on the scale it did in Rhodesia so as to force the native population into wage work out of sheer material need, the settlers did have labor needs which had to be met. The way these needs were satisfied along with the appropriation of land are at the crux of the anti-colonial movement in that country.

Angola was a relatively poor Portuguese colony on the west coast of Africa to the southwest of Zaire. During the 1950s it experienced a coffee boom which attracted a number of settlers into the northern part of the country. The nature of this settlement process is the proximate cause of the anti-colonial revolt in Angola. This started in 1961 and lasted until 1975 when Angola received its independence. The maps in figure 9.14 show this quite graphically. The only coffee growing area which did not generate what are called there national revolutionary events (worker revolts, murder of settlers) is to the south. The conditions under which coffee was produced there were quite different and, as we will see, this explains why there was no revolt in that region.

The settlers vigorously resisted the granting of independence to Angola. The reason for this is that they recognized that their success as planters depended on the reproduction of privileges that an independent state would not recognize. These privileges concerned access to land and to labor. As far as land is concerned the Portuguese colonial authorities operated a concessions system for the settlers in which supposedly unoccupied lands could be claimed and obtained for a nominal fee. The appropriation of land in the north

18 In the following two case studies I am greatly indebted to Jeffrey Paige's *Agrarian Radicalism* (1975).

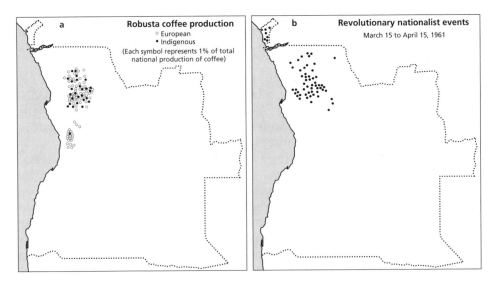

Figure 9.14 The national movement in Angola and the political economy of a settler society. **a**, the location of settler coffee plantations in Angola. Note that in the northern cluster these occurred alongside the production of coffee by indigenous peoples and this led to conflict over land as well as access to labor. **b**, the location of outbreaks of resistance aimed at the settlers and the Portuguese colonial government. There is a clear relation to the map in **a**, though also exceptions. The southern cluster of coffee producers did not experience these outbreaks of violence since production there was more mechanized, so there was less forcible drafting of indigenous workers. There was also less conflict over land owing to the fact that the indigenous peoples were not involved in the production of coffee.

Source: After J. Paige (1975) *Agrarian Radicalism.* New York: Free Press, pp. 231, 265.

by settlers, however, was in an area of maximum native population density, so it was inevitably at the expense of both small African coffee producers and traditional subsistence farmers. In fact, and despite the provisions of the concession system in which only unoccupied land could be claimed, a good deal of it was occupied by force with the support of the local colonial administrator. The latter also helped settlers clear "squatters" from their concessions.

On the other hand, the appropriation of land was never on a scale sufficient to solve the settlers' labor problem by creating a large stratum of landless Africans. Rather use had to be made of the colonial labor code. This asserted the fundamental moral obligation to work and the right of the state to enforce that obligation. Natives were therefore required to demonstrate that they were employed or face compulsory labor. Labor was also orderable for contravention of the criminal statutes or for failure to pay the native tax. Again the settlers depended on the colonial administrators to provide the requisite number of workers. Some of the workers were migrants from the south but the recruitment of these workers also depended on an implicit resort to the colonial labor code and its moral obligation to labor. In other words the settlers relied to a considerable degree on what could reasonably be called a system of forced labor.

To make matters worse conditions on the estates were far from appealing. Unlike on the estates further south lack of capital made mechanization difficult. In a context of increasing labor prices or falling coffee prices the only strategy available to the planter was to increase coercion: expanding the required daily task or tightening labor discipline on the estate.

On the other hand, and as Jeffrey Paige has argued, "without the special privileges of land concessions and compulsory labor the northern coffee estates could not exist. Since their privilege rested in turn on the distinction between the indigenous and nonindigenous status, the settler estates could not survive in a political system controlled by the African majority. Thus the estate owner found it necessary to resist any attempts to share political power with the African population even at the cost of an indefinite period of guerilla war" (Paige, 1975, p. 254).

The case of Vietnam

As part of a policy designed to control the colonized populations imperial powers often coopted some indigenous elements so that they might support the regime. They were granted privileges, economic and/or political. These privileges, like those enjoyed by settlers, placed them in an antagonistic relationship with those indigenes who did *not* enjoy those same privileges, but were rather their victims. These favored few often became an object of resistance. This also heightened resistance to the colonial regime since it was seen as protecting the privileges that the favored few enjoyed and at the expense of the vast majority.[19]

The history of Cochin China, part of French Indo-China, and what later became South Vietnam, is relevant here. Cochin China became a French colony in 1862. The French wanted to make the colony self supporting and towards that end they took steps to create a profitable economy. Their major vehicle for this was promoting the cultivation of rice for export. At the same time they had a problem of control. They solved both these problems, or so they must have thought at the time, by making large land grants – large enough to produce a surplus for export – to collaborators in the area of the colony most propitious for the cultivation of rice, the Mekong delta.

Production was carried on on these estates by share croppers working under conditions that were greatly to their disadvantage. Given the lopsided distribution of land resulting from the large concessions made to native collaborators – in Cochin China a mere 6,316 owners controlled 45 percent of the total cultivated land – peasants had little option but to submit to the demands of their landlords. Rents were very high and were collected regardless of the level of the harvest, often resulting in starvation for the peasant or making

19 This section relies heavily on Paige (1975).

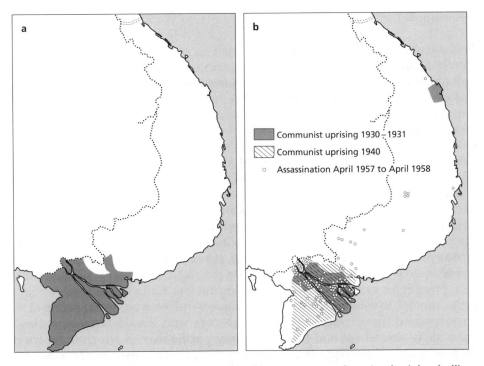

Figure 9.15 Rice exporting areas in (South) Vietnam, and anti-colonial rebellion, 1930–1958. Note that the rice-exporting areas **(a)** were the ones dominated by large landlords of Vietnamese origin and the sites of severe economic oppression. This helps account for the geography of revolutionary socialist events in that country **(b)**.

Source: After J. Paige (1975) *Agrarian Radicalism.* New York: Free Press, pp. 309, 322.

him an easy target for the money lending activities of the landlords. Exploitation was quite unusually intense.

This made the peasantry highly susceptible to anti-landlord propaganda and to organizing by the Communist Party. An initial revolt in 1930 was concentrated in the rice growing area dominated by the large landlords (see figure 9.15). Landlords and village notables were assassinated. The major objective of the revolt was not so much the overthrow of the French colonial regime but relief from tax payments. The revolt, however, was violently suppressed by the French military and taught the lesson that if the peasants were to deal effectively with the landlords then they first had to throw the French out.

Note finally the considerable parallels with the Angola case discussed above, with the difference that in that case it was settlers who benefited from the existence of the colonial regime. Instead of exploitation on the part of a favored indigenous group it was exploitation by a favored settler group which sparked off nationalist revolt.

Neo-colonialism

The essential impetus to colonial imperialism in the latter part of the nineteenth century was the expansion, under the stimulus of capitalism, of the economies of the Western European countries and of the United States. It was this which led them to look with fond eyes on the markets that less developed parts of the globe offered or on the raw materials that they could produce. Nevertheless, not all of those underdeveloped areas were ripe for the taking: so we confront a problem as to just how they might be, if necessary, subordinated to the needs of the economies of Western Europe and North America without assuming the status of colonies.

This problem became generalized after the Second World War since what had been colonies up until that time became, over the next twenty or twenty-five years, independent states. The period from 1945 to 1970 is one of retreat from empire so that today there are very few colonies left. On the whole this retreat proceeded remarkably peacefully. It was only in the earlier years, or where there were substantial colonies of white settlers, that it was more difficult. France, for example, fought a war in the late forties in Indo-China in a futile attempt to retain French colonies there. And later on, white settlers bitterly opposed independence in Algeria, Kenya, and Rhodesia. Nevertheless, the scale of the retreat and the dominantly pacific way in which it occurred do pose an interesting question which is similar to that posed by the existence of backward, but independent, states in the late nineteenth century: states like Siam and the states of Latin America. For if these areas were of major interest to the Western colonial powers why did they not subordinate them to the status of colonies; or where they did, why did they ever relinquish that control?

The benefits of colonial rule have to be calculated relative to its costs, of course, and this was undoubtedly a reason why the colonial powers in the post-war period were so willing to sit down and negotiate terms of independence at virtually the first sign of serious anti-colonial agitation.[20] This willingness was further galvanized by the communist bogy and the fear that if independence was not granted then the communists would move in and take over the independence movement. The growth of socialist parties in Western Europe also gave birth to an anti-imperial sentiment in the metropolitan countries on ideological grounds.

Yet could it be that the reason control was relinquished, or, in the case of other countries like those of Latin America, never taken up, was that there were other mechanisms apart from the bluntly colonial which could give the advanced capitalist countries of the West what they wanted? This is the burden of the neo-colonialist school of thought as well as that of so-called dependency theory, which has been concerned above all with the Latin American case. I treat both these arguments here, postulating neo-colonial

---

20   For the view that the benefits of empire for interests in the imperial country have been exaggerated see Strachey (1959).

mechanisms since mechanisms of dependency share a good deal with what is more narrowly termed "neo-colonial." I want also, however, to contrast an older with a newer neo-colonialism.

The old neo-colonialism

The essential argument here is that instead of a colonial government operating the economy on behalf of metropolitan interests, an independent government can do precisely the same thing – indeed, will have an incentive to do the same thing – under certain circumstances. Consider here the possibilities.

In the first place it may be that the state or the particular strata which form its social basis, elect or finance the government, are ones which benefit from patterns of trade and investment which, in turn, benefit foreign interests rather than the mass of the population. The investment of multinational corporations in less developed countries may have certain local multiplier effects which are valuable to some indigenous elements: distributorships, legal work, repair services, insurance services, some intermediate processing of products prior to export, trading in imports from overseas. Alternatively they may own the plantations – coffee, cotton, perhaps – which form the first stage in a chain which proceeds via the major global actors in commodity trade and processing and ends with Western European or North American supermarket chains.

In the second case, and perhaps alongside the conditions described in the paragraph above, what the local elite gets in exchange for policies friendly to foreign business and the foreign state is the military and financial aid to maintain its grip on a local population on the exploitation of which it depends for its own wealth. For in many of the more backward areas of the globe there is a chronic land distribution problem which runs in conjunction with highly repressive labor practices. This, for example, has long been the case in Central America where subsequent pressures for reform and political instability have led to US support for the elite, simply to prevent a communist takeover.

This led, and to some degree continues to lead in some instances, to a fairly predictable pattern of neo-colonialist politics: one in which the local elite could rely on a foreign power to protect their interests and those of the foreign investors. Historically this has been the pattern in Central America where the US has played the role of guardian angel to the local landed oligarchies, while they in turn have acted as local policemen for the business interests of the big US corporations. When the local elites have lost power the US has usually come to their aid.

Box 9.2 gives some examples of this but I also want to caution that I think this pattern is now on the way out. This is because the big corporations, the Western-based MNCs, have now found *new* ways of subordinating the production of LDCs to their own purposes without getting directly involved in production there and so without needing to call on the US to send in the marines or engage in dirty tricks such as those which brought down the Allende government in Chile in the early 1970s.

Box 9.2   *Old Style Neo-colonialism in Central America*

One of the most notorious and widely known examples involved Guatemala in 1954. The immediate context was a series of democratizing, liberalizing governments in Guatemala over the previous ten years. These had been generally pro-labor. The minimum wage had been raised and unions legalized. A start had also been made on land reform. These policy changes had serious effects for both the more privileged sectors of Guatemalan society and US corporations operating in the country. One of the major corporations involved was United Fruit which, in retaliation, ran down its banana exports from the country, seriously reducing its foreign exchange earnings; and also conducted a public relations campaign in the US intended to vilify the Guatemalan government as "communist."

Persuaded either that there *was* a communist threat – though the threat is extremely debatable – or that United Fruit's interests should be protected, the US government took steps to put the defenders of its interests in Guatemala back in power. World Bank loans were cut off, as was US military assistance. More crucially, however, a coup was engineered by the CIA. In the aftermath United Fruit and its allies in Guatemala got precisely what they wanted: United Fruit got its land back, the secret ballot was eliminated, and the "illiterate masses" were disenfranchised. The US government then took steps to consolidate the power of the new government through the provision of so-called "aid" (remember: "aid" can mean many things and it's not necessarily intended to "aid" the majority of a country's population).

The Guatemalan case brings to mind the more recent Nicaraguan instance. In Nicaragua a right-wing dictatorship (the Somoza dictatorship) that had been supported by the US was overthrown and a vigorously reformist, pro-working class, and pro-peasant government (the Sandinistas) took its place. The US government, as in the Guatemala case, sought to continue its alliance but this time in order to regain power for its Nicaraguan partners. Strategies included:

- Support for the Contras, the military wing of those Nicaraguan groups which wanted a reversion to the status quo. The Contras waged terror in their incursions into Nicaragua and did great damage to the country's economic infrastructure.
- Providing a base from which to pursue counter-regime activities. Originally this was the US itself, until the US procured bases for the Contras in Honduras.
- An economic embargo. This was particularly important since Nicaragua had depended so heavily on the US as an export market.
- The blocking by the US of all lending by such organizations as the IMF and the World Bank. Historically the influence of the US in those organizations has been preponderant.
- The endorsement of the opposition parties in elections. In particular the US hinted that it would lift the embargo if the Sandinistas were voted out of office.

The effect of these activities was to weaken the Sandinista regime, to deprive it of its social base, and ultimately to defeat it electorally. Thus, Nicaragua was burdened with huge defense costs in order to police its border against the Contras. Combined with the economic embargo and the physical destruction brought about by the Contras this created widespread economic deprivation. People also disliked the military draft. Indeed more than 15 percent of the population left Nicaragua during the war with the Contras. Some of these were people who disagreed with the regime but many were seeking to avoid the draft and/or economic hardship. Technical strata, the more educated groups, were overrepresented among those fleeing. It was among these that the Sandinistas were least popular, while the viability of the economy depended heavily on them. But the subsequent vote for "freedom" and "democracy" can only be interpreted with extreme skepticism: to say that people voted under an externally imposed duress is to put it mildly.

The new neo-colonialism

One way of conceptualizing the links between the less developed countries and the more developed, which lie at the heart of contemporary neo-colonialism, is in terms of the different stages involved in the production and bringing to market of the various agricultural and mineral products that are the characteristic focus of colonial relations (Murray, 1987).

There is an initial pre-production phase which tends to be dominated by major multinational corporations. With respect to agricultural products this may be a matter of genetic research, the provision of fertilizers and seeds, and the financing of production. In the case of minerals there are questions of geological survey and the assessment of deposits, trial drilling, stages which analogously tend to be dominated by the major multinational mineral concerns like Rio Tinto Zinc, Pechiney, Billiton-BHP, or Shell Oil.

There is then the actual production phase which, in the case of some agricultural products, may be through small peasants or locally owned and operated plantations. Indeed, in the case of some products the large Western companies have tended to withdraw from production altogether, dividing up their large land concessions and assigning title to them to small producers, though with certain provisions that ensure that they will continue to grow the product that the corporation trades in. In the case of minerals, likewise, it is not uncommon for the operation of the mines or oil fields to fall into local hands, usually through nationalization or simply through the rescinding of the leases on which operation of the mines by the major mining corporations depended.

Once produced in its raw form the product – coffee beans, raw cotton, crude oil, bauxite, sugar cane – then has to go through various stages of processing, transport, and distribution before it is finally sold to the consumer. Again, it is these stages which tend to be dominated by the major multinationals through facilities in the DCs, though native entrepreneurial elements may also share in this, as in the ownership of small sugar refineries or drying and storing sheds for coffee beans. It is these stages, moreover, where the greatest amount of profit lies simply because of the market power that the MNCs dispose of over the prices at which they buy and sell.

Reactions to these developments on the part of LDCs have been diverse. Among others they have included the formation of cartels through which to sell their products and so alter the balance of power between themselves as producers of raw materials and the big international buyers. The classic case is OPEC. The risks here include attempts to break the monopoly, perhaps by shifting investment to "politically safe" areas: some evidence of the importance of this is that it seems very unlikely that oil would have been developed in such environmentally difficult areas of the world as the North Sea and the north slope of Alaska without the emergence of OPEC as a significant player in the pricing of oil. This is because only at the monopoly prices set by OPEC was development in such areas at all economically feasible; with other products, like bananas, there are lots of places in the world where they can be produced and so elude cartels.

A second approach to dislodging the power of the Western MNCs has been movement, often state-sponsored, into the stages of production/processing where the greatest value-added lies. The risks here are ones of protection on the part of the big Western corporations. So if mineral producers try to enter the refining stages of production the large Western MNCs may simply protect their existing (DC) investments by requesting tariff protection from their host governments. Another risk is simply that global banks will refuse, on arguably commercial grounds, the credit necessary to move into those stages. Alternatively the big corporations will simply shift their supply lines in the direction of politically safer areas of the globe, playing one LDC off against another in order to do this. The consequence of this is a spatial division of labor which is to the advantage of the big MNCs, and to a lesser extent, to the advantage of the populations of the DCs. The pressures, moreover, are ones which work to the serious detriment of the workers in the LDCs who toil in the fields or in the mines. Market power gravitates to the agribusiness and mining MNCs which, accordingly, have the ability to dictate terms to the numerous producers of the commodities which they trade in and process. In producing for agribusiness, for example, just standing still may depend on the retention of repressive labor practices and on production from those large expanses of land which have been the target of land reform efforts. The threat to the landed elite, therefore, can be from two directions at once: from the MNCs which buy their product on highly competitive markets; and from the workers below who press for reforms.

> The power of this oligarchy does not exist in an international vacuum. It is part of the world market for agroexports and the international division of labor. While the oligarchy takes the first cut out of the income from agroexports, most of the profit goes to foreign corporations that control the trading, processing, and marketing of commodities. Central American growers and exporters only control a small portion of the coffee and other commodity-based industries. They have to accept the prices dictated by a world market in which they have very little influence. The fortunes of these oligarchs come not from their ability to set prices

and manipulate markets but from their access to cheap land and labor as well as their monopoly on credit and government assistance. . . . This international relationship benefits both the oligarchy and the foreign corporations. The corporations depend on the oligarchy to keep the laborers or peasants working for low wages. Those wages – less than one-tenth paid to workers in developed countries – guarantee the continued supply of low priced commodities for the international market. (Barry, 1987, pp. 48–9)

Dilemmas of Development in an Unevenly Developed World

There is remarkable continuity in the geography of development. Those countries which are the most developed today were for the most part the most developed fifty years ago. Even one hundred and fifty years ago the geography of gross national product per capita would be very similar to that which we observe today. It seems, in other words, that an early start bestows immense advantages. Capitalist development started in Western Europe. The institutions of private property, commodity exchange, and wage labor on which it depended were then transferred to North America and to the antipodes where a similar logic of development was unleashed. Japan is a different story, but in creating new top-down institutions through which to orchestrate the development process it created a model for a later generation of East Asian NICs.

The continuity of this geography is in part political. Economic clout provided military clout: the physical force through which empires could be created and subordinate roles, complementary to those of the major industrial powers, imposed on colonies. Market forces, however, have also played a role. Most investment today tends to occur within the charmed circle of the more developed countries simply because that is where most of the demand for new products, most of the skills necessary to producing them, along with states adept in their ability to regulate in a predictable, business-friendly manner, happen to be. And to the extent that market forces do not work in this way, to the degree that they are ineffective in preserving this degree of privilege for DC firms and populations, then the state will step in, protecting sunset industries or mobilizing international organizations like the International Monetary Fund (IMF) to make less developed countries develop along lines congruent with their own goals of maintaining an economic hegemony.

From the standpoint of less developed countries, eager to develop, and on behalf of whatever coalition of forces, this creates a dilemma. One of the approaches has been to use international organizations as platforms from which to exert pressures, largely of a moral kind, on the governments of the more developed countries to impose some sort of restraint both on themselves and on the MNCs headquartered there. Accompanying these demands has been a discursive construction of relations between DCs and LDCs which relies on images of territorial exploitation. I have called this set of beliefs Third Worldism and it is on this that the first part of the remainder of this section focuses.

Historically there have been other approaches to remediation, however. One of the more notable has been for states in their economic policies to try to isolate themselves from the world market, to severely limit and control their relations with other countries: to lock up investible money capital within national boundaries, to control the allocation of credit to the advantage of particular industries deemed vital to the national industrialization program. There are a number of examples of this. But by far the most radical and far reaching was the communist experiment, commencing with the Russian Revolution in 1917 and only finally breaking down some seventy or so years later. It is to a consideration of the communist alternative that I devote the second part of this section of the chapter.

Third Worldism

There is a belief, widespread among the less developed countries, that international trade and investment are from their standpoint, problematic; that the more developed countries intervene in these to their own advantage and to the disadvantage of the LDCs. In particular these interventions work to confine them to less advantageous positions in the geographic division of labor, and confirm the occupancy of the DCs in the more favored ones.

A major forum for these arguments has been UNCTAD (the UN Conference on Trade and Development). In particular UNCTAD has provided an arena for efforts to create what the LDCs call a New International Economic Order. Central to this are the resolution of major demands of the LDCs. These include:

1 A restructuring of world trade by which DCs would lower their tariffs on imports of such labor-intensive goods as shoes and textiles and on imports of technically unsophisticated products like steel. Such products are especially appropriate to the human resource base of the LDCs and so would allow them to industrialize in these directions. Table 9.1 indicates how world garment production has tended to shift towards LDCs. Nevertheless, the shift would in all likelihood be much more rapid still if the DCs did not maintain quite such severe import restrictions on these products.

2 A greater voice for LDCs in the management of the world monetary system, particularly in the operation of the IMF and the World Bank upon which LDCs rely heavily as a source of loans. The tendency in both these institutions has been for the number of votes held by a given nation to be proportional to its quota of the institution's subscribed funds; since the wealthier nations have the largest quotas they tend to have a preponderant voice in institution policy. An institution policy of particular concern has been that of "conditionality," which is discussed below. A major issue dividing DCs and LDCs has been that of the policies of the IMF. The concerns here are complicated, but a major one has to do with the leverage that control over international loan finance confers on the IMF for forcing

Table 9.1 Changing share of world clothing exports 1980–1995

	1980 (%)	1995 (%)
Less developed countries		
China	4.0	15.2
India	1.5	2.6
Indonesia	0.2	2.1
Thailand	0.7	2.9
More developed countries		
France	5.7	3.6
Germany	7.1	4.7
Italy	11.3	8.9
The Netherlands	2.2	1.8
United Kingdom	4.6	2.9
United States	3.1	4.2

Note the huge increase in China's share of world clothing exports. Among the more developed countries, only the US has managed to increase its share.
Source: After Dicken (1998, table 9.5, p. 291).

a particular model of economic development on those countries which go to it for loans: a model which emphasizes market forces and the importance of maintaining an economy open to global movements of commodities and finance. Not coincidentally this is a model that, to the extent it is indeed adopted by the LDCs, works to the advantage of DC corporations and to a lesser degree DC workers.

Typically when a country experiences balance-of-payments problems it can apply for a loan to cover the difference from the IMF. This has occurred particularly in those cases where private lending institutions regard the borrower as too risky. In exchange for the loan, however, the IMF has usually attached "conditions" designed to rectify the underlying cause of the borrower's balance of payments problem. These "conditions" typically involve a reassertion of market disciplines, with the ultimate goal of closing the gap between a country's *receipts* of foreign currency (from exports, from foreign investment in the country, from the wages repatriated by its emigrants, etc.) and its *needs* for foreign currency (to pay for its imports, to pay off foreign loans, to allow MNCs to repatriate profits, etc.). These conditions typically involve a retreat from intervention by the state into the economy and include:

- A contraction of government spending. The rationale here is that as government spending goes down, so the domestic market contracts. On the one hand imports decrease and on the other firms find they can no longer sell to the domestic market and so have an incentive to export. The balance of payments will then move into positive territory.

- The elimination of barriers to foreign investment. This, it is argued, will stimulate foreign investment and so increase the holdings of foreign currencies in the recipient country's central bank. This in turn will shift its balance of payments in a positive direction.

A problem here, however, is that by conceding ground to market forces in this way LDCs lose control of the development process. The reasons why the state is so involved in the economy in many LDCs are complex, and by no means is it the case that they all have to do with promoting the development process.[21] A major problem for newly industrializing countries is producing at a cost that makes firms there competitive with those from the DCs. Their firms and plants are often smaller and so lack economies of scale. Their social and physical infrastructures are much less well developed and workforces lack industrial skills. State interventions ranging from heavy investments in infrastructure to subsidies and protection for new industries are intended to facilitate moving firms to a position where they *can* compete.

A second problem is structuring the development process for national ends. One of the reasons the South Korean and Taiwanese states have tended to discourage foreign investment is that multinationals have their own agendas, and these do not necessarily include the upgrading of the industrial skills of the workforce, moving production on to higher value-added products, or investing profits in the country where they have been appropriated. Policies of both those states, as we have seen, have deliberately sought both those ends. But this has required considerable state control, including control over the allocation of credit.

The so-called structural adjustment policies of the IMF through which conditionality is imposed are antithetical to these goals. They require the retreat of the state, the opening up of countries to foreign investment, the withdrawal of state subsidies that have had developmental intent. Subsequent investment by Western MNCs, for example, can result in a loss of control over the developmental process. To take one example: in some countries banking is state-owned and this has facilitated the rationing of credit so that it can be channeled towards industries which the state believes have a chance of becoming competitive in world markets. It has also worked to limit the export of capital through (e.g.) investment in foreign bonds. So money capital has had to be invested in the country's economy. But opening a country up to foreign investment includes investment in banking and this undercuts that control of credit that can enhance a country's developmental prospects. It also creates a situation in which the country's infant industries are easy pickings for the MNCs, once again making national control of the development process considerably more difficult. In short, intendedly or otherwise, IMF policies have been the Trojan horse through which Western countries can impose their

---

21   Among other things, government controls may provide a vehicle for corrupt activities. Import quotas and licenses, for example, can be distributed to government officials or their relatives.

own development agenda for LDCs; an agenda from which their own corporations are likely to benefit but at the expense of the LDCs.[22]

The trade issue is also an extremely aggravated one. DCs place limits on
many imports from LDCs. LDCs may find a niche for some raw material
export that is in short supply in DCs: oil is a case in point. But when the producers of those raw materials try to shift to its processing into commodities
of higher value such as chemicals or refined minerals existing producers in
DCs are likely to oppose their importation, as indeed we saw earlier in this
chapter.

The trade issue, moreover, is tightly interwoven with that of the IMF. If
LDCs encounter obstacles in exporting their products to DCs then they may
very likely experience balance of payments difficulties and have to turn to the
IMF. Sometimes loans have been taken out to develop an export industry
which has subsequently been confronted with DC trade barriers. The Dominican Republic took out loans from DC banks to upgrade its sugar industry. The
US then reduced its sugar quota in order to protect US producers. This created
a debt problem which the Dominican Republic was unable to resolve.[23]

The communist alternative

> The world created by industrial capitalism remains a singularly unequal and
> divided one, yet what is striking is how states that wish to compete within it are
> forced, over time, to conform and converge. One can indeed speak here of the
> pathos of semi-peripheral escape: the repeated effort by states that are at some
> medium stage of the development process to accelerate this growth by adopting
> forms of political and economic strategy that circumvent the established norms.
> (Halliday, 1995, p. 219)

As Halliday goes on to point out, communism and the correlative central planning of the economy was one such strategy. Starting in Russia in 1917,
extended largely by force to Eastern Europe after the Second World War and

---

22   Opening up to foreign investment is not necessarily a consequence of IMF conditionality.
But Western pressure has often been part of the story. Gowan has provided a number of examples of just what the consequences can be, based on experience in Eastern Europe since the fall
of communism: "General Electric, after buying Tungsram in Hungary, closed the latter's production of vacuum equipment, electronic components, floppy disk and magnetic tape products
all of which were considered profitable by Tungsram management. The Hungarian cement
industry was bought by foreign owners who then prevented their Hungarian affiliates from
exporting; and an Austrian steel producer bought a major Hungarian steel plant in order to close
it down and capture its ex-Soviet market for the Austrian parent company" (Gowan, 1995,
p. 44).
23   An excellent example of the trade restrictions–indebtedness relation is provided by the case
of the Sudan. Sudan's difficulties started when the market for its principal export crop of cotton
– China – dried up. And the reason it dried up was that, prompted by the American textile industry, the US imposed restrictions on imports of Chinese textiles. Subsequently unable to pay back
its loans to government lenders like the US, the Sudan government had to appeal to the IMF for
a loan.

by popular revolution in China, North Korea, and North Vietnam, it represented, among other things, an attempt to break out of the developmental impasse posed by the world market and the types of neo-colonial mechanisms I have detailed above.

The institution of central planning mandated an isolation from wider, more global exchange relations since these could not be controlled by the plan. But it is precisely this isolation that some observers have identified as a necessary precondition for development in a world where other countries are significantly *more* developed, i.e. in a world of uneven development. Historically, therefore, countries like Brazil, Argentina, and South Africa attained their most rapid rates of economic growth during the world wars of the century when their manufacturers did not have to face competition from imports from Western Europe and North America. It should also be remembered that Germany and the United States made considerable use of tariffs on industrial products precisely as a boost to their own manufacturing growth. Indeed, it was a German, Friedrich List, who provided the theoretical basis for this top-down, regulative approach, one that has recently been taken up with vigor by NICs like South Korea and Taiwan.

This is not to argue, though, that the Soviet model had this as its only objective. Central planning also allowed control over the distribution of the product. Through this, a degree of equality could be achieved that has otherwise proven elusive in market economies. As Arrighi has pointed out, closure and central planning made a major difference to the well-being of the lower social strata. In terms of health, nutrition, and education the Soviet Union's lower strata did much, much better than those of Latin America, including Brazil (Arrighi, 1991, p. 57).

This points to the ideological underpinning of the Russian and Chinese revolutions. They were about a vision of humanity and human development essentially at odds with the capitalist model. As such they were always a threat to capitalism and, equally, the capitalist world was a threat to them. They wanted to spread revolution not just to bring about what they believed to be a better world but also to protect the revolution in their own countries. With respect to the latter this was because in both instances they had had to fight off forces either aided by the Western European countries and the US or, in the case of the Russian Revolution, actually from those countries. The revolution was always believed to be at risk, which also helps account for the desire of the Soviet Union to put a belt of communist states between itself and Western Europe after the Second World War. It should be recalled that it was Hitler's goal to bring low what he called the Bolshevik menace, and before the outbreak of war this was a goal widely shared in ruling class circles in the West.

At the same time their desire to spread communist revolution was a threat to various interests in the major capitalist countries. Quite why this was so is complicated. It was only in small part a fear that their own working classes would be infected and rise up to overthrow the status quo. For the most part the discourse of "freedom" had had its effect. Although the freedom that the capitalist class was concerned about was the freedom to make money, for most it had more to do with the freedom to choose a government, freedom of worship, freedom from censorship, freedom of choice in general. And indeed this was a telling point since communist countries were quite clearly highly authoritarian, even totalitarian, in their institutions; a frightening image for people who had experienced something different, however qualified in practice, even illusory, those freedoms might be. One result of this was that anti-communism could be an electorally useful policy for any government or government in waiting.

There were other concerns. Capitalism is inherently an expansionary form of economic system, including expansion in the geographic sense. To the extent that large areas of the globe became part of the communist bloc they were immunized against commodity exchange. They would no longer form part of the world market. Their consumer markets and their raw materials would be off-limits to Western corporations. This fed into the geopolitical imaginary of the domino effect: the concern that if one country fell to the communists, then its neighbors, and their neighbors, and so on, would quickly fall victim. It seems hard to understand something like the war in Vietnam and Cambodia except in these terms. This was something repeated in Central America subsequent to the Cuban revolution. Any peasant revolt could be interpreted as inspired from Havana and the US proved an easy prey as threatened ruling classes used the communist bogy to secure its help.

In the event the Cold War ended not with a bang but with a whimper. And from the standpoint of our interest in the dynamics of uneven development and its politics it is useful to reflect on precisely what happened. There is now general agreement that at the root of the crisis which ultimately led to the implosion of the communist regimes of the Soviet Union and Eastern Europe was the inflexibility of the planning system. Without the information provided by market prices it proved hard to coordinate the division of labor, so there were invariably shortages of some things alongside surpluses of others. In addition, revolutionary fervor, at least in the context of the revolution as it played itself out in the Soviet Union and its satellites, proved a poor substitute for personal incentives to make the plan work. The result was low productivity, poor quality products, and extremely inefficient distribution: getting agricultural produce to market and storing it was always a problem and attrition from rats incredibly high.

What brought the crisis to a head was a growing reliance on world commodity and financial markets at a time of general slowdown in the capitalist world. In contrast to their earlier isolationism as part of the logic of central planning, from the seventies on the communist countries started importing and later borrowing from the West in exchange for, primarily, raw materials. In the case of the Soviet Union it was largely food that was imported. In other cases high technology was imported in the hope of upgrading worker productivity. There was also a belief that exports would exercise some discipline on industrial managers as they were exposed to competition with capitalist enterprises: indicative of a general feeling of disillusion with the merits of central planning. The need for imports, however, was not matched by export growth. Falling commodity prices pursuant to the general downturn in the world economy from the early seventies on worked against the strategy. Goods originally intended for domestic consumption had to be exported, aggravating local hunger for consumer products. In addition foreign debt was incurred and as interest rates increased towards the end of the eighties this plunged the Eastern bloc into greater depths of crisis.

Undoubtedly the end of the Cold War has had implications for struggles over uneven development. The original inspiration for communist revolutions was the work of Marx and his critique of capitalism. Unfortunately Marx was less helpful when it came to providing blueprints for a classless society. Even so his work, often in highly garbled forms, provided an interpretation of their situation for oppressed classes everywhere, that could galvanize them to action, by identifying the nature of their exploitation, the character of colonialism, and the illusions that held people in thrall. The end of communism, at least for the time being, has served to diminish the credibility of the Marxist critique and enhanced the ideological power of capital and its defenders. The relation between Marxist theory and communist practice was always constructed as a tight one both by the communist regimes and by their antagonists, however unjustified. On the other hand, whatever the power of the Marxist critique, and it remains a very powerful one, the demise of central planning has disarmed, at least temporarily, those who would provide an alternative to capitalism or to the market.

Not only was Marxism an inspiration to peasant revolts, to national movements in places like South Africa, Zimbabwe, and Namibia, the communist regimes themselves were a source of serious material support. They educated cadres and provided *matériel* for insurgencies around the world, from Central America, through Southern Africa, to the Philippines. It is unclear, for example, that the African National Congress would have been able to sustain itself in exile without that support.[24]

Summary

"Development" is a complex category. When one refers to "geographically uneven development" it could be about variations in the productivity of workers, positions in a wider geographic division of labor, differences in income, or all three. So at the very least it is a term that must be treated with a good deal of care, and the relationships between the different meanings dissected. When the leaders of territorial coalitions talk about "development," however, it is typically a mix of the second and third meanings. What is aimed for is a more desirable position in geographic divisions of labor that will increase the value flowing through the place, region, country in question. And since not all can occupy the more desirable positions, and relegation to lower ones entails real material disadvantage, struggle around development and its inherent unevenness can be expected, and at a variety of geographical scales.

---

24   Though support from Norway and Sweden was also significant.

Within countries a common battle cry is "territorial justice." Through their policies central governments invariably have effects on the geography of development and on the abilities of localities and regions to "develop." As such they are the inevitable targets of claims for some sort of redress: for a "leveling of the playing field" on which the various growth coalitions compete; or for a reordering of their fiscal relation with the central government, for every local or regional government, every alliance of business-people with stakes in the health of a particular local economy, can find reasons why they should receive more in payments from the central government than they send in terms of taxes.

Over the past twenty years the politics of uneven development in Britain has slowly congealed around what has come to be known as "the North–South divide." This is associated with the policies of the successive Thatcher governments, though they are policies that began with Labour governments prior to her accession to power and have continued, largely unchecked by the displacement of the Conservatives by Tony Blair's (New) Labour Party. There has certainly been a tendency towards increasing interregional inequality in Britain compared with the slow convergence that was taking place prior to the mid to late seventies. There is also no doubt that this is associated with the policies that have become synonymous with Thatcherism: a pull back of the state, the rundown of nationalized industries prior to their privatization, the strict regulation of the money supply so as to wring out any remaining inflationary tendencies, the convertibility of sterling. Denationalization of industry, however, fell unevenly on the different regions, as did monetarism and the associated turn to full convertibility of sterling. Monetarism placed severe constraints on old line manufacturing, much of which was located in the North, while convertibility gave a strong fillip to the financial services industry centered in London and the Southeast. These changes also meant a squeeze on labor, especially in the North. Not surprisingly they have been reflected in party political support. The support for the Conservatives has strengthened in the South while Labour's heartlands have come to be defined more by the North. Significantly class and labor union membership are much more associated with voting Labour in the North than in the South.

The coalition of forces to which Thatcherism eventually came to appeal, therefore, was regionally based. Significantly much of what happened in that regard had no sense of top-down direction. The primary goal of Thatcherism was to impose a more competitive environment with the view that if that was provided then new possibilities in wider divisions of labor would open up; which they did, if one is to judge from the growth of the financial services industry and the emergence of Britain as a relatively low-wage industrial platform from which MNCs can serve the markets of the EU.

In turning to the politics of uneven development at the international scale I have chosen to concentrate on the colonial nature of the relations that link the more developed with the less developed countries. These define deep cleavages in development, and to be sure colonies came into being in order to serve the goals of imperial countries with respect to maintaining their edge in the international division of labor: they were to be, primarily, sources of the raw materials and markets required by the metropolitan country's industries. In some cases settlers served as part of the means through which this limited sort of development would be accomplished and as a result of the privileges they enjoyed they were to be a primary source of opposition to decolonization. But any development of the colonies also depended on maintaining civil order and in many countries the imperial powers found themselves forced to rely on local, indigenous leaderships. These too were to become associated with colonial rule and its privileges and, along with the settlers, to be primary targets of the dispossessed as they set about overthrowing it.

Decolonization has had very mixed results as far as reversing developmental disadvantage is concerned. Since independence new mechanisms have come into play for maintaining the subordinate roles of less developed countries in the international division of labor. To some degree these have required forcible intervention by the countries of the more developed world, though typically with a coalition of forces within the country in question that gains from servicing the needs of the MNCs, and which would also suffer from the serious social reform that would challenge their pre-eminence. This is not to argue that this type of intervention is always necessary. There has been some tendency to shift to neo-colonial mechanisms which rely more on market forces to do their work.

And market forces are a problem. The geography of development shows strong elements of continuity from one point in time to another: developed countries tend to remain developed countries partly because they are the most attractive places for a variety of reasons for investment. It is as if there is some gravitational pull which less developed countries must find ways of dealing with if they are to develop. There have in consequence been a number of experiments through which countries have tried to create a space for themselves in which they would be insulated from these wider effects; effects which tend to work against their development. The top-down policies of the East Asian NICs are one recent example, but the communist alternative embraced by the Soviet Union, and until recently by China, represents a more radical version of this. This latter obviously had goals in mind other than industrialization; there were also strong redistributional objectives, and one of the saving graces of the experiment was that they achieved a good deal of success in that regard. But markets have advantages of coordinating the sort of intense division of labor on which development depends and this proved to be the undoing of the communist alternative. To overcome the difficulties that were transpiring the Soviet Union opened up to greater levels of exchange with the West and that proved to be the slippery slope.

But even where market forces tend to work against the more developed countries, they can deploy other forms of power in order to retain their privileged positions in the geographic division of labor. These include protection of industries even where the goods in question can be produced more cheaply in the LDCs: labor-intensive industries like textiles and shoes, for example. At the same time the IMF has been mobilized to enforce a free market regime within those countries that makes orchestrating a top-down development process à la East Asian NICs all the more difficult.

REFERENCES

Arrighi, G. (1991) "World Income Inequalities and the Future of Socialism." *New Left Review*, 189, 39–65.

Barry, T. (1987) *Roots of Rebellion: Land and Hunger in Central America*. Boston: South End Press.

Freund, B. (1984) *The Making of Contemporary Africa*. Bloomington, IN: Indiana University Press.

Gordon, I. (1995) "Migration in a Segmented Labor Market." *Transactions, Institute of British Geographers* NS, 20(2), 139–55.

Gowan, P. (1995) "Neo-liberal Theory and Practice for Eastern Europe." *New Left Review*, 213, 3–60.

Halliday, F. (1995) "Interpretations of the New World." *Soundings*, 1, 209–22.

Hudson, R. (1986) "Nationalized Industry Policies and Regional Policies: The Role of the State in Capitalist Societies in the Deindustrialization and Reindustrialization of Regions." *Environment and Planning D: Society and Space*, 4(1), 7–28.

Hudson, R. and Williams, A. (1986) *The United Kingdom*. London: Harper and Row.

Johnston, R. J. (1987) "The Geography of the Working Class and the Geography of the Labour Vote in England 1983: A Prefatory Note to a Research Agenda." *Political Geography Quarterly*, 6(1), 7–16.

Murray, R. (1987) "Ownership, Control and the Market." *New Left Review*, 164, 87–112.

Paige, J. (1975) *Agrarian Radicalism: Social Movements and Export Agriculture in the Underdeveloped World*. New York: Free Press.

Savage, M. (1987) "Understanding Political Alignments in Contemporary Britain: Do Localities Matter?" *Political Geography Quarterly*, 6(1), 53–76.

Strachey, J. (1959) *The End of Empire*. New York: Random House.

FURTHER READING

On the facts of the North–South divide in Britain, see J. Lewis and A. R. Townsend (eds) (1989) *The North–South Divide: Regional Change in Britain in the 1980s* (London: Paul Chapman). On the politics of the North–South divide in Britain: J. Allen, D. Massey, and A. Cochrane (1998) *Rethinking the Region* (London: Routledge); M. Dunford (1997) "Divergence, Instability and Exclusion: Regional Dynamics in Great Britain," in R. Lee and J. Wills (eds), *Geographies of Economies* (London: Edward Arnold); R. Martin (1992) "The Economy," in P. Cloke (ed.), *Policy and Change in Thatcher's Britain* (Oxford: Pergamon Press).

A useful study of the contradictions of regional policy under Thatcherism as they applied to Wales is provided in J. L. Morris (1987) "The State and Industrial Restructuring: Government Policies in Industrial Wales." *Environment and Planning D: Society and Space*, 5(2), 196–213. Relatedly, on the story of another regional growth coalition in the British context, and one which resonates with issues of national identity, see P. Cooke (1983) "Regional Restructuring: Class Politics and Popular Protest in South Wales." *Environment and Planning D: Society and Space*, 1(3), 265–80. For an American case see W. J. Schiller (1999) "Trade Politics in the American Congress: A Study of the Interaction of Political Geography and Interest Group Behavior." *Political Geography*, 18(7), 769–90.

On the appropriation of the land in the colonies and the creation of labor markets, see: L. Callinicos (1981) *Gold and Workers 1886–1924* (Johannesburg: Ravan Press); J. D.

Overton (1990) "Social Control and Social Engineering: African Reserves in Kenya 1885–1920." *Environment and Planning D: Society and Space*, 8(2), 163–74.

On the politics of neo-colonialism, a classic study is Colin Leys (1975) *Underdevelopment in Kenya: The Political Economy of Neo-colonialism 1964–1971* (London: Heinemann). R. J. Barnett (1968) *Intervention and Revolution: The United States in the Third World* (New York: Meridian Books) has some good case studies. Kenneth Good (1976) "Settler Colonialism: Economic Development and Class Formation." *Journal of Modern African Studies*, 14(4), 597–620, is excellent on the politics of those settler societies where the settlers were in a minority. Barry, Murray, and Paige (see references) are also worth consulting. Also on the politics of colonialism and decolonization respectively see the entries under the same names in R. J. Johnston et al. (eds) (2000) *Dictionary of Human Geography* (Oxford: Blackwell).

Although now quite old, Cheryl Payer's (1975) *The Debt Trap* (Harmondsworth: Penguin Books) still remains a classic study of the relations between the IMF and the Third World. And on the New International Economic Order see P. W. Preston (1996) *Development Theory: An Introduction* (Oxford: Blackwell, pp. 242–4).

Chapter 10

The Politics of Globalization and Its Illusions

The bourgeoisie has through its exploitation of the world-market given a cosmopolitan character to production and consumption in every country . . . it has drawn from under the feet of industry the national ground on which it stood. All old-established national industries have been destroyed or are daily being destroyed. They are dislodged by new industries, whose introduction becomes a life and death question for all civilised nations, by industries that no longer work up indigenous raw materials but raw material drawn from the remotest zones; industries whose products are consumed, not only at home, but in every quarter of the globe. In place of the old wants, satisfied by the productions of the country, we find new wants, requiring for their satisfaction the products of distant lands and climes. In place of the old local and national seclusion and self-sufficiency, we have intercourse in every direction, universal inter-dependence of nations. And as in material, so also in intellectual production. The intellectual creations of individual nations become common property. National one-sidedness and narrow-mindedness become more and more impossible, and from the numerous national and local literatures, there arises a world literature. (Marx and Engels, *Manifesto of the Communist Party*, reprinted in Robert C. Tucker (ed.), *The Marx–Engels Reader*. New York: Norton, pp. 476–7)

Context

"Globalization" is a term that sprang suddenly into prominence in the late twentieth century, and despite the fact that, as in the quote above, Marx saw much of what has been attracting attention recently as patently clear in the second half of the nineteenth century. In its contemporary uses, however, it is somewhat more complex. For in addition to the sense that Marx and Engels drew on, the sense of the increasing internationalization of commodity exchange, the expansion of capitalism so as to draw everywhere in the world into its grasp, and some sense of its cultural consequences, there has been added an environmental dimension. Globalization in the cultural sense is clearly closely linked to globalization as an economic process, as Marx grasped in his reference to literature; it is, for instance, the cultural content of exports, of foreign investments like those of McDonald's, that is sparking fears in other countries of Americanization, Westernization, or whatever. The environmental problems are somewhat different, however, and have to do largely with the massive expansion of production that has occurred under the auspices of capitalist development and the way that has affected the air, the oceans, other sorts of water bodies, biodiversity, and so on. This in turn has generated pressure for policy solutions.

So globalization is political; it has a politics, and that politics is the primary focus of this chapter, though as I hope to demonstrate the politics of globalization is rather different from what many believe it to be. I am going to start out by discussing the various theses about that politics in its more economic incarnations: theses which tend to prevail not just in the media, where they are certainly very common, and not just in the pronouncements of politicians, but also in much of academe.

In the second major section I make some critical comments about this particular politics of globalization on what I call "its own terms." On its own terms I find the theses that prevail to be, and paradoxically, both over- and under-spatialized, and certainly far too pessimistic from the standpoint from which they tend to be written: that of the populations of the advanced capitalist societies. But the fact that most of the accounts tend to be written from the standpoint of the more developed countries should alert us to an issue that by and large has tended to elude those writing about the politics of globalization: that it has to do quite centrally with the politics of uneven development; a struggle between capital and labor, for sure, but also one that is refracted by the territorial concerns that we talked about in chapter 9.

In the third and final section of the chapter I take up the "other" politics of globalization; those that pertain to the environment and culture respectively. The environmental challenge is real. There are no illusions there. But in discussions of the cultural inflections of globalization some of the point seems to have been missed. And the points that have been raised tend to have be based on faulty inferences from what is happening around the world. The claims made about cultural homogenization and imperialism are cases in point. But that does not mean to say that they have not been drawn on, exaggerated, even erroneous as they might be, in struggles around the politics of globalization in its economic sense.

The Politics of Globalization Thesis

Consider first the dominant notion of a politics of globalization which has to do with the increasing globalization of the economic: the emergence of a global economy which increasingly challenges the ability of national governments to manage their respective economic affairs. Some of the classic indicators here include data on trade levels and foreign direct investment. Both of these have

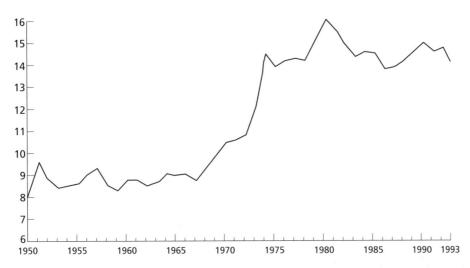

Figure 10.1 Exports and "globalization": exports as a percent of gross domestic product of Western developed countries (in current values). The major expansion of international trade, here indexed by exports, after about 1970 is often treated as a major expression of tendencies towards globalization. Note, however, how the upward tendency tends to level off after the late seventies.

Source: After figure 7.1 (p. 175) in P. Bairoch (1996) "Globalization Myths and Realities." In R. Boyer and D. Drache (eds), *States Against Markets*. London and New York: Routledge.

shown quite strong increases relative to gross national products since the late sixties. As far as the more developed countries are concerned (figure 10.1) exports, as a measure of trade, and as a proportion of gross domestic product took off from a level of around 9 percent in the late sixties and accelerated rapidly to reach a level somewhere between 14 and 15 percent. Foreign investment took off later, increasing rapidly during the eighties (figure 10.2). As far as trade is concerned, however, it clearly leveled off after 1980 and still at a rather modest relative proportion. And foreign investment – no more than 4 percent according to figure 10.2 – is still quite small relative to gross domestic products.

Nevertheless, and in turn, these changes are linked in significant part to the emergence of what some have called a New International Division of Labor. This is in contrast to an Old International Division of Labor and so needs some explanation in those terms. According to this argument, prior to the last third of the twentieth century, and certainly prior to the Second World War, the relation between more developed and less developed countries was one between manufacturing economies and those supplying raw materials to them. Trade between more developed and less developed countries, therefore, consisted of an exchange of manufactured goods for minerals, natural fibers, foodstuffs, particularly tropical ones.

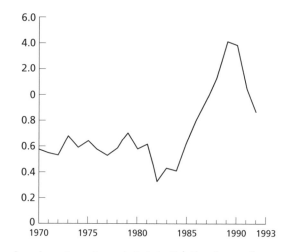

Figure 10.2 Foreign investment and "globalization": outflows of foreign direct investment of Western developed countries (in current values). This is not such a long time series as the one for exports. There was clearly a big increase during the 1980s but then a falling back. More generally, the year-to-year variation in foreign direct investment tends to be quite a bit greater than in the case of trade. Note also, however, how the percentages on the vertical axis are so much smaller than those on the vertical axis in figure 10.1.

Source: After figure 7.2 (p. 182) in P. Bairoch (1996) "Globalization Myths and Realities." In R. Boyer and D. Drache (eds), *States Against Markets*. London and New York: Routledge.

Since the late sixties, however, this has begun to change. Less developed countries have started acquiring a greater industrial presence. To a considerable degree, though not entirely, this has been mediated by the emergence of new divisions of labor within multinational firms. Firms based in the US, Canada, Britain, France, Germany, and Japan have started relocating the lower-skill parts of their labor processes to sites, often the free trade zones we discussed in chapter 3, in less developed countries: sites like those in Mexico just south of the US border. The geographic division of labor which emerges, therefore, is one in which the firm's offices and plants in more developed countries perform the roles of overall direction, research and development, and the more skilled parts of the labor process; while those in the LDCs are primarily engaged in the lower-skill parts and what they produce is shipped to the more developed countries where it will go through final processing/assembly stages. In this way multinationals have tried to reduce their overall costs of production, since wages in the LDCs are typically so much lower.[1] As a result countries which were once almost exclusively pur-

---

1   I am not arguing that MNCs necessarily own plant in LDCs. They may operate, rather, through subcontracting with locally owned and operated firms. This has been the pattern for a number of the footwear and clothing firms based in the DCs which then import them to be sold through the various retail chains.

veyors of raw materials to the more developed world now have growing industrial sectors. This would be true, for example, of Sri Lanka, which used to be primarily a source of tea, and Malaysia, whose major exports used to be tin, rubber, and palm oil.

Think and Learn

"Less developed countries have started acquiring a greater industrial presence." Recalling chapter 3, under what circumstances had industrialization previously taken place in some of those countries? Assuming that some of it had been through MNCs extending their divisions of labor into them, what sort of geographic division of labor was it? Were their plants in the less developed countries integrated into a labor process organized on a wider geographic basis or were they clones simply serving the domestic markets of respective less developed countries? How does your answer reflect on the conditions that have to be in place for MNCs to adopt one form of geographical organization as opposed to another?

This is not to argue that the MNCs have been the sole vehicles for industrialization in the less developed world. The story of the Far Eastern NICs, along with some elsewhere like Brazil, certainly cannot be reduced to one of investments by MNCs. In those cases there has been a top-down, state-driven orchestration of industrialization, and this has allowed state priorities rather than those of the MNCs to govern the process. Again, as we noted in chapter 3, the objective has been to progressively move up the hierarchy of positions in the industrial division of labor from the low-skill tasks that tend to prevail in those happy hunting grounds of the MNCs, the free trade zones, to more skill-intensive tasks. But this industrialization has also been export-driven, which means a further thrust to the process of globalization.

In the standard accounts of this process two preconditions are seen as significant. The first is that of the deskilling of labor processes. In order for the labor process to have low-skill aspects to it, in order for it to be possible to break it down into higher-skill and lower-skill stages, a deliberate process of taking the skill out of it has to occur. There is indeed a long history of precisely this under capitalism: breaking the labor process down into more and more fragments so that what is assigned to a particular worker may be no more demanding than tightening a particular screw, drilling a particular hole, or lifting parts from a dolly onto an assembly line (Braverman, 1974). The goal has been to lower labor costs, since typically the supply of unskilled workers is much greater than that of those with skills. In this way employers could take advantage of low-wage labor pools like women and immigrants.

Think and Learn

Is there anything necessary about the deskilling of labor processes, do you think? Is this the ultimate goal of business? Do you imagine that fifty years hence all workers will be reduced to mindless operations of pressing buttons or pulling levers? If so, why? If not, why not?

But these deskilled parts of the labor process can now be relocated to LDCs where the wages will be considerably lower than those that have had to be paid in the more developed world, even under sweatshop conditions. What has made the difference – and this is the second precondition – has been the development of new modes of transportation and communication. The fax, the electronic mail, the ability for computers to network with those elsewhere, the role that satellites have played in this, have greatly facilitated the task of coordinating the different parts of the multinational corporation: monitoring sales, inventories, shifting production at short notice from one plant to another as soon as shifts in demand or in costs occur.

Transportation has likewise benefited from technological changes and has been subject to long-term declines in real costs and, where needed, to a speed of delivery, through air cargo, unattainable in the pre-jet era. The container has become *the* mode of transferring products from one country to another. This has involved huge savings in terms of stevedoring since it dramatically reduces the need for labor to handle the products. Rather the container is simply transferred from a ship on to a truck or railroad for final delivery.

In some instances improved telecommunication has been the vehicle for the transportation of inputs. A number of American corporations have, consequently, been able to shift their data processing functions to lower-wage economies in the Caribbean and even as far afield as Ireland. This is especially attractive to firms with large data processing needs like credit card companies, but even airlines have found it important. In the Irish case data processing can take advantage of time zone differences: the data can be transmitted to Ireland at the end of the working day and be returned, duly processed, by noon the following day.

The other aspect of globalization in the economic sphere is financial. Trade with, and investment in, foreign countries requires foreign exchange dealings: purchasing dollars with pounds to invest in the US, changing dollars back into pounds in order to repatriate profits, for example. The past thirty years has, accordingly, been accompanied by a gradual elimination of the barriers to exchanging one currency against another. Without it, it seems unlikely that trade and foreign investment could have grown to the degree that they have. Moreover, these barriers used to be substantial and took many forms: limits to the amounts that could be repatriated by multinationals, the rationing of foreign currency to importers who needed it in order to pay their suppliers in foreign countries. This has changed very considerably and the major currencies, the dollar, the yen, the British pound, the Swiss franc, the German mark, the French franc, the Dutch guilder (these last three being replaced by the euro from early 2002), and so on, are now for the most part completely convertible into their foreign equivalents.

Think and Learn

If the barriers to convertibility come down then currencies can be easily traded one for another. Furthermore, that trade might in turn be the source of considerable profit for the trader. What sort of considerations might result in, say, a person selling yen in order to purchase dollars? If large numbers of traders were doing this, what might be the implications for the relative values of the yen and dollar? How might that, in turn, affect the prices of imports and exports for Japan and the United States respectively?

Consider now some of the implications of these trends for politics. Two related arguments have been made here. The first is that there has been a weakening of the state, a deterioration in its ability to manage the economy. The second is that globalization has resulted in an increase in the power of capital over labor. We consider each of these in turn.

The weakening of state regulatory power

This is the thesis of *territorial non-correspondence*: the events over which the state needs to exercise control in order to achieve its objectives are increasingly *outside* of its control. This can be appreciated through a consideration of the various mutations that state economic policy has undergone since the Second World War.

It is widely believed that, for a while, particularly during that period that has since been defined as "the Golden Age of capitalism," between about 1950 and 1970, states were quite effective in regulating national economies; effective, that is, in maintaining growth in incomes and full employment. In order to do this they engaged in a variety of policies designed to stimulate demand and so provide incentives for businesses to invest. Since about 1970, however, these policies have been decreasingly effective and the reason given for this is "globalization." For typically, in a world in which the movement of commodities is increasingly across international boundaries and in which there is a lively, and correlative, trade in currencies, attempts by individual states to use policy in order to maintain economic growth have, it is argued, generated other, quite unintended, and counteracting effects in the international economic environment.

Major weapons in the state policy armory have been monetary policy and fiscal policy. In monetary policy central banks work to counter declining economic growth by lowering the interest rate at which banks can lend money to customers. The principle here is that by lowering an important cost to businesses, they will be stimulated to invest in new machinery, so generating enhanced productivity and, ultimately, increased employment. At the same time the investment sets off demand for machine tools and construction, which adds to the demand in the economy and so works more directly to increase employment.

The argument is now made, however, that policies of this nature tend to be self defeating. The reason is that any state lowering its interest rate in a situation where other states do not risks setting off a devaluation of its currency. For moves of this nature can set up international differences in interest rates and trigger flows in so-called "hot money" in search of higher short-term yields on deposits. A multinational corporation or oil state with deposits in British banks, for example, is likely to react to a decline in British interest rates by withdrawing its money and depositing it in banks in a country – say Germany or France – where interest rates remain higher than they have become in Britain. The effect of this is to reduce the demand for pounds relative to the demand for euros. This will result in a devaluation of the pound relative to those other currencies. It is certainly true that this might have some *positive* consequences for the country trying to stimulate economic growth. It should, for example, make its exports relatively cheaper than those of its competitors. On the other hand it will make imports more expensive and to the extent that the country relies on imports that can work counter to increasing its rate of economic growth.[2]

Consider now, therefore, the potential of fiscal policy. The theory here is that in a period of sluggish demand a reduction in taxes may stimulate demand in the economy and hence investment and job growth. A well known example of such policies when they work are the tax reductions introduced by the Kennedy administration in the early sixties. These are widely credited with inaugurating a period of enhanced economic growth in the United States.[3] Such policies are now looked on with skepticism. It is believed that any reduction in taxes runs the risk of simply sucking in imports. And if increased demand is satisfied in this way then there are no benefits to domestic industry, no investment on their part, and no increase in jobs: merely a deterioration in the balance of payments which will ultimately trigger off a decline in the value of the national currency.

---

2   This might suggest the advisability to the state of making its currency inconvertible. The problem is, as we saw above, inconvertible currencies or, at least, limits on convertibility, impede trade and foreign investment. It can also limit the market for state bonds since foreigners will not want to hold them if they cannot convert the proceeds into their own currency or into one which they regard as "safe" (i.e. unlikely to experience a serious decline in value). So insulating oneself from international currency movements by strictly regulating trade in one's own currency is not regarded as a viable solution.

3   Perhaps erroneously, since the war in Vietnam and the requirements for *matériel* also boosted demand in the US economy.

Of course, it is possible that one could counter this by imposing trade restrictions on imports. The problem here, however, is that trade for most countries is a rapidly growing proportion, in value, of their gross national products. Many businesses depend on export markets. The imposition of tariffs on imported goods is likely to work counter to their interests by risking retaliation on the part of other countries. This retaliation may be due to the fact that if *their* imports are blocked by tariff restrictions then they are unable to earn the foreign currency to pay for exports from the country engaging in protectionist policies. Alternatively retaliation may be based on a view that protectionism simply threatens their overseas markets. The institution of protection by one country can also set off counter policies of a protectionist nature in other countries which in turn bring about a decline in global trade *in toto*. So what a country might gain in domestic demand through a stimulative fiscal policy it can risk losing by the drying up of international demand.

In sum, as a result of the enhanced globalization of economic relations, it is argued, states now have to confront a situation in which they are no longer "islands." Flows of money and of commodities across international boundaries have become so elaborated as to make unilateral attempts to manipulate respective national economies in the interest of enhanced growth increasingly difficult. The old macroeconomic tools are now stymied by a growing problem of territorial non-correspondence.

The capital–labor relation

In addition, however, it is now commonly argued that, as a result of "globalization," the labor movements of the more developed countries have suffered serious deteriorations in their ability to bargain for higher wages and improved welfare state benefits. There are several aspects to this. In the first place there is the claim that globalization has had certain direct effects on the capital–labor relation. This is linked to the increased international substitutability of sites and to the increased penetration of domestic markets by international competitors. As a result of the decline in real transportation costs and improved communication technologies the claim is made that the ability of firms to play one set of workers in one place off against another elsewhere has dramatically increased. This is not just a matter of the possibility of shifting production to a less developed country site, though this is important. There is also the matter of competition between workers in different parts of the more developed world. Wage rates can differ between these, though clearly not to the degree that they do between the DCs and the LDCs. But where it is a question of serving a particular market from a somewhat lower-wage site rather than from a higher-wage one, then there will be pressure to relocate to the former. This has been the source of much of the competition between workers in different member countries of the EU, for instance: with the elimination of barriers to trade between those countries firms have been rationalizing their locations and in this process relative wage rates have become important.

This argument is fortified by the evidence of "worker givebacks." These have received widespread publicity in the media and typically involve requests for a renegotiation of the contract with the workers' union local. The company, it may be argued, is having difficulties in competing with its rivals. It desperately needs to lower costs. One way in which it can do so is to renegotiate wage and benefit levels downwards; or alternatively to be able to hire new employees on a different, lower, wage scale. Whatever the particular form these negotiations take, the downward pressure on wages and benefits is clear. But even with renegotiation jobs are not necessarily saved. The firm may find that there is simply no alternative but to close down the plant and (e.g.) subcontract production to a firm in China.[4]

Think and Learn

The claim is made that the internationalization of production has exercised downward pressure on wages in the more developed countries. Another aspect of globalization is the explosion in international financial flows; the exchange of American dollars for Swiss francs, of Canadian dollars for the Japanese yen and so forth. Some argue that this has imposed a deflationary bias on macro-economic policy; states are afraid to boost demand for fear of igniting inflation. How do you think such a deflationary bias might affect wage and employment levels?

Even where changes in plant location between countries are not at issue, firms may still find, or at least argue, that the competition that they are facing in domestic and international markets has intensified as a result of an increased international presence. This has put downward pressure on prices and encouraged firms to shift to cheaper part-time workers hired perhaps from an agency and so not subject to union protections. Alternatively they may move production to small towns where unions are less likely to be present and workers straight off the farm perhaps, and so with a business mind set, are less susceptible to unionization. And if this doesn't happen, unionized firms may go broke and be replaced, perhaps reorganized (!) as non-unionized firms. The result is that employers pay cheaper wages and workers lose their collective ability, through labor unions, to enforce more advantageous terms.

There have also been certain more indirect effects on the capital–labor relation. These have worked through the state. The competitiveness of firms in the national economy is a matter of acute significance to states since it affects the ability of those firms to grow and so supply the state with the revenues out of which it can support its various obligations, not to mention its own extensive labor force of public employees. One of the ways in which this has

---

4   This appears to be the recent history of bicycle production in the US, for example.

worked has been pressure on the state to reduce its own costs of operation so that the burden of taxation on firms can be reduced.[5]

The watchwords have been privatization and marketization. In many countries, including the United States, Canada, the United Kingdom, and Germany, state agencies now have to put requests for services out to competitive tender. This has often meant a shift from the unionized public sector to non-unionized private firms that do (e.g.) the cleaning of government offices, printing for the government, the management of airports and docks, highway repairs, refuse removal, security services, construction, and the like. Similarly firms formerly in public ownership and kept afloat by government subsidy have been sold off and, as we saw in chapter 9 in our discussion of the North–South divide in the United Kingdom, this has often resulted in considerable, and in that particular instance localized, unemployment. The fact that most of these workers were highly unionized has dealt a further blow to the union movement. Parenthetically both privatization and marketization have increased the tendencies to globalization in the sense of the internationalization of investment. This is because they have increased the scope for private and therefore foreign investment.

Less clear is that there has been a move away from "welfare" to what is called "workfare." Eligibility for income supplements has been tightened up and the training of labor, and moving people into the labor force, have become new state emphases. In this way, it is believed, labor costs can be kept down and work readiness enhanced. Decreasing the magnitude and availability of income supplements reduces the time people are willing to spend out of work, increases the labor supply, and so becomes one more factor exerting downward pressure on wage levels.

Finally, an argument has been made to the effect that the international convertibility of currencies has resulted in governments pursuing broadly deflationary policies and these in turn have resulted in increased unemployment serving to further discipline the wage demands of workers (Albo, 1996). On the one hand, convertibility has increased the possibilities open to governments of borrowing in order to fund their operations. International investors and foreign banks purchase another state's bonds, for example, on the expectation that the national economy in question will grow sufficiently to provide a flow of taxes out of which they can be paid off. But convertibility also means that exchange rates become part of their calculus. Assuming that the bonds are denominated in the borrowing country's currency, any adverse shift in the value of that currency is likely to make them nervous. This means that borrowers have incentives to protect its value and, if necessary, see that it is enhanced. The borrowing country, in other words, has to be seen as having a

---

5   Clearly there is a tradeoff here: to what extent will reducing state expenditures and so its need for revenue spark growth in the private sector and so provide the increased revenue on the basis of which the state can once more expand? This was the crux of the so-called Laffer Curve, proposed by the American economist Arthur Laffer, and highly contested. According to it, as taxes came down so investment in the private sector would increase and state revenues would take off, so as to more than compensate.

stable currency or one on an upward trajectory in terms of its value. As a result, what has become seen as laxness in macroeconomic policy is any tendency to let inflationary pressures take over, any failure to increase the interest rate in order to forestall those pressures. But this tends to keep growth rates down and hence the demand for labor, with resultant adverse effects on the wages and benefits workers can negotiate.[6]

Critiquing the Politics of Globalization Thesis

The politics of globalization as I have set it out above clearly prioritizes relations over space: relations of trade, foreign investment, international financial flows. These are what it claims are creating problems of territorial non-correspondence and a shift in the balance of power between employers and employees. So space relations are central. But I am going to argue that, with respect to its political consequences, its true role has been misunderstood. As an argument or set of arguments it is a discourse that is both overspatialized and underspatialized. It is overspatialized in the sense that it attributes far too much causal power – indeed, it may have got the direction of causality wrong – to the increase in trade, foreign investment, increased international financial flows, and the creation of new international divisions of labor. As I hope to show, what has happened has been more complicated than that. The thesis of the politics of globalization, however, is also one that is underspatialized. This may sound highly paradoxical. But what I mean here is that the politics of globalization looks at space, at the politics of geographic arrangement, rather one-sidedly. Earlier in the book I emphasized the role of the variable embeddedness of activities at different geographical scales: the place or local dependence of various agents. This is a dimension that by and large is poorly developed in standard accounts of the politics of globalization. I treat these two issues in turn.

Overspatialization

According to the politics of globalization it is the extensification and intensification of the web of relations – trade, investment, corporate – connecting

---

6   From the particular standpoint of the present time (2001) this may seem a surprising claim to make. After all the US has enjoyed strong growth, tight labor markets, but stable wages for at least five years. Foreign investors have been eager to buy US bonds and the dollar has appreciated as a result. This it is argued is the result of the "new economy" which has allowed productivity increases sufficient to counter, through layoffs, the upward pressure on wages that would otherwise result. The "new economy," however, is a highly contentious subject and many of the claims made on behalf of it are quite suspect. We should also note that unemployment rates in Western Europe are still relatively high by the standards of what I earlier called "the Golden Age of capitalism." The same was true of the US prior to 1995. According to Kenworthy (1997), for example, the average rate of unemployment for OECD countries, including the US, for four successive periods was as follows: 1960–73, 2.2 percent; 1974–9, 4.1 percent; 1980–9, 6.8 percent; 1990–4, 8.2 percent. A rather compelling trend!

Table 10.1   The Golden Age of capitalism versus the long downturn

	Net profit rate		Output		Real wage		Unemployment rate	
	1950–70	1970–93	1950–73	1970–93	1950–70	1970–93	1950–70	1970–93
US	12.9	9.9	4.2	2.6	2.7	0.2	4.2	6.7
Germany	23.2	13.8	4.5	2.2	5.7	1.9	2.3	5.7
Japan	21.6	17.2	9.1	4.2	6.3	2.7	1.6	2.1
G-7	17.6	13.3	4.5	2.2	–	–	3.1	6.2

For private business as a whole. Statistics are averages for net profit rate and unemployment rate and rates of change for output and real wages.
Source: Brenner (1998, p. 5).

firms across international lines that is at the root of the problems confronted by states and by labor. What I am going to suggest here is that quite possibly the direction of causality is incorrect: that globalization is a strategy pursued by firms and states in response to something more basic, and that if we want to understand what has been happening in the world over the past thirty years it is that more basic condition that we need to identify and probe.

That more basic condition in my view is what has been called "the long downturn." To put this in a longer time frame: I identified the period from about 1950 to 1970 as what has been referred to as "the Golden Age of capitalism." During that period corporate profits in the advanced capitalist countries were buoyant, average incomes were increasing, rates of economic growth were strong, and rates of unemployment low. From the early seventies on, however, things changed quite dramatically. Corporate profit rates dipped, investment accordingly lagged, growth rates declined and unemployment increased quite dramatically. This is the period that in contrast to "the Golden Age of capitalism" has been dubbed "the long downturn" (Brenner, 1998) (see table 10.1). Whether or not the sharp uptick in the US economy since 1995 is an indication that this long relative decline is now over is unclear.

In those conditions it would be surprising if firms had not looked around for ways of re-establishing their profit rates, of reducing their costs of operation so as to increase net revenues, of exploring new niches in the division of labor that might afford relatively higher levels of profitability. The strategies were in fact diverse. One of these has surely been the decanting of relatively low-skill operations to sites in the less developed world so as to reduce labor costs. This has been a significant component of the New International Division of Labor (henceforth NIDL).[7] It resulted in increasing levels of foreign investment and also of trade as the parts produced offshore were imported

7   But not the only one. A number of the East Asian NICs, for example, have pursued top-down policies of industrialization that have relied more on domestic investment and less on that of the MNCs.

into Western Europe, North America, or Japan for further processing or assembly. The NIDL is by no means the complete explanation for rising levels of trade and foreign investment but it is part of it. Another has simply been the search for new markets overseas, perhaps supplied by branch plants there, as profitability rates in the home market declined.

But there have been other strategies as well. Firms have sought to lower their costs and heighten their profitability through technological change: by the introduction of machinery, new forms of industrial organization that enhance productivity, like the just-in-time arrangements pioneered by the Japanese and discussed below. There have been changes in labor market practices. Reserves of low-cost labor have been tapped in rural areas through relocation to small towns. The employment of women has increased markedly along with resort to the hiring of part-time workers who can be easily dismissed and for whom benefits can be limited.

The search for new areas of profitability has induced some shift in the sectoral composition of the advanced industrial societies and this has generated the distinction between "sunrise" and "sunset" industries. Many of the changes here are obvious: the relative decline of textiles, of iron and steel and footwear, of much of consumer electronics (TV and radio), and, in Western Europe at least, of coal mining; the rise of the computer industry, both hardware and software, and of related information technology firms, the emergence of the airplane industry and its suppliers, and the growth of that amorphous collection of activities, the "service" industries. These latter include: air transportation with all the employment that generates on the part of the airlines and airport activities; tourism and the development of retirement resorts; advertising; higher education; more highly specialized retailing; restaurants and fast food chains.

It is in part to these sectoral changes that we should also look if we want to understand why unionization levels in some countries at least have declined, so reducing the ability of workers to bargain for improved wages and benefits. Older, highly unionized industries like coal, the railroads, iron and steel, and shipbuilding have tended to show sharp declines in employment. As a result union memberships have been dented. Newer industries have yet to be organized. Furthermore, the problems of organizing (e.g.) back office workers with their highly feminized workforces, or hotel employees, fast food workers, office cleaners, in their dispersed workplaces, are different and require new strategies that have yet to be perfected, or in some cases even articulated, by the labor unions.

The long downturn has also put pressure on the state. Towards the end of the seventies, and long before there was talk of "globalization" and its implications, the welfare state was an important object of lay and academic scrutiny (for example, Offe, 1984). The talk then was of "the fiscal crisis of the state": an increasing gap between the demands being made on the state by citizens and the decreasing ability of the state to respond to those claims through government spending. As profit rates declined and unemployment mounted so the government found its revenues tapering off just at the time when it needed them in order to fund the growing need for various welfare programs. We now know that one of the responses was to encourage the development of "the competitive society": to privatize state activities, to deregulate, and to generally shift responsibilities away from the state back to the individual – ideologies of individual responsibility, and family values. These changes were especially apparent in Britain and the US under Thatcher and Reagan respectively.

Underspatialization

The picture of the world's changing economic geography depicted by those who emphasize the contemporary significance of globalization is one of the increasing mobility, even footlooseness, of capital. According to this view DC multinationals are hollowing out their operations and shifting them increasingly to less developed ones. Or, alternatively, they are setting up bridgeheads in foreign countries from which to invade their markets. Labor, on the other hand, is seen as immobilized in national spaces, as lacking the same ready mobility. So too does that apply to the state, which explains its concern for boosting respective national economies by making them attractive to outside investment.

But this image of mobile capital and immobile labor is a highly exaggerated one. For many firms, establishing branch plants in low-wage areas is simply not possible. I talked earlier about the preconditions for the NIDL as being twofold: the deskilling of labor processes and modes of transportation that are cheaper and/or faster. But many labor processes resist deskilling. This is partly a question of limits to standardization. Mass production is certainly a major, if not *the* major, form of production; but customization is still significant in both manufacturing and services. Customized products and services demand not just access to the customer, suggesting limits to the degree to which falling transportation costs can be locationally emancipating, but also abilities, honed by years on the job, to adapt skills to highly specific forms of application.

Similarly, for many industries and services cheaper/faster transportation is immaterial. As was briefly discussed in chapter 3 there is a whole category of goods and services that economists define as "non-tradables": these are ones which have to be close to the point of consumption. They include custom printing, the news and entertainment media, a variety of food processing activities ranging from baking to soft drinks, along with the obvious cluster of retail services. Others are less obvious. The upper end of the garment industry, for example, is one that requires close contact with customers so as to facilitate response to changing fashion.

In other instances locational substitutability is severely curtailed by the way in which firms are locked into relations with other firms. Examples of this sort of situation or different aspects of it abound. The successes of the computer firms in Silicon Valley are mutually dependent. The firms are locked into webs of suppliers and buyers with whom they can, so long as they remain in Silicon Valley, consult on the changes in inputs required for new products, how to design those inputs so as to minimize downstream assembly costs, etc. In some instances transportation costs may be important, particularly given the small sizes of many of the firms involved and their (consequent) inability to secure quantity discounts from trucking firms. The same sort of logic applies elsewhere: the financial services industries of London's City and New York's Wall Street, the fashion industry of Paris bringing together designers and small garment producers, the machine tool industry of Northern Illinois/Southern Wisconsin, and the optical products industry of Rochester, New York.[8]

Many of these agglomerations of interrelated firms are the outcomes of spontaneous development processes. Firms that later found themselves collaborating, benefiting from a common pool of similarly skilled workers, did not necessarily locate or develop there for those particular reasons. So these forms of local dependence often have a strong element of serendipity attached to them. That is not the case with the clusters of interrelated firms that have grown up around the American assembly plants of Japanese auto producers. These configurations have a significant planned element about them. This is because of the Japanese practice of so-called "just-in-time" production techniques. For a variety of reasons, partly having to do with keeping down inventory costs and partly to do with maintaining quality control, the components for the assembly plants are ordered "just-in-time": in relatively small amounts as and when needed.[9] This means that the component firms have to be clustered closely around the assembly plant, as indeed they are (figure 10.3). But planned or not, there is still a strong local dependence: in this case of the component firms on the assembly plant that they supply and vice versa.

Similarly the multinational character of MNCs can easily be exaggerated. There is a fairly common image of them as spread across the world: an image of having achieved the fortunate situation where they are dependent on no country in particular but have achieved a degree of footlooseness in their locations – a happy ability to substitute one location for another. This, however, is deceiving. Even though they are multinationals they all have *home bases*. They are clearly definable as (e.g.) American MNCs (e.g. GE, General Motors, Monsanto), Japanese MNCs (e.g. Nissan, Sumitomo Bank), or French MNCs (e.g. Michelin, Rhône-Poulenc, Péchiney). As such they have their headquarters in respective countries and their boards of directors will be almost entirely drawn from respective nationals. But there is much more to the home base than that. For despite their global operations most MNCs continue to have the bulk of their assets and their sales in their home countries (see tables 10.2 and 10.3), though there is some variation from one country to another.

In 1992 the percentage of assets in the US for American MNCs in the manufacturing sector was 70 percent, for France 54 percent, and for Japan a

8   Consider the specialties of the major "export" firms in Rochester: Bausch and Lomb, Kodak, and Xerox. These may not interact directly in the Silicon Valley sense but they certainly draw on very similar labor skills and it is therefore advantageous for them to cluster together.
9   For a highly informative discussion of "just-in-time" see Sayer (1986). And for an excellent study of "just-in-time" in practice as it applies to Japanese auto transplants in the US see Mair et al. (1988).

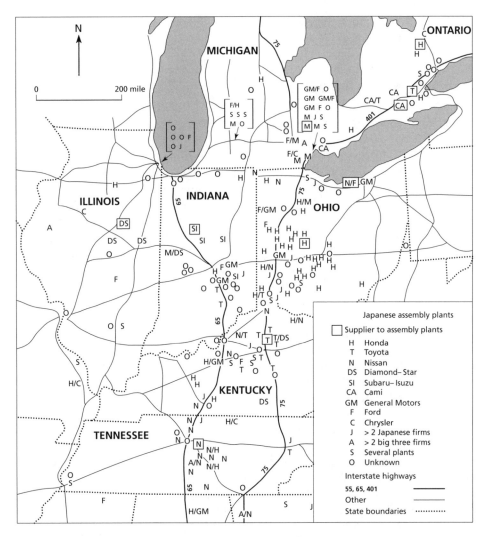

Figure 10.3 Japanese automobile production complexes in the Midwest and Southern Ontario. Note the way in which the suppliers cluster around the assembly plants – a function of the adoption of just-in-time production methods.

Source: A. Mair, R. Florida, and M. Kenney (1988). "The New Geography of Automobile Production: Japanese Transplants in North America." *Economic Geography*, 64(4), 364.

Table 10.2 Percentage of assets in home country for MNCs

	France	Japan	US	UK
Manufacturing	54	97	70	39
Services	50	92	74	61

Source: After Hirst and Thompson (1996, pp. 93–4).

Table 10.3 Percentage of sales in home country for MNCs

	France	Japan	US	UK
Manufacturing	45	75	64	36
Services	69	77	75	61

Source: After Hirst and Thompson (1996, pp. 91–2).

remarkable 97 percent, though for Britain the figure dipped to a much more modest 39 percent. The importance of the home market varied in a similar way: 75 percent for Japan, 64 percent for US multinationals, 45 percent for French, and 36 percent for those of the United Kingdom. In other words, and as Hirst and Thompson (1996, p. 98) affirm: "International businesses are still largely confined to their home territory in terms of their overall business activity; they remain heavily 'nationally embedded'."

We can speculate as to *why* MNCs continue to be dependent in these ways, though clearly the reasons will vary from one firm to another. Many of the reasons have already been covered in our discussion of the local dependence of industrial firms. Many multinational firms continue to rely on a core of industrial skills that can only be replicated elsewhere with difficulty. Virtually all research and development will be monopolized by the home base, for example, along with the more skilled parts of the production process. There is also the matter of relations with other firms, the need for collaboration, and the need to ensure quality control in components – something much more difficult to achieve when suppliers are located in other countries. Firms producing capital goods – machine tools, farm equipment, machinery for the restaurant business – also benefit from proximity to major consumers of their products and the feedback on product design and reliability that that facilitates.[10] There is also the question of knowledge of a particular national market. As box 10.1 indicates this has been an obstacle to the attempts of retail chains to establish themselves outside their home bases.[11]

10 So, and for example, Finland leads the world in the development of machinery for the lumber processing industry just as the most efficient pasta-making machinery comes from Italy.

11 Most businesses, of course, are not multinational. Although we *do* live in a world of giant global corporations with operating units in many different countries most businesses are much smaller and would have difficulty operating in other countries. Size is an important consideration since small firms do not have the deep pockets to finance an inevitably difficult adjustment period in a totally new and strange business environment. Who to appoint to run the overseas branch is often a major problem. The firm can, for example, transfer an existing HQ employee or hire a native. The existing employee knows the company's products well but not the overseas market; the native may know the market well but not the particular product the firm manufactures, particularly if it has features that are unique. On the other hand, few natives may want to apply to join a firm they have never heard of. Likewise, given the size of the firm there may be a very small pool of existing employees to draw on, and most of those may be only willing to do it at a substantial pay premium.

Box 10.1 *International Retailing and the Importance of Local Knowledge*

In an article entitled "Shopping All Over the World," *The Economist* (June 19, 1999, pp. 59–61) discusses some of the difficulties major retailing chains like Walmart and Boots have experienced in expanding outside of their home bases. Some indication of this is that returns on capital are often lower. This is true both for the French chain Carrefour which gets 44 percent of its sales from overseas, and for Walmart (17 percent). One of the reasons identified by *The Economist* is a deficient understanding of the market they are getting into: "multinational retailers need a fanatical attention to detail, and a willingness to do whatever local whim dictates" (p. 61). Some instructive examples are then given:

- In Brazil Walmart's stores were planned with insufficient parking space and with aisles that were too narrow. This was based on a misunderstanding of Brazilian shopping habits. Since most families only have one car they tend to shop at weekends. As a consequence Walmart stores could not accommodate the weekend rush.
- In Thailand branches of Boots, the British drugstore chain, found custom a little slow until they realized that their stores were too quiet for Thai tastes; at which they started playing pop music at full volume.

Think and Learn

In thinking about the politics of globalization as a politics of space, note how it tends to reduce it to a relation between the national and the international, as in the problem of territorial non-correspondence. But to what extent do these problems find parallels in the relation between, say, localities on the one hand, or the subnational state agencies which represent them and nation states on the other? And to what extent were the contemporary problems of the labor movement, as it struggles to cope with the task of organizing labor on an international scale, pre-figured by the history of the labor movement, or more accurately its historical geography, *within* nations?

Globalization and the Politics of Geographically Uneven Development

The politics-of-globalization literature has certain characteristic foci and, along with these, equally characteristic myopias. It concentrates on some aspects of what is happening and marginalizes the significance of others. Here I would identify three of its more significant features in this regard. Very crudely these emphases would be: cooperation (versus competition); class (versus territory); and space (versus time). Let me elaborate.

The notion of a regulatory deficit, of difficulties for the state as a result of a problem of territorial non-correspondence, poses the problem of coordination. The assumption here is that various agents consent to state activity, pay their taxes, accept its regulations, since they will be better off than they would be otherwise. This is the notion, therefore, of firms and workers cooperating with the state and with each other. There is certainly this cooperative aspect and it should not be ignored. But what this focus tends to do is to marginalize the importance of the, often very different, interests of those whose activities are to be regulated/coordinated. For them regulation is always a means to an end: it provides the possibilities of competitive advantages that would otherwise be denied. Any agreement around state policy, therefore, is inevitably a compromise that is good only as long as it facilitates the realization of individual interests. It is always a compromise for some rather than for others. When those ends are no longer realized, or when opportunities of competitive advantage outside those arrangements come into view, then the compromise breaks down and the search is on for a new one. In other words, regulatory deficits occur, state policies fail not because of some anonymous process of "globalization," but because as far as some of the agents are concerned, the existing arrangements no longer work to their advantage, or never did; and as circumstances change so they seek ways out.

We have also seen that the politics-of-globalization literature places heavy emphasis on the capital–labor relation. Recall here how the view is that the increasing mobility of capital, in the form of both direct investment and short-term financial flows, is exerting, in diverse ways, downward pressure on wages in more developed countries. Without the flux of short-term financial movements state policy could be less deflationary and this would facilitate economic expansion and so the demand for labor. The relocation of some parts of firm divisions of labor adds to the pressure by taking away significant elements in the economic bases of some communities. On the other hand, a major emphasis of this book has been the significance of territorial coalitions which bring together elements of both capital and labor in competition or conflict with similar territorial coalitions elsewhere. This should have been particularly clear from a reading of chapters 3 and 9, where a central concept was that of the growth coalition. Furthermore, as I have argued above, capital in the more developed countries has *not* been hollowed out by the creation of a NIDL. MNCs continue to remain highly dependent on their home bases and therefore concerned about labor market conditions and product markets there. Certain fractions of labor in particular are able to exercise some leverage over their (immobile) employers. The stance of the latter, therefore, has to be a cooperative one, though if deskilling should intervene and they could decant those tasks to some low-wage country they surely would.

The third and final emphasis is on space. The very term "globalization" prioritizes relations over space. This is apparent in the stress placed on the problems of territorial non-correspondence for states and of capital mobility for labor. But as I argued above, this is not the only way of looking at things. From the standpoint of "the long downturn," globalization is an effect rather than a cause: it is one, but only one, strategy that firms have resorted to in order to maintain profitability in increasingly difficult times. What gives coherence to all these different strategies is their neo-liberal caste: they all represent a return to market disciplines and, among some at least, the view that the market is the best regulator of all and that the role of the state should indeed be curtailed.

Let us turn next, therefore, to precisely this question of regulation: the question that has been raised in the literature by the supposed challenge of globalization to the state's ability to implement policies which will facilitate growth and full employment. For by no means is it the case that there is agreement on how regulation for those purposes is to be achieved. In particular, the emphasis given to state activity is no longer the consensus that it once was. There are many who believe that the balance between market and state in providing a steering mechanism for the national economy should shift more in the direction of the market: the belief, in other words, in the self-regulating character of markets. The state should significantly withdraw from its interventions, it should privatize, marketize, open the economy up to the drafts of competition from outside.

In some important respects this shift in policy discourse, to the extent that it is realized, represents a return to conditions as they were a hundred years ago. Just as today, currencies were regulated externally; not by the exchange of one currency for another as occurs now, but through adherence to the Gold Standard. At the same time the doctrine of free trade still ruled the roost, though as we saw in chapter 8 it was beginning to be challenged in Britain. Furthermore, the state had yet to acquire the accoutrements of broad welfare, and therefore revenue raising, responsibilities. It was still "an umpire state," quite small in the numbers of people it employed, and confined in its activities largely to defining and enforcing property rights and, of course, maintaining a far flung empire.

The foundations of the contemporary state were laid in the 1930s and 1940s. Depression and world war led to a retreat from the market and a major expansion of the state. The changes were complicated. The misery of the thirties certainly fueled pressures for a heightened measure of social security. It also led to a distrust of the market which in turn paved the way for the nationalization of major industries in Western Europe and increased regulation in the US. There was also a turning in, a retreat from trade, as states sought a greater measure of control over respective economies. This was encouraged after the war by the belief that only through a relative degree of closure could states effectively intervene to maintain growth and full employment. And the war itself entailed huge increases in taxation and state revenues; when the war was over, the money could be diverted to other purposes.

More recently, of course, the pendulum has moved in the opposite direction. The confidence in the state that appeared so justified in the so-called "golden years" has eroded. Both internally and externally the frontier between state and market has shifted. Firms have been denationalized. Industries, like the airlines in the US and more recently gas and electricity, have been deregulated. For their activities states now rely more on private firms, delegating tasks to the market. And as we have seen in this chapter, the barriers to currency exchange have been greatly reduced, paving the way for increased world trade and international investment.

But these shifts between state and market, both then and now, are fought over, contested. Behind *any* regulatory regime, any mix of state and market, the overthrow of an old one and the instantiation of a new one, there is a coalition of forces. These are invariably diverse, bringing together different capitalist interests in different branches of production, different fractions of labor, particular regions in a wide variety of combinations. So in Britain, and to draw on the example of the North–South divide discussed in chapter 9, the coalition in that case has been dominantly concentrated in financial services and in the South East, with an element of hi-tech. Manufacturing has not been so supportive, though there are clear exceptions, including those industries that produce major capital goods or inputs for them like jet engines, machine tools, and engineering services, which have found growing markets in the newly industrializing countries. The goal, then, is to promote their combined interests in growth and hence in defending, enhancing, finding new niches in a wider division of labor (not necessarily international), a project that they believe will be enhanced by that particular regulatory regime. But this means imposing it on others, both nationally and internationally, and perhaps with coalitions of forces in countries elsewhere who see advantages in such an imposition. In the latter regard the meeting of minds between the Reagan and Thatcher administrations is particularly pertinent.

Imposition occurs both discursively and through recourse to new modes of international practice, particularly the mobilization of existing international organizations and the formation of new, purpose-built, ones. It is at the discursive level that the notion of globalization has been particularly significant. Imposing a regime based more on the market than on the state is inevitably politically fraught. Privatization means that public employees lose their jobs. Denationalization and subsequent slimming down result in larger-scale unemployment which, if Britain is any indication, tends to be highly concentrated geographically. The effects of the deflationary environment ushered in by the convertibility of currencies include increased unemployment and insecurity. In this context globalization has proven a useful explanation, or even scapegoat. It is in order to gird for international competition, and maintain the confidence of international investors, or so we are told, that these changes must be made. What is happening has nothing to do with internal political struggles between one class and another, one branch of industry and another, one region and another. There is, as Mrs Thatcher said, no alternative.

This discourse clearly cannot occur in a material vacuum. There have to be events that can be referred to, cases that can be drawn on in order to bolster the case that is in course of being defined, ultimately to find its way into the media and be taken up by those others who see it working to their advantage. So, and for example, even those businesses that are immobilized and cannot take advantage of low-wage labor elsewhere in the world will join in the chorus; and in that case in the hope of securing shifts in government policy, a rewriting of labor law, a relaxation of corporate taxation which, they will argue, is to encourage inward investment, but from which they can also benefit. As Frances Fox Piven (1995, pp. 108–11) has written:

> The key fact of our historical moment is said to be the globalization of national economies which, together with "post-Fordist" domestic restructuring, has had shattering consequences for the economic well-being of the working class, and especially for the power of the working class. I don't think this explanation is entirely wrong but it is deployed so sweepingly as to be misleading. And right or wrong, the explanation itself has become a political force, helping to create the institutional realities it purportedly merely describes. . . . Put another way, capital is pyramiding the leverage gained by expanded exit opportunities, or perhaps the leverage gained merely by the spectre of expanded exit opportunities in a series of vigorous political campaigns.

In short, and from this standpoint, globalization functions as a discourse for imposing a particular regulatory regime: one in which the degree of globalization, the extent and intensity of its effects, is exaggerated.

But in this regard, what sense are we to make of the new levels of international organization that have emerged, like the EU and NAFTA, alongside the seemingly increased significance of older ones like the IMF, the World Bank and the World Trade Organization or WTO.[12] Are they indeed the answer to the need for new regulatory organizations to take care of the regulatory deficit that globalization has supposedly induced? Does their purpose lie elsewhere? Or could they be fulfilling a variety of purposes, and some more than others?

What immediately strikes one is that a coordinative function is only part of what they are about. Free trade areas like NAFTA and common markets like the EU have another purpose. This is widening the sphere of competition/substitutability so as to make the neo-liberal agenda more effective in its realization. For sure, part of that agenda is based on the belief that markets are self-regulating and barriers to their operation should be eliminated if they are to function properly. But this also has to be looked at in the context of the long downturn and the desire to restore profitability. Competition and the downward pressure it exercises on wage levels, on eliminating sunset industries and promoting sunrise ones, and generally weeding out the inefficient, is seen as the best way of accomplishing this purpose. In short, common markets and free trade areas are at one with the emphasis on deregulation, privatization and letting markets rip.

This is not to say that some form of state coordination is considered entirely out of court. The view that the state can perform some functions of a regulatory character still retains some importance and the emergence of consultative organizations like G-7 or ASEAN is obviously designed with that purpose in mind: providing forums for seeking consensus on what their economic policies should be so that they will complement rather than work against each other. This is a change from earlier in the century when, as table 10.4 shows, economies were at least as, if not more, open as they were in 1973. Then there

---

12   Formerly GATT, or General Agreement on Tariffs and Trade.

Table 10.4 Ratio of merchandise trade to GDP at current prices (exports and imports combined)

	1913	*1950*	*1973*
France	35.4	21.2	29.0
Germany	35.1	20.1	35.2
Japan	31.4	16.9	18.3
Netherlands	103.6	70.2	80.1
UK	44.7	36.0	39.3
US	11.2	7.0	10.5

Source: Hirst and Thompson (1996, p. 27).

were no such international regulatory bodies. So something seems to be at work other than the increased integration of national economies one with another, and one can only speculate as to what that is. One possibility is that it represents a changed internal political environment. In the earlier period of relatively intense trading activity the balance of political forces was somewhat different. The labor movement was politically weak, and typically represented by Labour and Social Democratic Parties that had yet to achieve a wide base of support. Today, on the other hand, at least seeming to do something about unemployment and incomes is politically unavoidable.

International organization, however, has also fulfilled a third function. This is to impose the new regulatory regime worldwide. There are several different aspects to this. The neo-liberal agenda is particularly dear to Anglo-Saxon hearts, which means in this instance the United States and its junior partner Britain. It is no accident that these two countries spawned Reaganism and Thatcherism respectively. One of the problems has been imposing their particular conception of what is an appropriate regulatory regime on other countries. The difficulties have been particularly aggravating in the Far East: not just the NICs there but also the country on whose economic policies many have modeled their own, Japan. As might have been evident earlier from the discussions in chapter 3, this is a model in which state orchestration of investment and trade has been strong. This orchestration has in turn meant some discrimination against foreign corporations, not just in terms of trade but also in terms of investment. The MNCs of other countries were looked at sceptically since their agendas might well not conform to those of the country in which they planned to invest.

Through US domination of the IMF and to a lesser degree of the OECD there has been a consistent drive to open up these economies to Western, particularly American, firms, seeking new markets and new field for investment. During the East Asian financial crisis of 1997 the IMF used its role as lender of last resort to impose particularly destructive regimens on countries like Thailand, Indonesia, South Korea, and Malaysia, forcing retreats from state subsidized industrial projects, like the national car in Indonesia, and opening up national banking sectors to foreign investment and ultimately ownership.

In this way they hoped to destroy at the same time the so-called "crony cap-italism" which went so much against the neo-liberal grain but which was also the means by which the NICs had managed to become NICs.

As far as the LDCs are concerned, however, it has been a case of all talk and little action. Free markets are espoused only to the degree that they work to the advantage of the more developed. This is something that we reflected on in chapter 9 in the context of Third Worldism, but the issues of protection are still there. The last big push towards global free trade was the Uruguay round of 1993 but in the course of those negotiations and in terms of the depth of the tariff cuts agreed to, the poorer countries conceded more than the wealth-ier ones. Since then the richer countries have found new ways of keeping out the sorts of products – footwear, textiles, steel, agricultural products – in which the poor countries have a comparative advantage. One of the major ones is the use of anti-dumping duties. These are the escape clause in World Trade Organization rules that allow more developed countries in particular to protect particular industries from the competition of the less developed. Tariffs cannot be raised since under WTO rules other countries could demand com-pensation or impose retaliatory tariffs of their own. But anti-dumping duties can be imposed if it can be shown that foreign goods are being sold cheaper than at home or below cost of production and when domestic producers can show that they are being harmed. What this misses, however, is that it is common for prices in international markets to be lower than in domestic ones simply because competition is greater. The bottom line, however, is that, according to one estimate[13] the average tariffs imposed by rich countries on manufactured imports from poor ones are four times higher than those on imports from other wealthy countries.

The results of this shift to a neo-liberal regime have been highly uneven, both in the more developed and in the less developed world and between them. In the more developed world there is evidence that interregional dis-parities have widened. The evidence for Britain was presented in chapter 9. This is entirely consonant with what one might expect given the sort of defla-tionary environment that has been characteristic of the past fifteen years or so. The weakest firms, both private and publicly owned, are weeded out and these tend to be concentrated in particular geographic areas. Similarly there have been widening income disparities in the US and in Britain. These have been less apparent in the rest of Western Europe, but instead unemployment rates have tended to be a good deal higher. Lower-skill workers are especially vulnerable to unemployment as a result of the deflationary bias that the inter-nationalization of financial markets has imposed. They are also the ones most likely to be displaced not just by imports from NICs but also by technical change as firms search around for cheaper ways of doing things. In the US and in Britain a burgeoning, but poorly paying, services sector has tended to take up much of the slack. But not in the remainder of Western Europe, where real wages have proven more resilient, so working against the creation

13   *The Economist*, September 25, 1999, p. 89.

of the low-wage labor force on the basis of which many of the service industries tend to thrive. Equally, when income distribution is examined on an international scale, the evidence is one of increasing disparities (Hirst and Thompson, 1996, pp. 69–71). This suggests, moreover, that the protagonists of Third Worldism do indeed have a point. Regulation is for some and not for others, and for some countries as opposed to the (much) greater balance of the world's population.

In sum, the long downturn induced a crisis of profitability across the world from which firms and the states which depend on them have been trying to extricate themselves over the past quarter-century or so. The strategies pursued have been diverse and can in no single-minded way be laid at the door of "globalization." There have also been ones of privatization, marketization, creating and developing new niches in wider geographic divisions of labor. Old regulatory strategies have been called into question. What the new strategies have in common is an emphasis on a return to the disciplines of the market. Through unleashing the forces of competition it is believed that the crisis of profitability can be resolved. Labor costs can be held in check, incentives provided for the emergence of new branches of the division of labor, and state regulation shown to be not just unnecessary when set beside the truths of the self-regulating market but an expensive execrescence that makes things worse.

Globalization of corporate operations, extension into new markets, has been one of the strategies pursued towards this end, but its material significance can be easily exaggerated. Its discursive significance, however, has been considerable. This is because as a construction that highlights some processes and marginalizes the importance of others, it has allowed states and the various forces into which they have entered in coalition to impose this new solution not just on their own peoples but on those elsewhere. The struggle around regulatory institutions *is*, therefore, about capital and labor since it is about profitability, but in the way envisaged here and not in the sense of mobile capital achieving enhanced leverage over labor *in toto*. And it is about the relation between state and market, but not in the narrow sense of a crisis brought on by territorial non-correspondence.

The "Other" Globalizations and Their Politics

In talking about the politics of globalization our discussion so far has emphasized the economic. From one standpoint this is as it should be, since that is where most of the popular interest has lain. As I mentioned earlier, however, there are other, environmental and cultural aspects to globalization. These tend to have been treated separately from the economic, compartmentalized. Whether or not that compartmentalization reflects reality is clearly something that we should now consider, along with a more detailed consideration of what is meant by globalization as applied to the environment and to culture.

"Global change"

In the debate about environmental change or what is sometimes called "global change," the globalization at issue is conceived in rather different terms and is only to a degree artifactual of the internationalized commodity exchanges which have come to define the essence of globalization as an economic process. For sure, the increasing distances over which people in places interact have received publicity. The increasing rate of air travel and the distances over which it occurs has been identified as one of the reasons why some diseases formerly confined largely to the tropics can occur in isolated outbreaks in northern latitudes; these have been found to be especially likely around airports and in summertime when conditions are most congenial for them.[14] But the most serious political contestation has occurred around issues that are the result not so much of economically mediated diffusion but of a physical diffusion. This is the politics of the atmosphere and the oceans considered as single systems in which the different elements seek an equilibrium.

The most obvious expression of this, and rightly so, is the case of global warming. The increasingly common view of climatologists is that average ambient air temperatures are indeed increasing[15] and that is due to the intensification of the so-called greenhouse effect. The latter is associated with the increase in greenhouse gases, particularly carbon dioxide, but also methane and nitrous oxide, which slow down the rate at which heat is radiated back to space by the earth. That the carbon dioxide and nitrous oxide content of the atmosphere has increased along with concentrations of methane is beyond doubt and this in turn is attributed to the combustion of fossil fuels and, to a lesser degree, of timber.[16]

Think and Learn

To what extent has the burning of fossil fuels increased as a result of the geography of late twentieth-century cities in the richer countries, the geographic form that urbanization has assumed there? And what does that suggest about the sources of opposition to, and of support for, plans to curtail the burning of fossil fuels?

---

14   For an excellent article on this topic see Cliff and Haggett (1995).
15   The latest forecast at the time of writing (January 2001) is that the average global temperature in one hundred years will be a colossal 10 °F (5.5 °C) greater than it is today.
16   Increases in rice culture and in cattle as world population increases, and as diets shift towards meat with increasing standards of living, at least in some parts of the world, also contribute to the generation of methane.

Accordingly, the release of greenhouse gases is highly concentrated in the more developed countries.[17] Energy consumption per capita there is huge, both for purposes of production and for consumption. But the resultant increase in carbon dioxide, methane, and nitrous oxide has effects that are felt worldwide. Precisely what the long-term effects will be is still unclear and some can be more readily detailed than others. But as glaciers and ice sheets melt there will undoubtedly be a rise in the level of the ocean. Given the fact that such a large fraction of the world's population is concentrated in coastal areas, this will be highly disruptive to say the least and very costly to mitigate. At the same time, zones of optimum agricultural production will shift. It is widely believed that these could have very adverse effects on the tropical zones, which include some of the least developed areas of the world. Infectious diseases hitherto confined to more tropical zones will also spread into more temperate ones. A major concern is malaria, which already kills 3,000 people per day. No vaccine is presently available and the parasites in question are becoming resistant to the drugs presently in use.

Global warming is therefore a serious matter. And despite the presence of contending views that it may be artifactual of other more transient causes rather than a long-term secular increase in greenhouse gases, it has clearly become an issue for international negotiation. Conferences have been held in an attempt to coordinate policy on the issue. Some of the more developed countries have policies. Others, notably the US, have dragged their feet.

The stakes are obviously huge. Recalling arguments made towards the end of chapter 4, the contemporary form of urbanization in the US, and to an increasing degree elsewhere in the world, is an energy-intensive one. It is the widespread use of the automobile that has allowed the low-density form of urban development that some define as sprawl. Consider here, and by way of example, the effects of a serious increase in gasoline taxes in the US.[18] Not only would the gasoline and automobile companies cry foul since it would threaten consumption of their products; so too would the lending agencies that hold the mortgages for residential property that would surely be devalued as people reordered their locational priorities. This is only a small part of the problem. If the same tax on fossil fuels extended to industry, to the generation of electric power, say, then impacts on profitability would be severe. Attention would shift to alternative forms of energy but the adaptation would be a costly one.

Having said that, it is patently obvious that the more developed countries are in a much better position to make these adjustments than the less developed. They are home to virtually all the MNCs with the huge resources that they have at their disposal. They also have the discretionary incomes which could be taxed in order to expedite the transition through public spending programs. Even so, there are hard political decisions. There seems an elementary territorial justice to the notion that since the DCs and their corpora-

---

17   Note, however, that China with its large population and heavy reliance on coal is the second-biggest contributor after the US.
18   Presently very low indeed by international standards, but invariably generating tremendous political heat whenever the question of raising them is broached.

tions are the ones that created the problem in the first place, they should carry most of the burden of reducing greenhouse gases and so, perhaps, reducing, even eliminating, the tendency towards global warming. Lending impetus to this line of argument is the fact that the less developed world is only just beginning the process of industrialization and to impose the same limits on them as on the DCs might well put a stop to it. If, on the other hand, the more developed countries were to accept stiffer limits one can see how that might accelerate development in the LDCs by increasing the attractiveness of production sites for the MNCs of the more developed countries. An examination of the tea leaves, however, suggests that this is unlikely to happen. Rather there will be a failure to agree, or the measures agreed to will be inappropriate to the scope of the problem. For the fact is, footdragging is a less costly option for the DCs. This is because they are the ones that will find it most easy to adapt to the consequences of global warming: to alter agricultural practices, to build dikes to keep the ocean out of especially vulnerable coastal locations (the Dutch, after all, have been doing it for years).

The connection to the politics of globalization in the economic sense, therefore, is relatively weak. The connection to what is driving the politics of globalization in both its real and imaginary senses, however, is huge. For it returns us to the Janus-faced facts of capitalist development and its logic. Its bounty has been immense. It has, as Marx predicted, created the world market, and knocked down barriers, real and metaphoric, to its expansion. Its technological feats are of an awesome magnitude. But so too have been its environmental effects, and these are much more real than some of the effects attributed to globalization in its economic guise.

And the cultural

As a cultural issue the debate about globalization has revolved for the most part around notions of homogenization and what has been called cultural imperialism. The growth of trade and direct investment has, it is argued, been the vector through which the cultural content of commodities ranging from blue jeans and fast food to TV programs and movies has spread from one country to another. This in turn has provoked two related sorts of response. The first is that this is eroding cultural differences, reducing the world to a condition of placeless-ness as the same forms of consumption crop up everywhere, the same chains of hotels and restaurants displace more local ones that expressed something of the distinctiveness of particular places. "Heritage" is being pushed aside and this is disturbing to those who cling to particular senses of difference and the national identities they underpin.

Think and Learn

We hear a great deal today about "national heritage." What precisely do you think it means? Who or what might have an enhanced interest in preserving "heritage" do you think, and why might that be?

The second response has been that of cultural imperialism: to draw attention to the national provenance of the new icons and the meanings that they convey. Most of the controversy here has revolved around Americanization, though there are variants. In France the concern has been more one of anglicization while in Iran it was what the mullahs termed West-toxification. But these differences also draw attention to the variety of concerns. In Canada and France the concern has been primarily that of what one might call the national cultural community: the TV program producers, movie producers, writers, national weeklies and monthlies. There have been in both countries attempts to regulate the local content of broadcasting and publishing, and in France, what is shown in movie theaters. In Iran, on the other hand, it had to do with the subversive cultural content of Western films and the images of (a relative) female emancipation that they conveyed.

Both of these critiques, however, that of homogenization and cultural imperialism, seem overdone. The importation of cultural influences from outside is nothing new. More importantly for the homogenization thesis, the people so "influenced" are not a blank sheet upon which the outside forces are to be impressed. Obviously they have their own cultural traditions, the ones whose durability the homogenization school worries about, and it is in terms of those traditions that the cultural imports are interpreted. It is less a process of imposition, blank acceptance, or even surrender than it is one of assimilation, reworking, the construction of new hybrid forms which go to form a new diversity, a new sense of difference, to replace the one that has been lost. We know that people get different messages from the same soap opera or movie, sometimes critical, sometimes not, and this difference should be enhanced when we consider the fact that people are socialized into different cultures.

Besides, culture does not exist in a vacuum. It gets refashioned in the context of changes in material practices. McDonald's, and other chains like them, are attractive in Western European countries for a good reason: they respond to a need for fast food that has resulted from, among other things, the same increase in the employment of women in the wage sector that we discussed in chapter 7. Culture necessarily changes as material practice changes, throwing up new needs, new desires, contradictions, and longings. Whether the response to these challenges is from inside or from outside a country is beside the point. It has to run the gauntlet of answering to the felt needs of people in particular places as those felt needs change.

Even if one accepts the cultural imperialism argument on its own terms it is a dubious one. McDonald's has been a lightning rod for this, as has Coca-Cola. But consider the ubiquity in the more developed countries of the Chinese restaurant, even in that self-conscious capital of gastronomy, France, of the Indian curry house on British High Streets, and the remarkable variety of "ethnic" restaurants that one finds in American cities, albeit with an American slant. Likewise the extent to which American TV programs and movies dominate is highly exaggerated. In Latin America the movies are much more likely to be Mexican or Brazilian, in India, Indian and in China, Chinese. And even when Indians live in Western Europe or North America, "Hindi-movies" are immensely popular. Similarly American movies and TV programs absorb/assimilate the foreign. One marvels, for instance, at the number of program ideas which originated in Britain to find a reincarnation in the United States.

So the significance of culture as it pertains to "globalization," and if it so pertains, lies elsewhere. And that I believe to be the case. In some cases it is almost transparently clear that fears about globalization and culture are being used to fight other, quite different battles, which even lack the strong cultural referents that one might expect from the way they are associated with globalization. Rather the battle lines are between big business and the smaller owner operators who fear displacement into the ranks of wage workers. Not surprisingly, France, where the small farmers, artisans, and shopkeepers – *le petit bourgeoisie* – have a long history of militant opposition to big business and the state which inadvertently or otherwise shifts the balance of advantage away from them,[19] has been at the center of these struggles.

In this regard an event which captured the imagination of wider audiences was the trashing of a McDonald's restaurant being built in the small town of Millau in Southern France. The leader, one José Bové, also leader of France's "Peasant Confederation," was arrested, his subsequent trial generating a good deal of publicity for both him and his cause. The headline in *The Economist* was "The French Farmer's Anti-global Hero."[20] The news magazine then went on to argue that "What matters [for M. Bové] is that McDonald's is a symbol of junk food, the product of an American led globalization threatening not just France's traditions but everyone else." This, however, misses much of the point. For sure Bové and his followers could harness France's culinary obsession, along with some anti-Americanism, in support of their cause. But as an earlier article in *The Wall Street Journal* had observed: "Every time he opens his mouth . . . he turns the talk from the anti-McDonald's crusade to the global struggle between agribusiness and the small farmer."[21] This means that his concern is equally with the rules governing competition between big business and *le petit bourgeoisie*, of which France's peasant farmers would be an important element. Accordingly Bové's wrath is turned more against agribusiness regardless of national origin than it is against McDonald's. McDonald's merely proved a convenient target through which to mobilize wider support. For example, a larger concern of his Peasant Confederation has been the subsidies which go to agribusiness in France – big grain farmers and some big meat companies – and which impinge on the ability of the small producer to compete (see also box 10.2).

Moreover, if cultural homogenization *was* occurring, and with or without domination from one particular culture hearth over the rest of the world, we should find that difference would be of decreasing importance in understanding the ebb and flow of world politics. What seems to be happening,

---

19   I am thinking here of Poujadism, a political movement with its own party in the 1950s, animated by small businesspeople hostile to a state which they believed was taxing and regulating them out of business.

20   *The Economist*, July 8, 2000, p. 50.

21   *The Wall Street Journal*, December 22, 1999, p. A16.

Box 10.2 *Cheese Wars*

Just how the struggle of the small producers against big business can mobilize sentiments of a cultural nature is also exemplified by recent controversies over the pasteurization of the milk going into cheese. France is a country of many cheeses, an important element in its culinary heritage, and many of these are made from raw milk. The use of unpasteurized milk runs the risk of consumers of the cheese experiencing the bacterial infection known as listeriosis.

A proposal before the World Trade Organization, and originating with the National Cheese Institute, the American industry's main lobbying group, seeks to make pasteurization the standard for all internationally traded dairy products. Clearly this proposal is double edged. On the one hand it can be presented as protecting consumer safety, even though the French cheeses are produced under impeccably hygienic conditions and rates of listeriosis are no greater in France than in the US. But equally clear is the fact that by keeping French specialty cheeses, and perhaps those of other countries, out of international trade, it would clear the way for the American producers to expand their markets, and once more drive out the small producer. Once again, however, in order to fight this proposal, the smaller producers have tried to mobilize wider cultural antagonisms. As one French producer claimed: "It's about whether we want a food landscape of 350 different cheeses or one, like the US, where you have the choice between cheddar, cheddar and cheddar."

Based on Amy Barrett (1999) "Why Defend Cheese that Smells Like Socks and Manure?" *Wall Street Journal*, May 27, p. A1.

however, is that far from being erased, old senses of difference, of cultural oppression in some cases, are being re-awakened, re-energized by the stresses and strains induced by the imposition of the neo-liberal agenda, including those various practices that have gone under the heading of "globalization." There have been increases in geographically uneven development. There has been a more general increase in the sense of material insecurity. In some instances this has paved the way for conflicts of a territorial sort, and these have been especially aggravated where historically sedimented senses of difference could be drawn on to marshall the troops and bolster material claims through appeals to "fairness." In other instances old senses of cultural oppression have been intensified. I conclude this section, therefore, with a brief examination of two recent examples of these processes at work: the breakup of Yugoslavia and the revolt in the Mexican State of Chiapas.

The breakup of Yugoslavia

Under communism the Yugoslavian federal state consisted of six republics and two autonomous regions. Of those six republics, four have become independent over the past ten years: Bosnia, Croatia, Macedonia, and Slovenia. All that remain of Yugoslavia are Serbia and Montenegro and the autonomous

Figure 10.4 The constituent republics and autonomous regions of the former Yugoslavia. All that remains effectively now of Yugoslavia is Serbia and Montenegro. Slovenia, Croatia, Macedonia, and Bosnia are independent states, though Bosnia is extremely weak. The fate of Kosovo is at the present time undetermined. It is occupied for the most part by a Western peace-keeping force and is contested by both Serbia and Albanian nationalist movements whose goals are unclear: unification with Albania or a separate statelet altogether.

region of Voivodina (figure 10.4). The future of the autonomous region of Kosovo, of course, is now in the balance. What led to this denouement is extraordinarily complex. The idea of being Yugoslavian was always weak. Primary identities were with various forms of sub-state nationalism, particularly with Serbia and Croatia and to a lesser extent with Slovenia and Bosnia. The hatreds between some of the constituent nationalities were intense, not least between Croats and Serbs. And there was a widespread suspicion of Serbs and fear of Serbian dominance throughout the federation. On top of this ethnic-cum-national mosaic and its associated tensions was superimposed a quite marked geographically uneven development, as table 10.5 indicates. Successive forms of the Yugoslavian state, the pre-communist[22] and the communist

---

22   Lasting from the establishment of a Yugoslavian state in 1919 out of the independent states of Serbia and Montenegro and the entrails of the Austro-Hungarian Empire, till the German invasion of 1941.

Table 10.5 Uneven development in Yugoslavia

	Income per head		Percentage unemployed		Percentage illiterate 1971
	1947	1981	1966	1981	
Slovenia	175	198	2.6	1.5	1.2
Croatia	107	125	6.1	5.6	9.0
Voivodina	109	117	5.4	12.5	9.0
Serbia	96	98	7.0	14.9	17.6
Montenegro	71	75	7.6	15.0	16.7
Bosnia-Herzegovina	83	67	5.3	14.1	23.2
Macedonia	62	67	16.4	22.3	18.1
Kosovo	53	30	21.0	27.7	31.5
Yugoslavia	100	100	6.9	14.9	15.1

Source: Alexandra Stiglmayer (1991) "Das Ende Jugoslawiens." *Informationen zur Politischen Bildung,* 223, p. 10.

one, had tried to redistribute income from the wealthier to the poorer republics and this had always been a source of tension among republics the peoples of which had less in the way of solidarity with one another than they did in mutual suspicion and even fear. In the event the economic proved to be the straw that broke the back of Yugoslavia.

During the seventies Yugoslavia made huge borrowings from foreign banks in order to fund an industrialization program. This latter was supposed to generate export earnings from which the loans could be paid back. Unfortunately things went wrong. Western markets went into a recession at the end of the seventies and this made it hard to sell the goods and so pay back the loans. Yugoslavia had to go begging cap in hand to the International Monetary Fund or IMF for further loans from which it could liquidate its existing debts. The IMF, however, also wanted assurance that Yugoslavia would be able to pay back the new loans for which it was the creditor and so imposed on the country an economic policy which greatly contracted demand there. This in turn served to increase unemployment and drive down wages. The federal government of Yugoslavia was obliged by the plan to devote something like 20 percent of all the country's earnings to paying off the debt. Real wages in Yugoslavia declined by 40 percent (!) between 1978 and 1983, with unemployment at about a third of the total labor force: hardly a recommendation for internationalizing the economy.

Significantly regional inequality worsened (see table 10.5), as it inevitably does as an economy contracts,[23] and each of the six republics saw itself as

---

23 Using the data from table 10.5, the degree of variation in income per capita across the republics and autonomous regions increased between 1947 and 1981 by almost one-third; the same was true of variation in unemployment between 1966 and 1981.

unfairly burdened by the austerity plan. Given the lack of strong national feeling – feeling for Yugoslavia as a nation, that is – this increase in tensions should not be surprising. Even then the situation might have been saved. But there were other things that had to be factored in. For as the eighties progressed so the Communist bloc, of which Yugoslavia was a member, if an errant one, went into terminal decline. This had two effects: (a) it increased pressures within Yugoslavia for greater marketization of the economy; (b) it enhanced the willingness of Western powers to interfere in the region, to expand their spheres of influence and to speed up the shift to a market economy.

With respect to marketization it is important to note that the pressures for it were geographically highly uneven. They were concentrated in particular in Slovenia and Croatia, the two most economically successful of the constituent republics (table 10.5). These were, significantly enough, the first of the two republics to make a move for independence. Serbia, on the other hand, was always opposed and preferred an economic policy characterized by a much higher degree of central state planning. With the implosion of the Soviet bloc a further stimulus to independence turned out to be the European Union. As socialism collapsed so the desire to join the West, and in particular the European Union, intensified. As Robin Blackburn (1993, p. 102) has argued, "Many Slovenes and Croatians became seduced by the notion that they could simply join the advanced West, with its enviable prosperity and liberality, allowing their more backward ex-fellow-countrymen to find their own level."[24] So in effect Yugoslavia was to be forsaken for another supranational union but one that would be much more decentralized than the Yugoslav federation.[25]

Think and Learn

I made reference in note 24 in this case study to some similarities with the situation created in Italy by the Northern League and its arguments for an independent state of Padania. How similar to or different from the Yugoslavian situation do you think it is?

---

24   Compare the case of Padania which we reviewed in chapter 1. One of the motives fueling the drive for separation there was to join the euro-zone: that part of the EU where the euro would be the common currency, so minimizing currency risk for exporters and importers. The concern of the Northern League and their supporters was that with Southern Italy in tow, the Italian state itself would be incapable of meeting the low national debt requirements that were the price of admission to the euro-zone (this has proved incorrect).

25   Even then the situation might have been saved and the descent into barbarism averted. But the Western powers had their own interests in the area, interests that were not necessarily aligned with the preservation of the Yugoslavian state. If, for example, the Western countries had withheld recognition from Slovenia and Croatia when they declared their independence it is quite possible that things could have turned out differently. But that was not to be. Germany, Austria, and the Vatican saw an opportunity to expand their respective spheres of influence and with the balance of political forces between the West and the Soviet bloc shifting in favor of the former they saw little reason for hesitation. Once they accorded recognition all hell broke loose, particularly in Croatia, where the Serbian minority was immediately made to feel very insecure indeed.

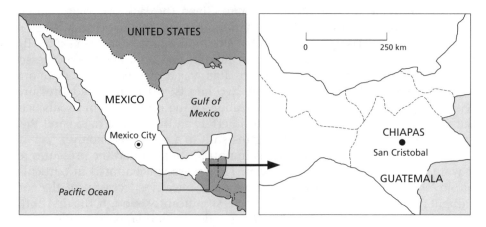

Figure 10.5   Chiapas in regional context.

The revolt in Chiapas

> Last year, the Zapatistas held a convention in the jungles of southern Mexico, titled "the Intercontinental Forum in Favor of Humanity and Against Neo-Liberalism." . . . The session ended with the Zapatistas doing a kind of drumroll and announcing the most evil-dangerous institution in the world today. To a standing ovation, the Zapatistas declared the biggest enemy of mankind to be the World Trade Organization in Geneva, which promotes global free trade. (Thomas L. Friedman, "Even Mexico's Zapatistas Sing in Chorus Against Global Trade." *Columbus Dispatch*, February 4, 1997, p. 7A)

On the first day of the New Year in 1994, significantly the day that the North American Free Trade Area agreement took effect, Indian peasants rose up in armed rebellion in the southern Mexican State of Chiapas (figure 10.5). Some towns were seized and contested with the Mexican army. There have been some land seizures on the borders of rebel strongholds. Indeed the ownership of land in the context of a castelike system of ethnic stratification in which the Indians were at the bottom was the major issue. The agents of harm against whom the revolt was directed were seen as the Mexican government, the local PRI (Institutional Revolutionary Party), which is the ruling party in Mexico, and large ladino landowners.

In Mexico's stratification system Indians are at the bottom. They are a despised group. They are the poorest of the poor. In many areas of Mexico the term *indio* is used to signify "lazy" and "uncivilized." In this way the majority of the population explains the poverty of Indians, at the same time affirming their own more meritorious attributes, though the history of the Indians in their relations with ladinos and mestizos suggests that this has nothing to do with it. Rather Indians live in the least productive parts of the country, the least agriculturally rewarding, not out of choice but because they were pushed there by those commanding greater physical force. But nowhere are Indians as despised as they are in the State of Chiapas. In the major center of the Chiapas highlands, San Cristobal de las Casas, and until about 35 years ago, Indians were not allowed to walk on the sidewalks, stay overnight, or look whiter-skinned Mexicans in the eye.

Land has been a major issue and the NAFTA agreement has made it even more so, deepening Indian hatred of ladino landowners and the Mexican government both. The land issue has numerous facets. The NAFTA agreement called for the end of the *ejido* or land redistribution program by the Mexican government but in Chiapas the program never really began. Many remain landless in a region of large estates. What land the Indians had has been vulnerable to seizure by the large owners by a variable mix of fraud and violence in which some Indians have disappeared: Chiapas is regarded by international human rights groups as the problem State in Mexico. In addition in recent years many Indians have been discarded by landowners who used to employ them as sharecroppers to grow corn. This has been done so that the land could be devoted to cattle raising. This is a form of use requiring much less labor. The abolition of subsidies for corn as part of the NAFTA agreement is likely to hasten this process and this also helps to account for the symbolism of the timing of the rebellion.

The Mexican state is seen as culpable not just because of NAFTA but because of its failure to do anything about the land distribution problem in the area. The PRI, moreover, works hand-in-glove with the ladinos of the area who are its support base. In return they have been allowed to ignore the needs of the Indians and override, often in the local court system, their rights. Large amounts of government money have been spent in the State – Mexico's poorest – but very large fractions of this have been siphoned off by corrupt local officials of the PRI who have used it to oil the local vote machine and buy off opponents. Indian hatred of the Mexican state, ladino landowners, and the onetime ruling party, the PRI, runs deep and the threats posed by NAFTA and its neo-liberal agenda have not only re-opened old wounds, they have rubbed salt into them.

Summary

As a concept with a specific set of meanings globalization is a relatively recent one. It is over the past fifteen years or so that it has insinuated itself into the media, the rhetorics of politicians, the concerns of academics. A major reason for this widespread interest is the way in which it has articulated with politics. And while the politics of globalization in the economic sense has perhaps received most of the publicity, there are also globalizations of the environmental and the cultural, each with their own distinctive politics.

As a concept in political economy it draws its strength from certain empirical observations and what are believed to be their political ramifications. Commodity exchange has become increasingly internationalized. Foreign investment has risen relative to gross domestic products. International financial flows have greatly increased. And there are clear signs of a New International

Division of Labor, one in which less developed countries acquire industries calling for lower degrees of skill and industrial experience. This has given renewed impetus to the industrialization of less developed countries. These changes are in turn attributable to two other conditions: the deskilling of labor processes so that some functions can indeed be decanted to less developed countries; and changes in transportation and communication which have allowed an extension of trade and the global production networks of MNCs.

From the standpoint of the states of the more developed countries and of labor there, these changes are believed to be problematic. For states there are regulatory problems. Due to the widening scope of the geographic division of labor, and the internationalization of financial flows, the old fiscal and monetary policies no longer work. States have lost the ability to create the conditions for steady growth and full employment. The problem for labor is partly a matter of the mobility of capital so that it can drive down wages and costs by threatening to exit to some site in the less developed world. It is also a function of the seeming reluctance of states, in a context of international financial flows, to pursue policies that have an expansionary bias.

A sense that this thesis might not be quite right, however, comes from an exploration of the way it draws on concepts of space. On the one hand it severely overestimates the mobility of capital. Jobs are *not* fleeing the more developed countries. Many firms find that it is only there that they can find the skills they need; or they require an access to markets which, despite cost and time reductions in transportation and communication, they cannot satisfy from a location outside the more developed world. On the other hand, the significance that the politics of globalization assigns to forces of internationalization of the economy – increasing trade, foreign investment, financial movements – is misplaced. What has been happening over the past thirty years or so needs so be set against the background of what has been called "the long downturn": a long-term decline in rates of profit, in rates of capital investment, in rates of growth in wages, and an increase in the unemployment rate. Firms along with states have sought a way out of this impasse, and foreign investment and the expansion of trade are indeed some of the strategies pursued. But by no means are they the only ones. Privatization, the marketization of government activities in order to induce greater competition, reorganizing geographic divisions of labor within the more developed countries themselves, deregulation, a renewed onslaught on organized labor, attempts to dismantle the welfare state are some of the others.

What sense, therefore, is one to make of this particular politics of globalization? The more fundamental condition that we should be examining if we want to understand the politics of our age is not globalization, but the attempt to impose a new regulatory fix through which firms and states can finally escape the grasp of the long downturn. This new approach to regulation is a neo-liberal one. It is an expression of a neo-liberal agenda of a return to markets and a retreat from state intervention. Through the discipline of markets, the competition they induce, it is believed, costs can be contained, and not least labor costs, and firms will have incentives to develop new products, new ways of doing things as a basis for renewed expansion. Moreover, there has been a shift of sentiment about the regulatory role of the state. Far from worrying about regulatory deficits induced by territorial non-correspondence, the view is that states try to do too much anyway and that markets can be much more effective in achieving growth with full employment than they have been credited with. And if there remains unemployment, then that is because states are still trying to do too much, engaging in wrong headed moves, like raising the minimum wage, which will simply provide employers with an incentive to replace workers with machines, and so increase the numbers of jobless.

The way globalization enters into this is certainly as one neo-liberal strategy among others. Accordingly states have been complicit in bringing it about: in shifting to convertible currencies, in joining free trade areas and common markets so as to widen competition. But what this overlooks is that globalization is also used discursively. Its significance is exaggerated as a way of pushing through the neo-liberal agenda, of making the medicine palatable to the broader electorate, of protecting the interests of those firms, embedded in DC locations, for whom a relocation to a low-wage country is just not on, for example, and of imposing the new regime, through organizations like the IMF and WTO, on less developed countries.

Some have clearly gained from this and some have just as clearly lost. The emergence of the North–South divide that we discussed in chapter 9 is part and parcel of what has been happening. There are widening income disparities in both the US and Britain. These have been less apparent on the European continent but unemployment has been a bigger bane there. Likewise the income disparities between rich and poor countries have widened. The coalitions of forces that have prevailed are typically place-based and include firms in growth sectors and some fractions of labor, particularly the higher-skilled echelons and the states of the more developed world that have orchestrated these efforts. It is through them that the new regime of competition is being enforced.

Meanwhile there is an environmental problem that is growing in its seriousness. The aspect of this that I chose to focus on was the implication of the effect of greenhouse gas emissions on global warming. This has been receiving sharply increased publicity of late because of the effect it will have on the level of the oceans and on the geographies of disease and of agriculture, all calling for expensive adaptation. The globalization that is at issue here is clearly a physical one. It is about the diffusion of gases throughout the atmosphere rather than directly about trade or foreign investment. And this returns us to the centrality of capitalist development. This is partly a matter of its expansionary logic resulting in the increased burning of fossil fuels. But equally, as negotiations are entered into, there are the huge investments that have been made that are dependent on the continued use of fossil fuel and which will act as an impediment to rational decision making on behalf of all the people on the planet through a shift to alternative forms of energy.

Cultural aspects of globalization have also attracted attention. The concerns here have been the related ones of homogenization, bringing with it the erosion of the cultural distinctiveness of regions and nations and a resultant sense of loss, and cultural imperialism. I suggested that these claims are exaggerated and we should give more credit to the ability of people to rework the various cultural artifacts and forces that they are exposed to as a result of what Marx referred to as "intercourse in every direction, universal interdependence of nations," and always in the context of changing material needs and possibilities. Difference is not disappearing and in fact, if anything, national difference and cultural difference have been given a renewed lease of life as geographically uneven development and colonial oppressions deepen in the context of neo-liberalism, engendering the sorts of violence that occurred in the wake of the breakup of Yugoslavia and the revolt in Chiapas.

REFERENCES

Albo, G. (1996) "The World Economy, Market Imperatives and Alternatives." *Monthly Review*, 48(7), 6–22.

Blackburn, R. (1993) "The Break-up of Yugoslavia and the Fate of Bosnia." *New Left Review*, 199, 100–19.

Braverman, H. (1974) *Labor and Monopoly Capital*. New York: Monthly Review Press.

Brenner, R. (1998) "The Economics of Global Turbulence." *New Left Review*, 229, 1–265.

Cliff, A. and Haggett, P. (1995) "Disease Implications of Global Change." In R. J. Johnston, P. J. Taylor, and M. J. Watts (eds), *Geographies of Global Change*. Oxford: Blackwell, chapter 13.

Hirst, P. and Thompson, G. (1996) *Globalization in Question*. Cambridge: Polity Press.

Kenworthy, L. (1997) "Globalization and Economic Convergence." *Competition and Change*, 2(1), 1–64.

Mair, A., Florida, R., and Kenney, M. (1988) "The New Geography of Automobile Production: Japanese Transplants in North America." *Economic Geography*, 64(4), 352–73.

Offe, C. (1984) *Contradictions of the Welfare State*. Cambridge, MA: MIT Press.

Piven, F. F. (1995) "Is It Global Economics or Neo-Laissez Faire?" *New Left Review*, 213, 107–15.

Sayer, A. (1986) "New Developments in Manufacturing: The Just-in-time System." *Capital and Class*, 30, 43–72.

FURTHER READING

The literature on globalization and its politics is immense. A good overview is provided by Frank J. Lechner and John Boli (eds) (2000) *The Globalization Reader* (Oxford: Blackwell). More focused and coherent are two highly stimulating books put together by faculty at The Open University and published jointly with Oxford University Press: John Allen and Chris Hamnett (eds) (1995) *A Shrinking World*; and James Anderson, Chris Brook, and Allan Cochrane (eds) (1995) *A Global World?* Peter Dicken's (1998) *Global Shift* (London: Paul Chapman Publishing) provides an excellent discussion of particular industries and how they have been affected by, as he puts it, "global shift." There are some excellent essays on a number of the topics treated in this chapter in R. J. Johnston, P. J. Taylor, and M. J. Watts (eds) (1995) *Geographies of Global Change*

(Oxford: Blackwell). Another useful compendium of essays is Ron Johnston and Peter Taylor (eds), *A World in Crisis* (Oxford: Blackwell). The book by Paul Knox and John Agnew (*The Geography of the World Economy*. London: Arnold, 1989) can also be perused with profit.

The original statement on the New International Division of Labor is: F. Frobel, J. Heinrichs, and O. Kreye (1980) *The New International Division of Labour* (Cambridge: Cambridge University Press). On the deskilling of labor processes the key reference is H. Braverman (1974) *Labor and Monopoly Capital* (New York: Monthly Review Press).

More advanced, critical discussions of the issues set out in this chapter can be found in: P. Hirst and G. Thompson (1996) *Globalization in Question* (Cambridge: Polity Press); L. Kenworthy (1997) "Globalization and Economic Convergence." *Competition and Change*, 2(1), 1–64; and R. Boyer and D. Drache (eds) (1996) *States Against Markets* (London and New York: Routledge).

On the relation between globalization and uneven development there are some good books written for a lay audience, including: J. Bennett with S. George (1987) *The Hunger Machine* (Cambridge: Polity Press); and T. Alexander (1996) *Unravelling Global Apartheid* (Cambridge: Polity Press).

On the plight of Indian agriculture in a Mexico that is subordinating national fortunes to the free trade dictates of NAFTA see T. Barry (1995) *Zapata's Revenge: Free Trade and the Farm Crisis in Mexico* (Boston: South End Press).

Conclusion

At the heart of political geography are the ideas of territory and territoriality. These bring together the ideas of space, and therefore "geography," and of politics, and hence "power." Power and space in their interrelation are therefore what we have been concerned with in this book, but from a particular angle: that of the world as it has changed over the past fifty years or so, generating insights not just into itself but also into the political geographies of the past. It is, in short, an approach to the political geography of the contemporary world.

To talk about the contemporary world, however, commits us to a view that emphasizes the institutions and processes specific to that particular world; not the world of antiquity, therefore, or of the Middle Ages, but the world of capitalist development, the nation, and the nation state. At the same time it commits us to some of the changing forms thrown out by those economic processes and institutions as they have been transformed during the capitalist epoch: to the politics of identity, for example, to the politics of colonialism, and more recently to that of neo-liberalism and its associated politics of globalization.

Capitalist development is a highly territorializing process. It requires fixity and it requires movement. Money has to be invested in fixed facilities, building up supporting physical and social infrastructures if value is to be produced, and people and firms acquire interests in their continued viability. But once value is produced and converted back into its money form there is no necessity that it be invested in a way that protects those investments in place. The result is the wide variety of territorial practices that have been at the center of this book's argument: exclusionary zoning, tariff protection, a spatial restructuring to make an area more attractive to inward investment, the social definition of others as alien and undesirable, and the assertion of claims for "territorial justice."

The dilemma of fixity versus movement, and the resultant territorializing impulse, have been central to the discussions in this book. Without it the territorial form of the contemporary state, its centralized character, the polic-ing of movement across its boundaries, make only limited sense. The state is the organization through which are expressed interests in places at diverse geographic scales. And while the state's internal scale division of labor between more local and more central branches mirrors to some degree scale variation in the place-specific interests that seek protection or realization through it, there is no one-to-one relation between the geography of interests and the particular branch of the state that people seek to mobilize. For those with highly localized interests what may work in terms of leverage is an alliance with other "locals" and an approach to the central state; in fact this accounts for much of the logic of the political parties as well as of the labor movement – alliances of people with interests in particular places.

The state that emerged under capitalism was a specific type. Apart from its high degree of centralization it has also been a nation state. The nation is a very modern idea. It is, furthermore, a highly potent one in the name of which a good deal of violence has been perpetrated. It is a major way in which people differentiate themselves from others: part of the politics of difference. Quite why it has had such an allure is not an easy question to answer, but the sheer universality of the idea suggests that there is something quite fundamental in the social pressures of living in a bureaucratized and commodified society which gives it resonance: an island of seeming stability in a world where, to borrow a phrase from Marx, "all that is solid melts into air."

This is by no means to argue, however, that in understanding the political geography of the contemporary world we can recite a set of structures like capitalism, the state, or even the different political parties, talk about the way in which people might be formed by them, and leave it at that. People may act under conditions that they might not otherwise choose, but they do act. They are agents who develop ideas, test them out, and sometimes, through them, give sharp impetus to transformation over both time and space; though again, whether their ideas resonate, take root, depends on conditions they do not control. Moreover, they never do this alone. Their acts depend on webs of interdependence with both contemporaries and people in the past who bequeathed ideas, resources that can be later used in different ways. They are, in short, *social* acts. People construct their world in and through each other. It is a socially constructed world. Capitalism, the state, the nation are not exempt from this dictum, nor are the countless changes in concrete form that each of these has exhibited over time.

What gets constructed, how they get constructed, who constructs whom, is always a matter of struggle. Capitalist development is a tension ridden process that generates profound anxiety about the future; about having a job, about the viability of one's business, about a level of personal security which the state can only contribute towards and not ensure. The most central axis of tension around which conflicts tend to congeal is that of class: the division between workers on the one hand and those owning the means of production on the other. This is signified, among other things, by the way in which the dominant cleavage in the party political systems of the Western democracies tends to be along class lines, with left-wing parties typically enjoying close

relationships with respective national labor movements. There are other frac-
tures, however, and these tend to divide both business and labor. Firms
compete with one another, and workers likewise, so that the class tension can
often coexist with other forms of conflict which threaten to displace it from
the consciousness of the protagonists and replace it with other forms of awareness, other constructions of the world, and their place in it.

And to be sure, a common axis around which conflicts between businesses
and conflicts between workers tend to break out is the geographic. To repeat:
firms and workers have interests in particular places. Under conditions of
uneven development, conditions which are inevitably produced under capitalist forms of development, though whose precise geography we can never
anticipate, these tensions can be severe: workers and their employers both,
for example, fear being undercut by cheaper products from elsewhere where
cheaper wages prevail and/or there are firms exhibiting a higher level of technical prowess. Even without uneven development there is always the prospect
that it will occur; that "our" city will be deserted by major investors for cities
elsewhere, for example; and as a result of our embeddedness in it, our local
dependence, we will not be able to move to where the major investors are creating the jobs or the boom conditions for us as retailers, developers, home
builders, and the like. In these circumstances, and not surprisingly, various
forms of cross-class alliance are likely to come into being: the core of the
growth coalitions we discussed in chapter 3.

It bears emphasis, however, that these alliances, the particular cleavages
that dominate politics, are always constructed. They don't just happen. People
have to articulate positions, seduce would-be allies, debate with their antagonists, form breakaway labor unions or new political parties, fund candidates,
get media exposure. And to the extent that public opinion starts to shift in
their direction, certain views will start to assume the form of the "taken-for-
granted" and the constructors can rely on others to do their work for them.
This process of social construction, moreover, can assume all manner of forms.
It can be as conscious as the actions of the Taiwanese government as it
tries to undo former layers of sinicization in favor of a Taiwanese sense of
nationhood. It can be as visceral as the reactions of those South Africans that
we discussed in chapter 5 who defined in the most derogatory terms they
could imagine the outsiders, including Poppie, whom they saw as threatening their housing and job possibilities: "Bushmen." Or, of course, it can be as
cynical as a US Senator who, up for re-election in a tight contest, plays the
race card.

In that construction process, stories are told, facts selectively drawn on,
some events embellished, the significance of others marginalized, and language used in order to dramatize, evoke, stir emotions so that rhetoric
becomes a central element in social construction. Whether explicitly or not,
geography will be drawn on in these accounts. Accordingly maps are an integral part of the process, whether the maps the South African apartheid government published showing white settlers from the southwest meeting,
somewhere in the east of the country, black settlers coming from the north;
or those deployed in the US to show the way in which federal resources are

being redistributed from Coldbelt to Sunbelt States. It is so often, therefore, and to give an old aphorism a new target, a case of "lies, damned lies, and maps."

Quite what the constructions will be, which will be tried, which will work, is inevitably a contingent matter. We know that difference will be constructed and that the tensions inherent in a class society, the insecurities of both capitalist and worker, will underlie the impetus to differentiate, to justify positions of advantage, to argue for "rights." But the degree to which people identify with class, the degree to which alternative forms of identity along racial, ethnic, gender, or simply territorial lines assume importance, depends on the circumstances of time and place. Social construction, in other words, is context specific. What happens in one place, the particular combinations of territory and class that emerge, for example, will be different from what happens elsewhere.

But context is no *deus ex machina*. It too is socially constructed and gets reproduced by virtue of people's activities, activities which could change over time and often do. But how malleable it is, how resistant to change, is highly variable. Capitalism is part of the context into which we are all born. We are brought up in the expectation that someday we will have to seek employment and school is urged on us as a necessary preparation for "getting on in life." Through our actions, through (e.g.) showing up at work, moreover, we *re-produce* capitalism just as every time we spend the money in our pay packet. Buying (and sometimes selling) seems to be the unalterable horizon of our lives from a very early age, and to be sure if anything is taken for granted it is the durability of capitalism. This sense is fortified by the fact that when overthrown in revolution it has later – much later in the case of the Soviet Union – been restored to its former role as arbiter of life chances and the engine people look to as the means of raising their material standards of living. Communist experiments, whether those of the Soviet Union or of China,[1] proved no match for capitalism's ability to raise material standards even though they achieved considerable success in terms of equalizing life chances and securing a minimum in housing, healthcare, and nutrition: levels of achievement that have still to be attained in the most prosperous countries in the contemporary world, and a success for which they are given scant credit by the Western media and politicians.

Even so, even though the institutions of a capitalist society seem set for the indefinite future, hard to imagine as something we could live without despite their clear weaknesses, we should recall that they were themselves socially produced. Capitalism was born in Western Europe as a result of a quite inadvertent juxtaposition of events whose identity is still not conclusively determined. But the revival of trade in the Middle Ages, the English enclosures of land for sheep and the abrogation by feudal lords of the land rights of peasants, the confiscation of monastic lands and their distribution to court favorites, all contributed to the separation of immediate producers from the means of production, particularly land, and their reuniting with it through the

---

1   Prior, that is, to China's embrace of the market.

money of the few, so creating that commodification of labor power which is capital's essential precondition. Once in place it acquired a momentum which deepened its grip, provided incentives for its deliberate extension into other areas. Recall, for example, the way in which the mining companies went about creating a labor market, and hence imposing capitalism, in Southern Africa: the thoroughly intended dispossession of the vast majority of the population so that they would have to seek wage work.

I do not want to be misunderstood here, however. In understanding the social construction of capitalism we should never forget Marx's famous dictum about people making history but not under conditions of their own choosing. The various social experiments that emanated from late fifteenth- and early sixteenth-century England – and in retrospect it seems reasonable to define them in those terms – could never have succeeded without certain conditions of a material kind being in place. Long periods of human history have to have elapsed before those conditions are present. It seems fair to say that that history has been characterized by a gradual improvement in people's ability to convert naturally occurring substances using natural forces into items of use. Only on the basis of a certain level in the development of the productive forces can capitalism come into being. This is because it depends on the availability of a surplus product over and above the needs of the immediate producers: a surplus product which, once produced under the command of the capitalist, can be invested in *new* means of production and allow that cycle in which the owners of capital accumulate on an ever expanding scale.

All of our other social relations partake of the same fundamental explanation: they are socially constructed, reproduced, transformed. But, and unlike the case of capitalism, in many of these other instances we can observe their transformation so that they seem more malleable. We experience change in our life times. That is certainly the case for gender relations in the Western societies and it is true of race in that most racialized of societies, South Africa. But, and again, the material conditions have to be right. The drafting of increasing numbers of women into the wage labor force gave them an access to the social power of money that they had not had before and this proved a major force for liberation. In South Africa the racialized institutions of apartheid created a skills shortage which gave impetus to the demands of white business for some dismantling of such age-old policies as job reservation by race. And the growth of large numbers of blacks in the cities created those thresholds through which organizations could form and bring pressure to bear on the white government.

Moreover, to refer to South Africa in this way alerts us to the role of specifically national contexts; the institutions of apartheid in that instance. At various points in this book I have referred to differences between the US and Western Europe (and there are certainly differences within the latter that we could have explored). They are both capitalist in their modes of economic organization, but there are significant differences in their institutions. Western European countries, for example, have tended to be more spatially redistributive. The state tends to redistribute money from areas where there are higher tax revenues to those where tax revenues are lower and where otherwise public provision would be lacking; as a result the sorts of appalling differences in educational provision that one encounters in the US, often over very short distances, tend to be greatly attenuated. Likewise there has been a history of redirecting employment away from areas of labor shortage to areas where unemployment rates have tended to be stubbornly high. There are other differences. National parks in the United States are on land owned by the state; in Western Europe they tend to consist of private land but with fairly stringent conditions on how that land can be used. And of course there are reasons for this difference rooted in part in a history of large-scale land ownership by the state in the US that goes back to the country's colonial origins and also in an ideology of private property that is less circumscribed by notions of social obligation. Again, these are matters of social construction. Political histories are different. The colonial experience is one: and remember that the US empire was always within North America – not a question of "the white man's burden" but of "manifest destiny." Similarly, the greater strength of the labor movement in Western Europe resulted in stronger central states than in the US case and a stronger redistributional ethos.

We can observe change within national contexts. The EU is having definite effects on the territorial dilemmas of Western European countries. As a result of the sweeping away of barriers to trade in Western Europe firms locate with reference to the Western European market as a whole and this has brought the member states into a competition one with another to attract in the necessary investments that is redolent of a territorial competition long characteristic of the American States. The EU is a long, long way from the federal structure of the American state but there has been some closing of the gap.

Context then is a matter of place *and* period, time *and* space, and it is with some remarks about the conjoining of these two great universal dimensions of the framework of human experience that I wish to conclude. In the title to one of his books Eric Hobsbawm has referred to the "short century" from 1914 to 1991 as *The Age of Extremes* and in part what he wanted to capture with it were the immense changes characteristic of it. But if we concentrate on those changes rather than the more dramatic events of world war, communist revolution, the end of the Cold War, and so forth, the full hundred years seems more appropriate. The beginning years of the century and the closing ones were seemingly very different. There is surely a veritable catalog of these differences but I want to emphasize here four of them. In the first place at the beginning of the century most of the world was divided up into empires. The most obvious of these were the overseas empires of the Western European nations: Britain, France, Portugal, Italy, Germany, Belgium, the Netherlands. Their empires were still in full bloom and not like the aging hulks of the, nearer to hand, Austro-Hungarian and Ottoman Empires. In addition Japan too had an overseas empire and there were also the continental empires of Russia and the United States. Empire was legitimated, moreover, by racist ideology: *la mission civilisatrice*, and the "white man's burden" that we talked about in chapter 7. And it was the white *man's*. Not only were ideas of racial superiority and inferiority taken for granted, so was the superordinate role of the male. Bear in mind that before the First World War women enjoyed the franchise nowhere. By the end of the century, however, empire had virtually vanished to be replaced by a multiplicity of nation states, though in many instances the "nation" qualification has been a little ambitious. From the dominance of empire, therefore, to the dominance of the nation state. Along with that has gone a reassessment of racial and sexist doctrine, though I would not want to link the fortunes of sexism too closely with those of empire.

The second contrast I want to point out concerns the relation between state and market. At the end of the twentieth century and continuing beyond the millennium we have grown accustomed to the idea of markets pushing back the state. In chapter 10 I talked about the implementation of a neo-liberal agenda and the transfer of various functions hitherto assigned to the state to private firms or even private citizens: a shift of the public–private boundary, in other words, in favor of the private. This has been associated also with a dismantling of state barriers to the market in the international sphere so that trade and foreign investment have certainly expanded over the past twenty-five years or so as a proportion of the global product. But at the beginning of this century the tendency was in the other direction. Markets were starting to shrink in the face of an expanding state. The very early beginnings of the welfare state – an old age pension was introduced in Britain in 1907 – were beginning to appear. Labor unions were pushing for an expansion of their rights rather than resisting efforts to dismantle them. And while world trade and investment were on a relatively greater scale than they have yet to attain today, the forces of change were working in the opposite direction: economic policies of a more autarchic nature were clearly attaining more favor.

Third, the environmental problem, in all its varied manifestations, is something that we have become accustomed to hearing about. Dark warnings of the effects of greenhouse gases, the deterioration of groundwater supplies, loss of biodiversity, pervade the media. But one hundred years ago these issues were not even on the horizon. And when the environment, or more accurately our relations with it, did start to become an issue, it was defined as a much more local problem: soil erosion in the American dust bowl in the thirties, air pollution in British cities in the fifties, overpopulation in India from mid-century onwards. Only quite recently have we become sensitized to the global dimensions of local acts, acts that include the burning of fossil fuels and biological reproduction. As a result the environment is now part of the international agenda. Nation states, it is believed, need to coordinate more in their respective policies if the earth is to continue to be a habitable place.

Fourth, and finally, geographic divisions of labor and of consumption have been transformed. These changes have been quite complex. It is not just the shift from the Old International Division of Labor to the New (from the OIDL to the NIDL): a shift, that is, from a division between industrial countries on the one hand and raw material producing ones on the other; to a division between countries engaged in the more skilled aspects of industrial processes and those engaged in the less skilled. Apart from the fact that this is highly overgeneralized, that the NIDL has been superimposed on, rather than replaced, the OIDL, there have also been changes in the organization of firms.

The NIDL has in significant part been mediated by the multinational corporation with its own internal geographic division of labor. Similar forms of multilocational organization are now characteristic within as well as between countries. As we saw in chapter 3, there are branch plant towns, headquarters cities, places specializing in R&D, and so forth. Again, this type of firm organization would have been the great exception at the start of the twentieth century and would have been almost entirely confined to international mining companies.

Likewise, geographic divisions of consumption, in Western Europe and North America at least, have changed. By 1900 "the suburbs" were being identified as such. But as yet there was no "inner city" and none of the differentiation that we now experience in the geography of the living place. People of all generations lived alongside one another, aging parents often living with married children; the retirement community or the "starter home" developments were a long way in the future. Similarly, most people still lived in sight of their workplace – the factory, the mine, or the docks – and walked to work. Between then and now, therefore, there have been vast changes in personal mobility, in choice, and therefore in the magnitude of the income stream on which those particular changes have been predicated.

Much of what we take for granted in our contemporary political geography has depended on these changes. Not least, the competition for more advantageous positions in geographic divisions of labor has been conditional upon the division of the corporation into its various functionally related, but spatially separable, parts. Similarly the competition that goes on between local governments in metropolitan areas, in the United States at least, for fiscally enhancing land uses – upmarket residential developments, retirement communities which come without children to educate – depends upon those particular forms being possible.

In between these two periods, approximating more to Hobsbawm's "short century," were the intense political convulsions that, along with the growing rapidity of social change, help justify the idea of the "age of extremes." Of primary significance in understanding these was the rise of the labor movement and the nationalist reaction which, in significant part, it engendered. By 1920 union membership had increased to about half the labor force in Austria, Belgium, Britain, Denmark, Germany, and Italy.[2] There was also a rising socialist vote. By 1925 over a third of the total vote in Austria, Belgium, Britain, Germany, Norway, and Sweden was going to socialist parties of various complexions, and by 1945 this had swollen to about half the total vote in Austria, Britain, France, Norway, and Sweden, though with some slight tendency to fall off thereafter. There were, however, counterattacks of which nationalism was a major though not the only one. In its most extreme manifestations this reaction, designed to define leftist leaders as traitors to the country and to bind the working class to a national mission, merged with anti-semitism. This was

2   On the labor movement in twentieth-century Europe and the various attempts to obstruct and derail it, see Michael Mann (1995) "Sources of Variation in Working-class Movements in Twentieth Century Europe." *New Left Review*, 212, 14–54.

thoroughly contradictory since Jews were identified with *both* the class enemies: bolshevism *and* international capital. But the message was the same: Jews, in diverse ways, subverted the nation.

Obviously there is far more to the "age of extremes" than this. Without the First World War, without the depression of the thirties, things might have been different. Likewise without capitalism's "Golden Age" in the fifties and sixties there might never have been that period of compromise between organized labor and business, for the general expansion of the global economy during that period made it easier for capital to accept labor's demands and for the welfare state to expand. We should also note that the years since the early seventies have been different again, with labor on the defensive as capital seeks to find a new basis for continued accumulation in the context of long-term declines in profitability worldwide: what we referred to in chapter 10 as "the long downturn."

Likewise, without the Second World War the process of decolonization might have taken much longer. For sure, the fact of socialist-led governments in a number of the Western European countries made it easier ideologically for independence to be granted: empire had always generated ambivalent feelings among socialist leaderships. But the Second World War not only exposed the imperial powers as vulnerable to those not of the white race: Japanese success in Asia was especially significant in undermining the British, Dutch, and French presence there. It also left the imperial powers economically exhausted and in a less than strong position from which to confront the growing nationalist movements in the colonies with a determined show of force. It did happen, as in Algeria and French Indo-China, but the lessons were on the whole, and apart from the Vietnamese War, quickly learnt.

But having said all this, having pointed to the dramatic changes that have occurred over the past hundred years, there are also striking elements of continuity between what the world was like at the beginning and what it was like at the end. In formal terms colonialism is clearly dead. Nevertheless, arguments about neo-colonialism, which continue to resonate, suggest that the underlying conditions for it are not. Those underlying conditions are ones of uneven development and uneven development is something that is endemic in capitalism. Not all places develop at the same time. Those with an early start tend to get privileged in the future competition for more attractive positions in the geographic division of labor. Firms tend to cluster there so they also gain from increasing returns to scale consequent on agglomeration. This is so regardless of the geographic scale one is talking about: localities, regions, nations, even whole continents. But maintaining these advantageous positions is also dependent on places elsewhere in the world continuing to be the hewers of wood and drawers of water; confined to the less advantageous roles in the geographic division of labor. The new neo-colonialism leaves this up to market forces, in contrast to the political interventions characteristic of the older version. And while the old racist understandings of colonial relations have fallen out of fashion they have been replaced with new codes of superiority and inferiority: First World versus Third World or even more developed versus less developed countries. Change, therefore, but also continuity.

Likewise, the centrality to social life of the capital–labor relation continues unimpaired. The balance of advantage may ebb and flow, but there is no sign that labor's present retreat is terminal; that workers have agreed that what capital believes is best for them, is indeed best. The notion of a retreat of the labor movement is overgeneralized anyway. US and British experience in terms of declining union memberships has not been duplicated in many of the advanced industrial countries. In some, union membership was higher in 1990 than in 1970, including Belgium, Canada, Finland, Italy, and the Scandinavian countries.[3] Moreover, the three "lacks" or insufficiencies, that I identified in chapter 2 – insufficiency of work, insufficient purchasing power, lack of a sense of significance – continue to be ones to which workers are vulnerable, so why, therefore, should antagonism and conflict diminish? Moreover, the particular forms in which class antagonisms and the struggle for leverage express themselves vary. Direct confrontation is only one of them. Class antagonisms combine with territorial interests in odd ways, so that class conflict can yield to conflict between territorially organized coalitions, as in the case we referred to in chapter 3, where the issue of workers' compensation – an employer–employee issue, if ever there was one – was converted into an issue of territorial competition. Or in other instances workers try to achieve some advantage with respect to their employers by excluding workers from other countries.

Which brings us, logically enough, to that clarion call of the hour, globalization. In contrast to much of the media rhetoric, we have already seen that globalization is nothing new. Its form may have changed, as in the rise of the multinational corporation, but trade and foreign investment were, if anything, more significant at the beginning of the twentieth century than today. What seems to be different, however, is the way in which it is being used as a new way of asserting domination over the South and over workers. Much of this is discursive, constructing stories, based on anecdotal evidence and some carefully massaged statistics, whose moral is what will happen to workers in the North, what happens to whole countries in the South, if they do not attend to the new global "realities": without renegotiation of labor contracts the jobs of the workers will disappear; without the adoption of Northern standards of currency convertibility, free trade, and the like countries in the South will be deprived of capital. And so it goes.

But again, there is nothing particularly new about this. The goals remain the same: goals of restraining the pretensions of labor and of those countries which think they can manage capital and trade flows to *their* advantage and to the disadvantage of the major multinationals and the countries where they are based. So with respect to the latter, globalization as a discourse becomes one more means of asserting a colonial relationship over the South, channeling its development in ways consistent with the needs of the multinationals. With respect to workers in the North it is ironic that while in the thirties the nation was seen by business as the bulwark against the demands of labor, the

3 L. Kenworthy (1997) "Globalization and Economic Convergence." *Competition and Change*, 2(1), p. 36.

means through which workers could be weaned away from the promises of socialist parties and the like, today it is the *inter*national which has been grasped as the solution.

This seems an appropriate point at which to bring, in a manner of speaking, this conclusion to its own conclusion. This book has provided an approach to the political geography of the contemporary world. As the term "political geography" implies it is a particular window on the world. But talk of globalization, if only as an idea that has developed out of proportion to the material reality it is designed to describe, brings home the *necessary* character of geography in understanding our world and its politics. The threat of capital mobility is a potent one because people have commitments, material and symbolic, as in the nation, to particular places. And to the extent that that contradiction between mobility and fixity attacks vital interests it will generate various forms of resistance, various strategies to turn geography to local advantage. In these struggles the power of the state will become a focus, and so therefore the state's structure, including, necessarily, as I hope to have shown, its geographic structure.

We can learn about these things every day, simply by reading the newspaper. Journalists are – have to be – political geographers as they report on (e.g.) plant closures, rezoning disputes, the doings of the World Trade Organization, though they are not necessarily very good as political geographers. In earlier versions of this text used in my lecture courses I relied a great deal on extracts from newspapers and accompanying critical discussions. At the very least, therefore, what I hope to have accomplished is to sensitize the reader to that vast flow of news which confronts us everyday and which is inevitably about places, their interrelations, and how those are reflected in politics. But not just *any* sensitivity: it has to be a *critical* one.

Index